MULTIPARAMETER
EIGENVALUE
PROBLEMS
STURM-LIOUVILLE THEORY

MULTIPARAMETER EIGENVALUE PROBLEMS

STURM-LIOUVILLE THEORY

F. V. ATKINSON

UNIVERSITY OF TORONTO
TORONTO, ONTARIO, CANADA

ANGELO B. MINGARELLI

CARLETON UNIVERSITY
OTTAWA, ONTARIO, CANADA

CRC Press
Taylor & Francis Group
Boca Raton London New York

CRC Press is an imprint of the
Taylor & Francis Group, an **informa** business
A CHAPMAN & HALL BOOK

CRC Press
Taylor & Francis Group
6000 Broken Sound Parkway NW, Suite 300
Boca Raton, FL 33487-2742

First issued in paperback 2019

ISBN-13: 978-1-4398-1622-6 (hbk)
ISBN-13: 978-0-367-38322-0 (pbk)

Library of Congress Cataloging-in-Publication Data

Atkinson, F. V.
    Multiparameter eigenvalue problems : Sturm-Liouville theory / F.V. Atkinson, Angelo B. Mingarelli.
        p. cm.
    Includes bibliographical references and index.
    ISBN 978-1-4398-1622-6 (hardback)
    1. Sturm-Liouville equation. 2. Eigenvalues. I. Mingarelli, Angelo B. (Angelo Bernardo), 1952- II. Title.

QA372.A843 2010
512.9'436--dc22
                                                                    2010038203

Visit the Taylor & Francis Web site at
http://www.taylorandfrancis.com

and the CRC Press Web site at
http://www.crcpress.com

Preface

A literary executor messes with the literary legacy of the author
whose works he was meant to protect with unintended consequences.[1]

The task of resurrecting a manuscript written by the hand of one of the foremost
experts in differential equations in the school of Hardy-Titchmarsh, Frederick
(Derick) Valentine Atkinson, has been a daunting one. Derick died on the 13th
day of November 2002, in Toronto, Canada, after a long illness that kept him
confined to a bed and unable to work for some 10 years.

He was born in Pinner, Middlesex, on the 25th of January 1916 of a father
who was a well-read journalist and a mother who was a homemaker. We would
certainly be led astray if we were to confine ourselves at this point on further
biographical items. The interested reader may of course consult the elaborate
obituary[2] or the brief note to the Royal Society of Edinburgh[3] for further
biographical and professional details.

Unlike Kafka and Hemingway, Atkinson left no written record forbidding the
publication of any unpublished works after his death. Thus, soon after the
funeral services were held, the task of publishing his "opera omnia" became
the order of the day, a seed that had been germinating since at least 1972.
The family held that a complete work existed by the end of July 1992 (one
month prior to his debilitating stroke) and that this manuscript was ready for
publication. The final document was etched on as many as 8 x 3.5 in. IBM-DOS
formatted diskettes. Unfortunately, neither the manuscript nor these diskettes
have ever been found.

We review here, for the record, the events that led to this publication. On many
an occasion during the winter of 2003, the second author was always greeted
warmly by the Atkinson family during his frequent visits to Toronto. He was

---

[1] www.complete-review.com/reviews/moddeut/krugerm.htm, July 1, 2008.
[2] *A glimpse into the life and times of F.V. Atkinson*, Math. Nachr., **278** (12–13) (2005),
1–29
[3] http://www.rse.org.uk/fellowship/obits/obits_alpha/Atkinson_fdv.pdf.

encouraged to search for the missing disks through the wealth of material left behind by Derick. This body of work consisted mainly of notes, books, scraps of paper, even napkins bearing equations, many three-holed binders, 5.25 in. floppy disks, and two personal computers (an old Amstrad and an IBM PC-XT) with antiquated dot-matrix printers. At the suggestion of the family, we contacted Prof. Paul Binding (Calgary) who, sometime in the spring of 2003, kindly returned some disks he had obtained in the past and a stack of some 300 manuscript pages of what appeared to be an old, but definitely not final draft of what would lead to this work.

The problem of sewing this quilt of a book together would become the most challenging task. At our disposal we had many 5.25 in. floppy disks with files labeled *.T3V and dated anywhere between January and July 1992, as many as four manuscript "Table of Contents," and four essentially different but frequently overlapping typewritten/printed drafts of the book (with a total of about 1500 pages). After some forensic work, we managed to decipher that *.T3V indicated files that were "T3 volumes" and so had been written in a now defunct software called Scientific Word© made by TCI Software Research, a one-time precursor of the modern-day Scientific Workplace© produced by MacKichan. Fortunately for this adventure, my own father had left behind an old and still working IBM PC-XT running IBM-DOS 5.0 and equipped with dual floppy drives (one wide and one short). It was perfect. After having copied the old Atkinson files for the sake of posterity, the task was now to find a way to "read" them.

So it appeared that we should secure a working copy of Scientific Word. It came as a surprise when we realized that Derick actually kept the required software in a dusty old box on one of his basement office bookshelves. Installation was not successful at first since we needed an old monochrome display screen as well, and, in addition, the printer drivers were absent and impossible to find for the old printers we had at our disposal. Managing our way around this problem by finding an old monochrome monitor the installation then succeeded, and the ancient language was mastered anew. After reading those old DOS help files, the files were opened so that we could view the final days of the work that Atkinson left behind, unfinished, unpolished, and we suspect only some 65% of the way to a complete text. It was clear that many sections were missing and others were incomplete. Occasionally, paragraphs were cut short in the middle of a sentence, and the excitement quickly turned to a mild, though not complete, state of despair. After all, at least we had *something*, and this coupled with the hardcopies that were floating around should allow one to "complete" the book. Of course, this was all theoretical until we embarked upon the actual task of assimilating all the material into a coherent whole. In the end, we found and compiled at least four manuscripts of this book, which we name individually herewith: the *Tracing Paper* version, the *Black Binder* version, the *Daisy Wheel* version, and the *Green Binder* version.

The Tracing Paper version consists of many loose (unnumbered) leaves of tracing paper filled with mathematical equations and theorems scattered about the Atkinson Nachlass. Taking into account the fading and almost complete disintegration of the paper, the watermark, the hand-inserted symbols, and definite use of a typewriter as opposed to either an IBM selectric or PC, we would identify this document as the oldest of all; estimating its creation at or about the early 1970s, possibly even the late 1960s.

The Black Binder version is familiar to the second author as this is the version Atkinson carried with him and kept in his office in the mid-1970s when he was Derick's doctoral student. We observed that this black binder version consisted of eight chapters whose headings agreed almost exactly (save for the final chapter) with the *Contents of Volume II* on p. ix of Atkinson's already published Volume I (Academic Press, 1972). Hence this version is essentially in complete agreement with what Atkinson *expected* to submit to the publishers back in 1972, when he had agreed to publish Volume II of his *Multiparameter Eigenvalue Problems* (i.e., the book that led to the present text). However, it is now clear that he was dissatisfied with the contents of this version and so did not submit the manuscript for publication although he did release copies of parts of this volume to colleagues via documented private communications.

The next version, dubbed the Daisy Wheel version, is so-called because it was printed off a Daisy Wheel printer; it is of later date, perhaps from the early to mid-1980s when such printers were fashionable. It differs from the Black Binder version essentially in its contents; new ideas permeate the text, some sections are omitted and new ones are inserted. Yet the corresponding work in our possession is still incomplete relative to the table of contents for this specific version.

Now comes the Green Binder version (GB); we thought at first that this may have been the final version since it agreed almost verbatim with the contents of the 5.25 in. diskettes dating from January to March–April 1992 with a few files dating to July 1992 (containing mostly references). Printed off a high-quality dot matrix printer we used this version as the working version for the publication of this monograph. It consisted of 12 chapters with titles that differed from what he had expected to submit in 1972. Indeed, this was a substantially different book.

In retrospect, one thing became clear and very sad. Atkinson was definitely struggling fiercely with the completion of the book during the spring and summer of 1992, months before his collapse in August of the same year. We can surmise this because the Green Binder version makes reference to equations and sections that sometimes appear in the previous versions (Daisy Wheel, Black Binder, etc.) and this with different numbering! In addition, one gathers from the floppies that some files were saved in the early hours of the morning and this on many consecutive days. As a result, GB became filled with hundreds of misprints and typographical errors as one had to refer to the older versions constantly for what he had intended to say. Like the writers of the New Testament in Christian doctrine, we suspect that Atkinson had some older manuscripts before him when he wrote up the Green Binder version and not unlike those writers he lifted the material from his older texts and fused them into the GB. Yet, like Schubert's major symphony, GB remained "unfinished." Still, in agreement with the family's views, we believe that there must have been another (later) set of diskettes other than the ones in our possession. In some places in GB, there are lines of text that do not appear *anywhere* in the disks that we found. Thus these pages in GB must have been printed off a later (but certainly not final) set of diskettes. It is unfortunate that we, and history, will never see what Atkinson had in mind as a final manuscript. This work can only be a good approximation at best.

Now for the layout of the book: Because this book never had an actual title we decided to give it the natural name appearing on the cover as it was to be considered a continuation of Atkinson's Volume I. However, the "Volume II" was dropped by the publisher. The table of contents (ToC) used here is mainly drawn from GB version's ToC with some sections omitted as they were either incomplete or missing entirely from either the diskettes and/or the hard copy. No Preface/Introduction was ever found for any of the versions; all existing drafts consisted solely of solid mathematics with preambles to the sections being addressed. More details on the differences between this text and the documents referred to can be found in the Notes at the end of every chapter. On this subject of "Notes," we observe that Atkinson was fond of the Notes idea. Indeed, here he could use his encyclopedic knowledge of the field to be free to outline his reviews and impressions of completed work and introduce lines for future research. Unfortunately, we cannot even begin to address this issue in his style and in this manner. This tragedy leaves one with another approximation; the Notes in this text, written by the second author, were reconstructed from Atkinson's own original notes for the few chapters for which he actually had germs of notes. We have also added a section on Research Problems and have illustrated the examples using Maple©. Finally, the bibliography has been updated to the end of 2008, although we cannot make any claim as to its thoroughness. We have tried to be faithful to the first author's wishes for the inclusion of references prior to 1992. Thus references to pre-1992 documents are drawn directly from Atkinson's GB bibliography, and only a few additional ones that appeared to be relevant to the material presented are included (see the notes and the bibliography for more information related to this item).

This text would certainly have had more impact had it appeared in 1992 or 1993 when it was originally expected to appear (Atkinson had already contacted another publisher at that time). The intervening years have seen more than 80 papers in the subject of multiparameter spectral theory, a body of work that supersedes this text in breadth and generality. Still, one of the triumphs of this book lies in its exposition. It is written in the solid style typical of his texts, filled with ideas and techniques that have inspired researchers from far and wide. It can be considered a lasting tribute to his contributions to this subject over a span of more than 40 years.

The book is intended for senior-level graduate students and researchers in differential equations alike. It can, however, be read by strong senior undergraduates with an honors background, yet it may be slow-going but the rewards are high. I trust the reader will learn from this book and then contribute to the area as a result: This would have been Atkinson's wish, and it is certainly mine. Any remaining typographical and other errors are the responsibility of the second author so, if you get the chance, please write so that we can update this by means of a web-based erratum.

I am indebted to a number of people for assistance in this project: To Dusja Atkinson and her children Leslie and Vivienne for their recollection of the man that we all knew and admired and for their insights into the final stages of this book. In particular, my profound gratitude goes to Vivienne Atkinson Chisholm without whose constant encouragement over the past six years this work would likely have never been completed. For technical support, I am grateful to Antonio Rodriguez, his assistant Francisco, and to the Department of Mathematics of the University of Las Palmas in Gran Canaria for building me and providing me access to a pure MS-DOS 5.0 machine with a wide floppy disk drive so as to gain passage to critical original documentation related to the task at hand. As well, I acknowledge, with thanks, Professor J. M. Pacheco of the University of Las Palmas for an invitation as visiting professor there in the years 2006–2008, during which time this reconstruction became a reality. I also thank George Pearson of MacKichan Support, MacKichan Software, Inc., for providing critical updates of Scientific Word and T3 to TEX software bundled in its package. I must acknowledge with thanks a grant in partial support of this project from the Office of the Vice-President of Research and International at Carleton University for the years (2002–2005). I also express my deep appreciation to my partner Karen for her unfailing support through these past few years, and to my daughter Oliviana–both of whom were there caring for my younger children so that I could pursue this venture uninterrupted by the daily requirements of raising a family while overseas.

Finally, I wish to thank the following reviewers for their time: Professors Afgan Aslanov and Larry Turyn for reading over the first few chapters and providing helpful comments, suggestions, and correcting numerous typographical errors. My most profound gratitude goes to Professor Hans Volkmer who painstakingly read the entire manuscript over a period of a mere few months thereby correcting interminable errors both mathematical and grammatic.

Angelo B. Mingarelli
Sardina del Norte, Gran Canaria, Spain
amingare@math.carleton.ca

Contents

Chapter 1

Preliminaries and Early History

1.1 Main results of Sturm-Liouville theory

Motivated by problems of periodic motion in continuous media, such as the periodic flow of heat in a bar, Sturm and Liouville were led in 1836 to identify a class of problems in second-order differential equations that have inspired much of modern analysis and operator theory, and continue to do so. Their original formulation, already fairly general, called for the study of the non-trivial solutions of

$$(p(x)y')' + (\lambda w(x) - q(x))y = 0, \quad a \le x \le b, \qquad (1.1.1)$$

with homogeneous boundary conditions at $x = a$, $x = b$, of the form

$$Ay(a) + Bp(a)y'(a) = 0, \quad Cy(b) + Dp(b)y'(b) = 0, \qquad (1.1.2)$$

where A, B are not both zero, and likewise C, D. Here all quantities are real, the prime $(')$ signifies d/dx, the functions $p(x)$, $w(x)$ are positive, and the interval (a, b) finite; we discuss these restrictions in more detail shortly. In their work, one meets, perhaps for the first time, the notion of an "eigenvalue" (or characteristic value), being a value of the parameter λ for which (1.1.1) has a non-trivial solution (the "eigenfunction") satisfying the boundary conditions (1.1.2). The noteworthy propositions in these early papers include

(1) the eigenvalues are all real and form an increasing sequence

$$\lambda_0 < \lambda_1 < \lambda_2 < \ldots \tag{1.1.3}$$

without finite limit-point,

(2) if $y_n(x)$ is the eigenfunction corresponding to λ_n, then $y_n(x)$ has precisely n zeros in the open interval (a,b), (Sturm oscillation theorem)

(3) the eigenfunctions are mutually orthogonal, in the sense

$$\int_a^b y_m(x)y_n(x)w(x)\,dx = 0, \quad m \neq n \tag{1.1.4}$$

(4) the eigenfunctions form a "complete set," in the sense that an "arbitrary" function $f(x)$ from some general class can be expanded in some sense in the form

$$f(x) = \sum_{n=0}^{\infty} c_n\, y_n(x), \tag{1.1.5}$$

where the coefficients c_n can be determined with the aid of the orthogonality relations (1.1.4),

(5) incidental lemmas, such as the Sturm comparison and separation theorems, estimates for eigenvalues, etc.

1.2   General hypotheses for Sturm-Liouville theory

At the time when Sturm and Liouville wrote, many basic notions of real analysis, such as continuity and convergence and their refinements, had been little developed. Then, as now, an explicit solution in finite terms of the general second-order linear differential equations was not available; theorems on the existence of solutions, and their continuous dependence on initial data or on a parameter, were to come later. This lack of a rigorous basis for the theory was noted, for example, by Klein, and led to a series of papers by Bôcher in 1897/8. It is in the first place a matter of providing a secure or extended basis for the original results of Sturm and Liouville, and secondly of noting significant possible variations in these hypotheses that may lead to distinct results.

A common formulation, not the most general one, requires:

(i) the interval (a, b) is finite,

(ii) $p(x)$ is continuous and positive on the closed interval $[a, b]$,

(iii) $w(x)$ is continuous and positive on $[a, b]$,

(iv) $q(x)$ is continuous and real on $[a, b]$,

(v) $p'(x)$ exists and is continuous on $[a, b]$.

This last requirement makes it possible to make the identification

$$\{p(x)y'\}' = p(x)y'' + p'(x)y', \tag{1.2.1}$$

so that a "solution" is an element of $C^2[a, b]$. However, (v) can be dropped if we use the notion of a "quasi-derivative," in which y and py' are both differentiable, without y having to be twice differentiable.

Problems satisfying the above conditions are sometimes termed "non-singular," or "regular," and those that fail to satisfy one or more of them "singular." In particular, the latter term is often used if the interval (a, b) is unbounded in one or both directions, if $p(x)$ vanishes at an endpoint, or if $q(x)$ or $w(x)$ are not integrable at an endpoint. In such cases, one may speak of a singular endpoint.

In the regular case, boundary conditions such as (1.1.2) make sense. In singular cases, y and py' may have to be interpreted in a limiting sense. The matter can be settled by transformation to a regular case, when that is possible.

The classification into "regular" and "singular" needs qualification in two regards. The first is that the conditions can be considerably weakened in respect to continuity and differentiability, without impairment of the main results. The second is that changes of variables, dependent and independent, may transform a problem from singular to regular, or *vice versa*.

We achieve a broader analytic framework if we do not demand that a solution be continuously twice differentiable, but rather that y, py' be locally absolutely continuous. We postulate in this setting

(a) $p(x) > 0$ *a.e.* in (a, b) and $1/p \in L(a, b)$,

(b) $w(x) > 0$ *a.e.* in (a, b) and $w \in L(a, b)$,

(c) $q \in L(a, b)$.

It is not required that the interval (a, b) be finite.

Also of interest will be departures from these assumptions that actually affect the results. The most important of these will be that in which $w(x)$ is allowed to change sign.

1.3   Transformations of linear second-order equations

Suppressing for the moment the parameter λ, we list two special forms of the second-order equation, namely, the formally self-adjoint form

$$\{p(x)y'\}' + q(x)y = 0, \quad a < x < b, \tag{1.3.1}$$

its specialization

$$d^2y/dt^2 + h(t)y = 0, \quad c < t < d, \tag{1.3.2}$$

and the general form with middle term

$$y'' + f(x)y' + g(x)y = 0, \quad a < x < b. \tag{1.3.3}$$

In the sequel, we shall rely largely on the form (1.3.2), with the restrictions

$$-\infty < c < d < \infty, \tag{1.3.4}$$
$$h \in L(c, d). \tag{1.3.5}$$

and so we must consider the possibility of reducing the others to it.

A simple way for getting from (1.3.1) to the form (1.3.2), effective in some cases but not all, is given by taking as a new independent variable some primitive of $1/p$, as in

$$t = \int_{\xi}^{x} du/p(u) \tag{1.3.6}$$

for some $\xi \in [a, b]$. Here it must be assumed that $1/p(x)$ is almost everywhere positive and is at least locally integrable; the device is not available if $p(x)$ changes sign. We then get the form (1.3.2) with

$$h(t) = p(x)q(x), \tag{1.3.7}$$

where x is a function of t given through (1.3.6). Checking now the desiderata (1.3.4)–(1.3.5), we see that these are, respectively, equivalent to

$$1/p \in L(a, b), \tag{1.3.8}$$
$$q \in L(a, b), \tag{1.3.9}$$

in view of (1.3.6).

If $1/p(x)$ is, as before, positive almost everywhere in (a, b) and locally integrable, but not necessarily in $L(a, b)$, we can have recourse to a more powerful device. In one such case we determine a new independent variable t by

$$\tan t = \int_\xi^x du/p(u), \tag{1.3.10}$$

for some $\xi \in (a, b)$, so that now

$$c = \arctan \int_\xi^a dx/p(x), \quad d = \arctan \int_\xi^b dx/p(x) \tag{1.3.11}$$

and $[c, d]$ is in all cases finite, being a sub-interval of $[-\pi/2, \pi/2]$, so that (1.3.4) is ensured. We introduce also the new dependent variable

$$z = y \cos t \tag{1.3.12}$$

and find after some calculation that z satisfies the equation

$$d^2z/dt^2 + \{1 + p(x)q(x) \sec^4 t\} z = 0. \tag{1.3.13}$$

Since (1.3.4) is assured, it remains to consider (1.3.5), or in this case whether $\{1 + p(x)q(x) \sec^4 t\} \in L(c, d)$, as a function of t. This may be reduced to

$$\int_a^b |q(x)|\{1 + (\int_\xi^x du/p(u))^2\} dx < \infty. \tag{1.3.14}$$

For a similar criterion, we refer to (Kaper, Kwong, and Zettl (1984)).

To bring the general linear form (1.3.3) to the form (1.3.2), we can begin with a transformation to the formally self-adjoint form (1.3.1), for example, by multiplication by $\exp\{\int f \, dx\}$. We then get the form (1.3.2) on taking a new independent variable of the form

$$t = \int^x \exp\{-\int^v f(u)du\}dv, \tag{1.3.15}$$

where the lower limit of integration is arbitrary but fixed. More generally, one may start by replacing the dependent variable in (1.3.3) by some $z = y\eta(x)$, where η is at our disposal; a standard choice is

$$z = y \exp(\int^x f(u)du/2). \tag{1.3.16}$$

1.4   Regularization in an algebraic case

The problem of regularization may be partly clarified by discussion of a special case. We cite the equation, a modification of the hypergeometric equation

$$y'' + \{\gamma/x - \delta/(1-x)\}y' + \lambda y/(x - x^2) = 0, \qquad (1.4.1)$$

with real γ, δ, taken over the interval $0 < x < 1$. To get the formally self-adjoint form, we can proceed as suggested at the end of §1.3, multiplying by

$$p(x) := x^\gamma (1-x)^\delta \qquad (1.4.2)$$

to get

$$\{p(x)y'\}' + \lambda w(x)y = 0, \quad 0 < x < 1, \qquad (1.4.3)$$

where

$$w(x) = x^{\gamma-1}(1-x)^{\delta-1}. \qquad (1.4.4)$$

The transformation $dt = dx/p(x)$ yields a finite t-interval if

$$\gamma < 1, \delta < 1, \qquad (1.4.5)$$

while the requirement $w \in L(0,1)$ will be satisfied if

$$\gamma > 0, \delta > 0. \qquad (1.4.6)$$

Thus the equation is regularizable by this method if $\gamma \in (0,1)$, $\delta \in (0,1)$.

The more general method of (1.3.10)–(1.3.13) allows some relaxation in (1.4.5). Assuming (1.4.6), the requirement (1.3.14) reduces to

$$x^{\gamma-1}(1-x)^{\delta-1} \cdot |\int_c^x dt/p(t)|^2 \in L(0,1)$$

for some $c \in (0,1)$. Since $\int_c^x dt/p(t) = O\{x^{1-\gamma} + (1-x)^{1-\delta}\}$, the requirement (1.3.14) is satisfied if $\gamma < 2$, $\delta < 2$.

1.5   The generalized Lamé equation

We go now to the early history of multiparameter spectral theory and its beginnings in the work of Klein and others in the 1880s. This has a rather different flavor to the original work of Sturm and Liouville, sketched in §1.1, being concerned with differential equations of the general form

$$y'' + \left\{ \sum \gamma_j/(x - e_j) \right\} y' + q(x)y/\Pi(x - e_j) = 0. \qquad (1.5.1)$$

Here the e_j are various real numbers with $e_1 < e_2 < \ldots < e_p$, the γ_j are also real and non-zero, and $q(x)$ is a polynomial of degree $\leq p - 2$. This framework encompasses the equation (1.4.1) if $p = 2$, that of Lamé if $p = 3$, and Heun's equation if $p = 4$. Of special interest is the Lamé case, where all the $\gamma_j = 1/2$ when a wide variety of the special functions of mathematical physics are covered (Whittaker and Watson, §10.6, and Chapter XXIII.)

An eigenvalue problem results if we treat one of the coefficients in $q(x)$ as a disposable or spectral parameter, to be chosen so that (1.5.1) should have a non-trivial solution that vanishes, or satisfies some other boundary condition at the ends of some real interval. A multiparameter eigenvalue problem results if we treat $k \leq p - 1$ of the coefficients of q as spectral parameters, so that for each of k intervals (a_r, b_r), $r = 1, \ldots, k$, there should be a non-trivial solution satisfying similar conditions at the endpoints of several intervals. Problems of this nature, k boundary-value problems, each involving the same k parameters seem first to have been discussed by Klein in 1881, in the case $p = 3$, with the $\gamma_j = 1/2$, in connection with the boundary problem of potential theory for a region bounded by confocal quadric surfaces to be discussed shortly.

A boundary value problem for (1.5.1) over an interval (a, b) can be seen as regular or singular, in respect of possible involvement of the singularities at the points e_j, or of the points $-\infty$, ∞. If the interval $[a, b]$ is finite, and does not contain any of the e_j, it will be regular, and the standard Sturmian boundary conditions can be used. However, other cases have interest, particularly that in which (a, b) coincides with one of the intervals (e_j, e_{j+1}); in such a case, we can ask whether regularization is possible. Sturmian type boundary conditions for the regularized version can then be transported back so as to complete the original version of the problem.

In an alternative approach to the question of boundary conditions in singular cases, we use Frobenius theory and characteristic exponents to devise suitable boundary conditions. This was the method used in the period under discussion. Thus with the singularity e_j of (1.5.1) we associate the roots 0, $1 - \gamma_j$ of the indicial equation. Then, at least if γ_j is not an integer, there will be a pair of solutions, one analytic at this point, and another behaving as $(x - e_j)^{1-\gamma_j}$.

Thus in the case (1.5.1) with $p = 3$, so that $q(x)$ is linear, with two coefficients or "accessory parameters," one asks whether they can be chosen so that the solution is either regular at the e_j or behaves as $\sqrt{(x - e_j)}$. More precisely, the desired solution should be either a polynomial, or a polynomial times one or more of the $\sqrt{(x - e_j)}$, $j = 1, 2, 3$; such e_j would of course be branch-points of the solution when considered in the complex plane. "Klein's oscillation theorem" would assert that such a problem has a series of solutions, one for each way of assigning zeros to the two intervals (e_1, e_2) and (e_2, e_3). This topic and similar questions were taken up by Klein in a series of papers, often couched in somewhat different terms, starting with the paper (Klein 1881a).

1.6   Klein's problem of the ellipsoidal shell

Klein's original memoir contains what is perhaps the first multiparameter Sturm-Liouville problem of a general nature. He considered the solution by separation of variables of the Dirichlet problem for Laplace's equation in an ellipsoidal shell. In view of its historical importance, we sketch the details. Starting with his definition of ellipsoidal coordinates, we set up the equation

$$x^2/\lambda + y^2/(\lambda - k^2) + z^2/(\lambda - 1) = 1, \qquad (1.6.1)$$

where $0 < k^2 < 1$. Considered as a cubic equation in λ for fixed real x, y, z not all zero, this has three real roots μ, ν, ρ, with

$$0 < \mu < k^2, \ k^2 < \nu < 1, \ 1 < \rho < +\infty. \qquad (1.6.2)$$

For fixed $\lambda = \mu \in (0, k^2)$, (1.6.1) determines a hyperboloid of two sheets, for fixed $\lambda = \nu \in (k^2, 1)$ a hyperboloid of one sheet, and for $\lambda = \rho > 1$ an ellipsoid. A shell is then defined by (μ, ν, ρ) regions,

$$0 < \mu_1 \le \mu \le \mu_2 < k^2, \ k^2 < \nu_1 \le \nu \le \nu_2 < 1, \ 1 < \rho_1 \le \rho \le \rho_2 < +\infty, \qquad (1.6.3)$$

together with sign-restrictions on x, y, z. Klein's problem calls for a solution of Laplace's equation in the interior of this shell, taking assigned values on its six faces.

A preliminary step is to break up the problem into three sub-problems, in which we ask for a solution vanishing on two pairs of opposite faces, for example, given by

$$\mu = \mu_1, \mu_2, \ \nu = \nu_1, \nu_2, \qquad (1.6.4)$$

and taking the assigned values on the faces $\rho = \rho_1$, $\rho = \rho_2$; this sub-problem can be further broken down into the cases where the solution vanishes on one or the other of the ρ-faces and takes the required values on the other. The case of a solution vanishing on the faces (1.6.4) is amenable to the method of separation of variables.

The coordinates can be transformed by the map

$$t = \int_0^\lambda d\lambda / \sqrt{\{\lambda(\lambda - k^2)(\lambda - 1)\}}$$

to yield alternative coordinates u, v, w, in which Laplace's equation takes the form

$$(\nu - \rho)\partial^2 \Phi / \partial u^2 + (\rho - \mu)\partial^2 \Phi / \partial v^2 + (\mu - \nu)\partial^2 \Phi / \partial w^2 = 0.$$

Separation of variables is possible and we find that it is satisfied by a product $\Phi(u, v, w) = E_1(u)E_2(v)E_3(w)$, where the E_j are solutions of the equation

$$d^2 E / dt^2 = (A\lambda + B)E. \tag{1.6.5}$$

Equivalently, in terms of the coordinates μ, ν, ρ, we have a solution

$$F_1(\mu)F_2(\nu)F_3(\rho),$$

where the F_j are solutions of

$$\begin{aligned} F'' + (1/2)\{1/x + 1/(x - k^2) + 1/(x - 1)\}F' \\ + F(Ax + B)/\{x(x - k^2)(x - 1)\} = 0, \end{aligned} \tag{1.6.6}$$

which is a form of Lamé's equation (see Whittaker and Watson, Chapter XXIII). One then observes that this equation is regular in the intervals $(0, k^2)$, $(k^2, 1)$. The problem for the case (1.6.4) calls for A and B to be determined so that (1.6.6) has non-trivial solutions F_1, F_2 over the intervals (μ_1, μ_2), (ν_1, ν_2), respectively, vanishing at the endpoints. In this original case of "Klein's oscillation theorem," it is claimed that a unique pair of values A, B exists, so that F_1, F_2 have any assigned numbers of zeros in their respective intervals.

1.7 The theorem of Heine and Stieltjes

Extending the earlier work of Heine (1881) on Lamé-type equations Stieltjes (1885) proved one of the earliest and most elegant results in multiparameter theory. It relates to the equation

$$A(x)y'' + 2B(x)y' + C(x)y = 0, \tag{1.7.1}$$

where $A(x)$, $B(x)$ are given real polynomials, of degrees $p + 1$ and p, and $C(x)$ is an unspecified polynomial of degree $\leq p - 1$. The problem is to find the possible polynomials $C(x)$ so that (1.7.1) has a non-trivial polynomial solution. Supplementary assumptions are

(i) The coefficients of the highest powers in A, B are positive

(ii) The zeros of $A(x)$, $B(x)$ are real and distinct, and separate one another.

The theorem is then that the solution-polynomials $C(x)$, $y(x)$ exist and are characterized uniquely by any assigned distribution of the zeros of $y(x)$ between the zeros of $A(x)$. This result is discussed by Szegö (1939), who notes that Heine considered the existence and the number of solutions $C(x)$ with y of preassigned degree, while Stieltjes (1885) completed the full result just stated. Szegö himself gives a proof related to that of Stieltjes, involving the minimization of a certain logarithmic potential.

The problem is equivalent to that of finding solutions of (1.5.1), which are polynomials of degree n, where as before $e_1 < \ldots < e_p$, $q(x)$ is an unspecified polynomial of degree $p - 2$, and

$$\gamma_j > 0, \quad j = 1, 2, \ldots, p. \tag{1.7.2}$$

1.8   The later work of Klein and others

Extensions of Klein's work on the ellipsoidal shell, that is to say, boundary-value problems that stay clear of singularities, seem due in the first place to Bôcher, in a series of three papers (Bôcher 1898–1900), in which he proves the natural extension to k-parameters and k intervals, using the method of induction over k. This is reproduced in Ince (1954, Chapter 10). Bôcher also treats the case of an endpoint at a singularity, but restricts the γ_j to the range $(0, 1/2]$.

The question of the appropriate range of values for the γ_j in (1.5.1), if the e_j are to be the endpoints for boundary-value problems, remains of interest. The range $(0, 1)$, and of course the Lamé value $1/2$, permit transformation to a regular problem. In the Stieltjes result of §1.7, there is no upper limit. The question of oscillation theorems "beyond the Stieltjes limits" occurs in several places in Klein's writings (e.g., Ges.Werke, Bd.2, p. 597). A paper on related one-parameter problems by Haupt and Hilb (1923) should deserve more study.

It should be mentioned that Klein's interests leaned more to the conformal mapping properties of the ratios of two solutions, considered in the upper half-plane, with the singularities being branch points. These ratios describe "Kreis-bogenpolygone" on the Riemann sphere; again one can pose a sort of spectral problem, requiring the parameters to be such that this circular polygon should have special metric features. Among Klein's interest was the connection with the "fundamental theorem of automorphic functions."

1.9 The Carmichael program

By the early 1920s it had become clear that the scope of multiparameter theory extended beyond differential equations of generalized Lamé type, or even differential equations altogether. Perhaps the first recognition of this appears in the work of Anna J. Pell (1922) on linked integral equations. One can go beyond this and consider linked equations of rather arbitrary different kinds. This was proposed by R. D. Carmichael, who has three papers (1921a, 1921b, 1922), skimming over the whole range of eigenvalue problems of various types, including algebraic, differential, integral, and difference equations; these offer many ideas for research and display many intriguing formulae. In particular, he calls attention to the most basic case, that of linked matrix problems, and gives it some preliminary treatment.

This algebraic problem deserves interest on its own account, though it can also be studied for its suggestive value for other problems and, as Carmichael noted, for its possible use as a basis for approximation. As it had been treated elsewhere, we mention here only a few details. It is a question of an array of square matrices

$$A_{rs}, \quad r = 1, \ldots, k, \; s = 0, \ldots, k, \tag{1.9.1}$$

in which $A_{r0}, \ldots, A_{rk}$ are of the same order n_r. The ordinary notion of an eigenvalue may then be adapted provisionally to mean a set of k scalars such that the k equations, with y_r an n_r-column vector,

$$\left\{ A_{r0} + \sum_1^k \lambda_s A_{rs} \right\} y_r = 0, \quad r = 1, \ldots, k, \tag{1.9.2}$$

all have a non-zero solution. Equivalently, such an eigenvalue is a common zero of the polynomials

$$\det \left\{ A_{r0} + \sum_1^k \lambda_s A_{rs} \right\}, \quad r = 1, \ldots, k. \tag{1.9.3}$$

Since these polynomials are in general of degrees $n_1, \ldots, n_k$, the number of such common zeros ought to be, again in general, the product $\prod n_r$ of these degrees. If we think of the matrix in the r-th row of (1.9.1) as acting on a space G_r, of dimension n_r, then $\prod n_r$ is also the dimension of the Kronecker or tensor product

$$G = G_1 \otimes \cdots \otimes G_k. \tag{1.9.4}$$

It is thus reasonable to anticipate that the eigenvectors, Kronecker products of the y_r, might span this space. Carmichael introduced the dual equations to (1.9.2) and set up a bi-orthogonality with a view to the eigenvector expansion.

There are some difficulties here. The existence of a complete set of eigenvectors is not quite trivial even in the self-adjoint or hermitian case, while in the non-self-adjoint case there is the possible complication of the presence of root-vectors in addition to eigenvectors. Carmichael (1921a, p. 77) however considered it unnecessary at that point to develop the theory for exceptional cases in order to benefit from the heuristic value of this algebraic case.

There seems to have been only limited follow-up to Carmichael's papers; possibly readers concluded that the problems mentioned had been essentially solved. Nevertheless, firm results for the algebraic case are needed in applications, particularly if this case is to be used for approximations. Reference may be made to Atkinson (1964a, 1972), and, for recent results on the hermitian case, to the book by Volkmer (1988). While we do not rely greatly on this topic in what follows, the completeness of eigenvectors in the hermitian case does provide one way of considering completeness for multiparameter difference equations, and also in turn one way of attacking the completeness problem in the differential case. We proceed to list some typical problems in the multiparameter area.

Notes for Chapter 1

We start with some historical remarks dealing with the intended layout of the text: As pointed out in the Preface the Atkinson manuscript at our disposal for the preparation of this book is certainly not the final one as conceived by Atkinson even though the existence of such a final manuscript has been ascertained by his family. Unfortunately, a complete thorough review of his *Nachlass* failed to turn up this final version. Judging from the dates on the 5.25-inch floppy disks used by Atkinson for storage purposes we can estimate the date of the manuscript under review to January 1992, during Atkinson's 76th year. The final version should have been dated sometime in July 1992, just prior to the onset of his final illness which, unfortunately, made it impossible for him to do any more mathematical work.

The manuscript drawn from Atkinson's *Nachlass* had indentations for additional subsections in this first chapter; material with the headings *Mathieu's equation, The spheroidal problem* along with some additional references (e.g., the author name "Schäfke" by itself) not specifically linked to any particular subheading or subject but perhaps an afterthought. In any case, neither these manuscript pages nor copies and/or disks containing this information have ever been found, hence any possible guess as to the presentation of this material is left out. The same applies to some of the subsequent chapters, and this will be addressed as the book unfolds.

Section 1.7 is missing from the files under review although a marginal note in the table of contents in Atkinson's hand notes that he had forgotten to include it in the manuscript. The "spheroidal problem" alluded to above may have been slotted for this location, but this is only a guess. Thus, we have renumbered the sections accordingly. This spheroidal problem had been considered earlier by Lamé (1854) and independently by Thomson (1863). (See also Thomson and Tait 1873). A complete exposition of the results in §1.6 may be found in Whittaker and Watson (1927, Chapter XXIII, §23.3).

Section 1.8 as presented here is followed, in the original manuscripts, by a section entitled "General multiparameter problems" equivalent in content to the present §1.9. The section "Heading to the present" §1.9 is the same as the heading used in another (undated) version of §1.9 only to be found in manuscripts but not in digital form.

The transformation (1.3.16) normally requires f to be differentiable, in which case (1.3.3) becomes

$$z'' + h(x)z = 0$$

with $h(x) = g(x) - f(x)^2/4 - f'(x)/2$.

At the end of §1.7 Atkinson added the references "We refer to the papers and discussion in Klein's collected works" and the partly cryptic references, "... and to the papers of Hilb, Flackenberg [sic], ... (ref. Thomson and Tait, Theoretical Physics, Natural Philosophy 1867, 1879 Phil Trans. 1863, Lamé ??, W-W chapters??)" some of which we have deciphered and added in the references at the end. We note that Hilb's interest in this general area was, no doubt, due to Klein's influence in Munich (where Klein had taught until 1880). Hilb's 1903 dissertation in Munich was entitled *Beiträge zur Theorie der Laméschen Funktionen*. It was directed by C. L. F. Lindemann, who had been a student of Klein's. Those papers by Hilb (some co-authored by Otto Haupt of nondefinite Sturm-Liouville theory fame) dealing with Lamé functions and Klein's program in this vein can also be found in the references.

The exposition in §1.9 can be used for motivational purposes to complement Chapter 6 of the companion text, Atkinson (1972). For a survey of the case of a single Sturm-Liouville equation with two parameters, see Binding and Volkmer (1996).

1.10   Research problems and open questions

1. It is well-known that (1.3.3) may be reduced to (1.3.2) by means of a normalization and a change of independent variable (Ince [1954]). For example, if $f \in L(a, b)$, then, (1.3.3) may be cast in the form of (1.3.1) with $p(x) \equiv F(x)$,

$q(x) \equiv g(x)F(x)$, where $F(x) = \exp\{\int^x f(s)\,ds\}$. Since $F(x) > 0$, (1.3.6) may be used to bring this to the Liouville normal form (1.3.2).

Consider now (1.3.3) as a matrix system of differential equations, that is, y is a (real) square matrix of unknown functions, f, g are given matrices of the same size as y and $f, g \in L(a, b)$ (in the sense that all their entries share this property). One may even assume with a slight loss of generality that the coefficient matrices f, g have smooth entries. *Does there exist a suitable transformation of variables that reduces the matrix system (1.3.3) to the form (1.3.1) or even (1.3.2)?*

Remark Observe that the "exponential device" referred to in the opening paragraph is unavailable in this setting unless there are major (and undesirable) commutativity assumptions on the values of the matrices f, g. One way of circumventing this difficulty consists in rewriting the *vector system* (1.3.1) as a *canonical system* (so now $y \in \mathbb{R}^n$ is a column vector, say)

$$u' = JB(t)u,$$

where $u = \mathrm{col}\,\{y, py'\} \in \mathbb{R}^{2n}$,

$$J = \begin{bmatrix} 0 & I_n \\ -I_n & 0 \end{bmatrix},$$

and

$$B(t) = \begin{bmatrix} Q & 0 \\ 0 & P^{-1} \end{bmatrix}.$$

See Chapter 9 in Atkinson (1964a) for more information on such systems and Chapter 11 in Hartman (2002), or Reid (1971, 1980) for details on the connection between vector and matrix systems of the form (1.3.1).

2. Another line of research would be to consider the results of this chapter (e.g., transformation theory, two-parameter linear problems) in the more general setting of *dynamic equations on time scales*. The study of two-parameter problems on time-scales would be a noteworthy contribution to the understanding of the relationship between discrete and continuous boundary problems. This approach would be in the same spirit as the original ideas in Atkinson (1964a) then taken up by Mingarelli (1983) that is, to find a framework that could be used to unify discrete and continuous problems, which are differential and difference equations simultaneously. For a survey of this relatively new field, see Agarwal *et al.* (2002).

3. As mentioned in (2), a different viewpoint at unification of discrete and continuous problems was undertaken by Atkinson (1964a) and others using Volterra-Stieltjes integral equations, a topic that dates back to the Russian school of Kreĭn with *Stieltjes strings* being the main thrust. In the West, this

study became abstracted through the work of the probabilist W. Feller (1950s) (cf., *Feller derivatives*, see Mingarelli [1983] for further references).

There appears to be no systematic study whatsoever of even two parameter boundary problems for either a Stieltjes string or for dynamic equations on time scales.

An obvious advantage would be the immediate unification of the continuous and discrete second-order, two-parameter case without having to resort to more specialized discrete settings. Thus, for example, it would not be out of the question to see the pioneering work of Billigheimer (1970) in the study of boundary problems for five-term recurrence relations (or fourth-order difference equations) as being a consequence of such a general theory. Indeed, one may even contend that this book in its entirety may become a special case of such a *grand unified theory* since as the reader will note that different techniques have to be used to handle the discrete and continuous multiparameter cases, especially when treating expansion problems.

Chapter 2

Some Typical Multiparameter Problems

2.1 The Sturm-Liouville case

We now present formally the special case of primary interest, the multiparameter version of classical Sturm-Liouville, in its simplest and most classical form. We suppose given k real and finite intervals

$$I_r = [a_r, b_r], \quad r = 1, 2, \ldots, k, \tag{2.1.1}$$

and real functions satisfying the usual continuity requirements

$$p_{rs}(x_r) \in C[a_r, b_r], \quad 1 \le r, s \le k, \tag{2.1.2}$$
$$q_r(x_r) \in C[a_r, b_r], \quad 1 \le r \le k. \tag{2.1.3}$$

In addition, we need $2k$ real numbers α_r, β_r, $r = 1, \ldots k$, to specify the boundary conditions; without loss of generality, we make the usual restriction that

$$0 \le \alpha_r < \pi, \quad 0 < \beta_r \le \pi, \ r = 1, \ldots, k.$$

We ask for non-trivial solutions of the k differential equations

$$y_r{}''(x_r) - q_r(x_r)y_r(x_r) + \sum_1^k \lambda_s p_{rs}(x_r) \, y_r(x_r) = 0, \tag{2.1.4}$$

17

$a_r \leq x_r \leq b_r$, $r = 1, \ldots k$, which satisfy the respective boundary conditions

$$y_r(a_r)\cos \alpha_r = y_r'(a_r)\sin \alpha_r, \ r = 1, \ldots, k, \qquad (2.1.5)$$

$$y_r(b_r)\cos \beta_r = y_r'(b_r)\sin \beta_r, \ r = 1, \ldots, k. \qquad (2.1.6)$$

We have here taken advantage of the remarks of §1.2, in using the simple form y'' for the second-order term in (2.1.4). Naturally, we are interested only in the situation that none of the $y_r(x_r)$ vanishes identically. A set of values of $\lambda_1, \ldots, \lambda_k$ for which this is possible will be termed an "eigenvalue." Eigenvalues are thus again k–tuples of scalars. An eigenvalue is thus an inhomogeneous object.

An alternative concept is to consider a $(k+1)$–tuple $\lambda_0, \ldots, \lambda_k$ of scalars, not all zero, an eigenvalue if the equations

$$\lambda_0\{y_r''(x_r) - q_r(x_r)y_r(x_r)\} + \sum_{s=1}^{k}\lambda_s p_{rs}(x_r)\, y_r(x_r) = 0, \qquad (2.1.7)$$

$r = 1, \ldots, k$, likewise all have non-trivial solutions satisfying the boundary conditions. This may be termed an eigenvalue in the homogeneous sense. For the present, we use the previous inhomogeneous sense.

The term "eigenfunction" is used for the product

$$y = y(x_1, \ldots, x_k) = \prod_{r=1}^{k} y_r(x_r), \qquad (2.1.8)$$

where the $y_r(x_r)$ are solutions of (2.1.4)–(2.1.6). We then have

$$\{\partial^2/\partial x_r^2 - q_r(x_r)\}y + \sum_{s=1}^{k}p_{rs}(x_r)\lambda_s y = 0, \quad r = 1, \ldots, k, \qquad (2.1.9)$$

and the boundary conditions

$$y \cos \alpha_r = (\partial y/\partial x_r)\sin \alpha_r, \ x_r = a_r, r = 1, \ldots, k, \qquad (2.1.10)$$
$$y \cos \beta_r = (\partial y/\partial x_r)\sin \beta_r, \ x_r = b_r, r = 1, \ldots, k. \qquad (2.1.11)$$

It will appear later that certain sign restrictions, which we term "definiteness," must be imposed, if we are to obtain results similar to those for the standard one-parameter case. These will be introduced in the next two chapters; they correspond to restrictions usual in the one-parameter case.

In the absence of these definiteness conditions, we can only say that the problem is at least formally well posed, in that the eigenvalues $(\lambda_1, \ldots, \lambda_k)$ are the roots of k simultaneous equations. We denote by $y_r(x_r \ ; \lambda_1, \ldots, \lambda_k)$ the solution of the r-th equation (2.1.4) such that

$$y_r(a_r; \lambda_1, \ldots, \lambda_k) = \sin \alpha_r, \quad y_r'(a_r; \lambda_1, \ldots, \lambda_k) = \cos \alpha_r, \qquad (2.1.12)$$

so that (2.1.5) is automatically satisfied. Then for (2.1.6) we need that

$$y_r(b_r; \lambda_1, \ldots, \lambda_k) \cos \beta_r - y_r{}'(b_r; \lambda_1, \ldots, \lambda_k) \sin \beta_r = 0, \ r = 1, \ldots, k. \quad (2.1.13)$$

The eigenvalues are thus the common zeros of k entire functions of the k variables $\lambda_1, \ldots, \lambda_k$ (cf. [2.1.4]). Of course, we cannot without further assumptions say much about the existence or distribution of such common zeros. We take this up in Chapters 5–8, using oscillatory arguments.

The set of eigenvalues is naturally termed the "spectrum" of the problem (2.1.4)–(2.1.6) (at least in this case, although we point out that, strictly speaking, the spectrum can generally contains other types of points, but these do not appear in this context).

2.2   The diagonal and triangular cases

In the general case of (2.1.4)–(2.1.6), the values of $\lambda_1, \ldots, \lambda_k$ occurring in the spectrum are inseparably linked; we'll see later that they do not run through sequences of values independently. However, there is one case in which this does happen. Although this case is degenerate from the point of view of the general multiparameter theory, we note it for its illustrative value, to say nothing of its importance in applications. In this and the next section, we may assume that the coefficients (2.1.2)–(2.1.3) satisfy the minimum integrability requirements, $p_{rs}(x_r), q_r(x_r) \in L(a_r, b_r), 1 \le r, s \le k$.

The case is that in which each of the λ_s occurs, with a non-zero coefficient, in just one of (2.1.4). Renumbering the parameters or the equations, if necessary, we may suppose that

$$p_{rs}(x_r) = 0, \quad r \neq s, \quad\quad\quad\quad\quad (2.2.1)$$

We may term this the "diagonal" case, since the matrix formed by the $p_{rs}(x_r)$ is then diagonal. The situation is then that we have k separate one-parameter Sturm-Liouville problems, without parameter linkage. The spectrum of the multiparameter problem is given by letting each λ_s run independently through the corresponding set of eigenvalues.

A slightly less degenerate situation is likewise of great importance in applications. We term this the "triangular" case, for its resemblance to the inversion of a triangular matrix. We suppose that, after any necessary re-numbering, we have

$$p_{rs}(x_r) = 0, \quad r < s. \quad\quad\quad\quad\quad (2.2.2)$$

Then (2.1.4) with $r = 1$ takes the form

$$y_1'' - q_1 y_1 + \lambda_1 p_{11} y_1 = 0, \quad a_1 \le x_1 \le b_1, \qquad (2.2.3)$$

and this, together with (2.1.5)–(2.1.6) with $r = 1$, is a standard one-parameter Sturm-Liouville problem. This gives an admissible sequence of values for λ_1, at least if $p_{11} > 0$. We then pass to (2.1.4) with $r = 2$, and so to

$$y_2'' - q_2 y_2 + (\lambda_1 p_{21} + \lambda_2 p_{22}) y_2 = 0, \quad a_2 \le x_2 \le b_2. \qquad (2.2.4)$$

The procedure is then that for each λ_1 admissible in (2.2.3), we substitute in (2.2.4) and obtain together with (2.1.5)–(2.1.6) with $r = 2$ an eigenvalue problem for λ_2. We then take admissible pairs (λ_1, λ_2) so obtained and use them in (2.1.4) with $r = 3$, to yield an eigenvalue problem for λ_3, and so on. Thus, at any rate so far as the spectrum is concerned, the multiparameter problem reduces to a sequence of one-parameter problems. In general, of course, such a reduction does not take place.

These reductions, when possible, do not entirely dispose of the problems of such degenerate cases. It is still of interest to study the completeness of the eigenfunctions obtained in this way.

2.3   Transformations of the parameters

It is often convenient to make hypotheses concerning the signs of functions, parameters or operators appearing in a multiparameter problem; this raises the problem of when such sign-hypotheses can be brought into being by a simple transformation. We take first the case of a translation or change of origin in the parameter-space.

For any real numbers $\mu_1, \ldots, \mu_k$, we may introduce new spectral parameters $\lambda_s^\dagger$ by

$$\lambda_s^\dagger = \lambda_s + \mu_s, \quad s = 1, \ldots, k. \qquad (2.3.1)$$

Substitution in (2.1.4) yields the equation

$$y_r''(x_r) - q_r^\dagger(x_r)\, y_r(x_r) + \sum_{s=1}^{k} \lambda_s^\dagger p_{rs}(x_r)\, y_r(x_r) = 0, \qquad (2.3.2)$$

where

$$q_r^\dagger(x_r) = q_r(x_r) + \sum_{s=1}^{k} \mu_s\, p_{rs}(x_r). \qquad (2.3.3)$$

and as before $a_r \leq x_r \leq b_r$, $r = 1, \ldots, k$. The boundary conditions are unaffected since, in this case at least, they do not involve the spectral parameters. While not altering the problem in any essential way, such a transformation may serve the purpose of arranging that the eigenvalues or the potentials $q_r(x_r)$ have special sign-properties. This device is sometimes used in the one-parameter case to remove zero as an eigenvalue.

Consider next the effect of a homogeneous linear transformation, in which new parameters $\lambda_s^\dagger$ are introduced by the relations

$$\lambda_s = \sum_{t=1}^{k} \lambda_t^\dagger \gamma_{st}, \tag{2.3.4}$$

where (γ_{st}) is a non-singular k-by-k matrix of constants. Inserting this into the differential equation, we have

$$\sum_{s=1}^{k} p_{rs}(x_r) \lambda_s = \sum_{t=1}^{k} p_{rt}^\dagger(x_r) \lambda_t^\dagger, \tag{2.3.5}$$

where

$$p_{rt}^\dagger(x_r) = \sum_{s=1}^{k} p_{rs}(x_r) \gamma_{st}. \tag{2.3.6}$$

The linear transformation of parameters thus induces a dual transformation of the coefficient functions. Again, it may be possible to choose the γ_{st} so as to give some special form or sign-properties to the $p_{rt}^\dagger(x_r)$. We take this up in Chapter 4.

2.4   Finite difference equations

With the y_r defined as in (2.1.4)–(2.1.12), for an unbounded sequence of integers $m \to \infty$ we form finite-difference approximations to these functions, denoted

$$z_r(x_r) = z_r^{(m)}(x_r), \quad a_r \leq x_r \leq b_r, \ r = 1, \ldots, k \tag{2.4.1}$$

the superscript $^{(m)}$ will often be omitted. To form these we first divide the intervals $[a_r, b_r]$ into m equal parts, of lengths

$$h = h_r^{(m)} = (b_r - a_r)/m, \tag{2.4.2}$$

with nodes $x_{rj} = x_{rj}{}^{(m)}$ such that

$$a_r = x_{r0} < \ldots < x_{rm} = b_r. \tag{2.4.3}$$

For simplicity, we take m to be independent of r.

As in the one-parameter case, the z_r are initially defined at these nodes x_{rj}, starting with the initial data

$$z_r(x_{r0}) = \sin \alpha_r, \quad z_r(x_{r1}) = \sin \alpha_r + h_r \cos \alpha_r. \tag{2.4.4}$$

We then continue the definition at the nodes with the aid of recurrence relations

$$h_r{}^{-2} \Delta^2 z_r(x_{r,j-1}) + z_r(x_{rj}) \left\{ \sum_{s=1}^{k} \lambda_s \, p_{rs}(x_{rj}) - q_r(x_{rj}) \right\} = 0, \tag{2.4.5}$$

for $j = 1, \ldots, m-1$; here Δ is the forward difference operator, $\Delta z_r(x_{r,j-1}) = z_r(x_{rj}) - z_r(x_{r,j-1})$ and as usual $\Delta^2 z_r(x_{r,j-1}) = \Delta(\Delta z_r(x_{r,j-1}))$.

Finally, we complete the definition of z_r between the nodes by linear interpolation; z_r will be continuous and piecewise linear, $z_r{}'$ being constant except at the nodes.

The functional dependence is expressed in full by

$$z_r{}^{(m)}(x_r) = z_r{}^{(m)}(x_r; \lambda_1, \ldots, \lambda_k).$$

At the node x_{rj}, this will be a polynomial in $\lambda_1, \ldots, \lambda_k$ of degree at most $j-1$. Corresponding to the boundary condition at b_r we impose the condition

$$z_r(x_{rm}) \cos \beta_r - h_r{}^{-1} \{ z_r(x_{rm}) - z_r(x_{r,m-1}) \} \sin \beta_r = 0, \quad r = 1, \ldots, k. \tag{2.4.6}$$

Here the left is a polynomial in the λ_s, of degree at most $m-1$. The eigenvalues, k-tuples $\lambda_1, \ldots, \lambda_k$, will be common zeros of the equations (2.4.6). We denote them by $\mu_n{}^{(m)} = \mu_{1n}{}^{(m)}, \ldots, \mu_{kn}{}^{(m)}$, where n runs through some index set. We have an eigenfunction orthogonality in the sense

$$\sum_{s=1}^{k} \sum_{j=0}^{m} (\mu_{sn} - \mu_{sn}{}') p_{rs}(x_{rj}) z_r(x_{rj}, \mu_n) z_r(x_{rj}, \mu_n{}') = 0, \quad r = 1, \ldots, k,$$

and so, if the eigenvalues are distinct, we have

$$\det \sum_{j=0}^{m} p_{rs}(x_{rj}) z_r(x_{rj}, \mu_n) z_r(x_{rj}, \mu_n{}') = 0, \tag{2.4.7}$$

and so (cf. [4.2.4])

$$\sum_{u_1=0}^{m} \cdots \sum_{u_k=0}^{m} \det p_{rs}(x_{ru_r}) \prod_{r=1}^{k} z_r(x_{ru_r}, \mu_n) \prod_{r=1}^{k} z_r(x_{ru_r}, \mu_n{}') = 0. \tag{2.4.8}$$

These constitute orthogonality relations between the eigenfunctions, which are now products

$$\prod_{r=1}^{k} z_r(x_{ru_r}, \mu_n) \qquad (2.4.9)$$

defined on the k-dimensional grid

$$(x_{1u_1}, \ldots, x_{ku_k}), \quad u_r = 0, \ldots, m, \ r = 1, \ldots, k. \qquad (2.4.10)$$

2.5 Mixed column arrays

We give now a brief overview of the formalities associated with systems of equations in which the coefficients are linear operators, ranging from the most basic case in which these operators are representable by square matrices of finite order to the situation of main interest here, where they will be ordinary linear differential operators. The general theory of such systems follows to some extent the theory of a very special case, namely, that of ordinary linear simultaneous equations. With a $(k+1)$-by-$(k+1)$ array of scalars

$$a_{rs}, \quad 0 \leq r, \ s \leq k, \qquad (2.5.1)$$

we may associate a system of equations

$$\sum_{s=0}^{k} a_{rs} x_s = 0, \quad 0 \leq r \leq k. \qquad (2.5.2)$$

As is well known, non-trivial solutions are possible iff the determinant

$$\det a_{rs} = 0. \qquad (2.5.3)$$

In the operator analogue, we cannot proceed so rapidly. We now suppose given an array of linear operators

$$A_{rs}, \quad 0 \leq r, s \leq k, \qquad (2.5.4)$$

Here, for each r, the $k+1$ operators $A_{r0}, \ldots, A_{rk}$, have a common domain-space G_r and map into a common range-space H_r. The space G_r may be the same as H_r, but need not be, and either may include the other as a subspace. However it is in no way requisite that these should be inter-related for different values of r. Usually, these linear spaces will be of the same general nature, but even this is not necessary. However, two operators from different rows of the array

(2.5.4), say, A_{rs}, $A_{r's'}$ when $r \neq r'$, do not act on the same objects, though they may act on separate copies of the same object, The composite $A_{rs}A_{r's'}$, is not defined; indeed, it need not be defined when $r = r'$ either.

The fact that the operators (2.5.4) do not have a common domain means that we do not have an immediate analogue of the simultaneous equations. However, several analogues do indeed exist. In the simplest of these, we set up an eigenvalue problem. We ask for a set of λ_s, not all zero, such that

$$\sum_{s=0}^{k} A_{rs}\lambda_s y_r = 0, \quad r = 0, \ldots, k, \tag{2.5.5}$$

for some set of elements y_r, $x_0, \ldots, x_k$, none of them zero. In other words, the set of operators

$$\sum_{s=0}^{k} A_{rs}\lambda_s, \quad r = 0, \ldots, k, \tag{2.5.6}$$

must all be singular, or more precisely must have non-zero kernels. The "unknowns" x_s in (2.5.2) have thus been replaced by spectral parameters λ_s.

The structure is still deficient, however, in that we have no means of eliminating the λ_s and the y_r between the equations (2.5.5) so as to produce an analogue of (2.5.3). This is where the tensor-product apparatus comes in. For the general theory, we need to introduce a tensor product space

$$G = G_0 \otimes G_1 \otimes \cdots \otimes G_k; \tag{2.5.7}$$

this is generated by "decomposable" elements

$$y = y_0 \otimes y_1 \otimes \cdots \otimes y_k; \tag{2.5.8}$$

where $y_r \in G_r$, $r = 0, \ldots, k$. In addition, with the A_{rs} we associate "induced" operators, denoted $A_{rs}^{\dagger}$; roughly speaking, $A_{rs}^{\dagger}$ is given by making A_{rs} act on the factor y_r in (2.5.8). Moreover, $A_{rs}^{\dagger}$ can act on any tensor product containing G_r as a factor; we do not introduce a fresh notation for such actions.

With the aid of this machinery, we can pass from (2.5.5) in the first place to the equations

$$\sum_{s=0}^{k} A_{rs}^{\dagger}\lambda_s y = 0, \quad r = 0, \ldots, k; \tag{2.5.9}$$

the question is now whether $\lambda_0, \ldots, \lambda_k$, not all zero, can be found so that the equations (2.5.9) have a common solution $y \in G$, not zero. Second, we can set up the equations

$$\sum_{s=0}^{k} A_{rs}^{\dagger}u_s = 0, \tag{2.5.10}$$

in which $u_0, \ldots, u_k \in G$. Again the question is whether (2.5.10) have a solution $u_0, \ldots, u_k$ in which the u_s are not all zero. The last forms perhaps the truest analogue of (2.5.2) in that it has the property that set of solutions of (2.5.10) forms a linear space, a subspace of

$$G \oplus \ldots \oplus G,$$

the direct sum of $(k+1)$ copies of G. On the other hand, it is weaker than (2.5.9); a non-trivial solution of (2.5.9) yields one of (2.5.10) but the converse conclusion is in no way immediate.

From either of (2.5.9), (2.5.10), we can deduce an analogue of (2.5.3). This involves the "operator-valued determinant"

$$\Delta = \det \ A_{rs}^{\dagger}. \tag{2.5.11}$$

This, an operator from the tensor-product G to the tensor product

$$H = H_0 \otimes H_1 \otimes \ldots \otimes H_k,$$

is formed by expanding the determinant in the usual way as a sum of signed products of its entries. The products are to be in the sense of operator-composition; the requisite products do have sense, the order of the factors being immaterial. The conclusion from (2.5.9) is that if the λ_s are not all zero,

$$\Delta y = 0. \tag{2.5.12}$$

Summing up, we have as trivial propositions that the existence of non-trivial solutions of (2.5.5) implies the same for (2.5.9)–(2.5.10). The reversal of such implications was taken up in Atkinson (1972) in the finite-dimensional case. We take up here similar questions in certain differential operator cases.

2.6   The differential operator case

We now fill in a few of the details in the situation of relevance to this volume. Using the conventional letter for differential operators, we assume we are dealing with an array

$$L_{rs}, \quad 0 \leq r, s \leq k, \tag{2.6.1}$$

where L_{rs} is an ordinary linear differential operator, acting on functions of an independent variable x_r. For example, we may have

$$L_{rs}(y_r) = \sum_{m=0}^{m_{rs}} q_{rsm}(x_r)(d/dx_r)^m y_r(x_r); \tag{2.6.2}$$

forms also occur in which the coefficients q_{rsm} come under the sign of differentiation, along with derivatives of y_r. It is not excluded that L_{rs} might be of order zero; that is to say, it might be the operation of multiplication by some function of x_r.

The spaces G_r will be, in this case, spaces of functions defined on respective intervals I_r. The functions will of course have to be suitably differentiable, with appropriate restrictions of continuity or of absolute continuity on the derivatives. For the purposes of the general theory, the independent variables $x_0, \ldots, x_k$ occurring in these function-spaces are considered as unrelated; there is no objection to them all having intervals of variation that overlap or coincide.

In this situation, a formal introduction of tensor products would need considerable preparation. Fortunately, an explicit construction is immediately available. We take in the role of the tensor product G a space of functions on the cartesian product of the various x_r-intervals, that is to say, on the set

$$I = \{(x_0, \ldots, x_k) |\ x_r \in I_r,\ r = 0, \ldots, k\}. \qquad (2.6.3)$$

The induced operators $L_{rs}^\dagger$ are simply the L_{rs} interpreted as partial differential operators; on a function $y(x_0, \ldots, x_k) = y$, they act according to

$$L_{rs}^\dagger(y_r) = \sum q_{rsm}(\partial/\partial x_r)^m y \qquad (2.6.4)$$

The functions in G must be such that partial derivatives of the requisite orders exist and commute. "Decomposable elements" of G, corresponding to (2.5.8), will be functions of $x_0, \ldots, x_k$, which are products of functions of these individual variables.

Corresponding to the constructions of § 2.5, we have

(i) the system of ordinary differential equations,

$$\sum_{s=0}^{k} L_{rs}\lambda_s y_r = 0, \quad r = 0, \ldots, k, \qquad (2.6.5)$$

linked only by the parameters,

(ii) the system of simultaneous partial differential equations

$$\sum_{s=0}^{k} L_{rs}^\dagger \lambda_s y(x_0, \ldots, x_k) = 0, \quad r = 0, \ldots, k, \qquad (2.6.6)$$

(iii) the system of partial differential equations

$$\sum_{s=0}^{k} L_{rs}^\dagger u_s(x_0, \ldots, x_k) = 0, \quad r = 0, \ldots, k, \qquad (2.6.7)$$

(iv) the determinantal partial differential equation

$$Ly = 0, \qquad (2.6.8)$$

where formally

$$L = \det \ L_{rs}^{\dagger} \qquad (2.6.9)$$

and the determinant is to be expanded as a sum of products or operator composites in the usual way.

Assuming that all functions concerned are sufficiently differentiable, it is immediate that products of solutions of (2.6.5) yield solutions of (2.6.6), that solutions of (2.6.6) yield solutions of (2.6.7), and that the latter yield solutions of (2.6.8). For example, to deduce from (2.6.7) that

$$Lu_0 = 0, \qquad (2.6.10)$$

we apply to the typical equation in (2.6.7) the partial differential operator, which is the co-factor of $L_{r0}^{\dagger}$ in the determinant (2.6.9), and sum the results.

2.7 Separability

A partial differential operator L in the variables $x_0, \ldots, x_k$ which admits representation in the form (2.6.9) may be said to be "separable." More generally, we say that it is separable if it can be written as

$$L = \mu \ \det \ L_{rs}^{\dagger}, \qquad (2.7.1)$$

where μ is a non-zero function of $x_0, \ldots, x_k$, and as before L is a partial differential operator, involving x_r only, both as to the derivatives and as to the coefficients. Such a representation, if possible, will not be unique. If L is separable, the equation $Ly = 0$ possesses a considerable stock of particular solutions, given by products of solutions of the ordinary differential equations (2.6.5). Here the $\lambda_0, \ldots, \lambda_k$ may be any set of numbers, not all zero. For certain purposes, solutions of this nature, together with linear combinations of them, are sufficient.

In this connection, the numbers $\lambda_0, \ldots, \lambda_k$ are termed "separation constants"; the number of them, $k + 1$, is equal to the number of independent variables. Frequently, it is convenient to go over to an inhomogeneous formulation, by fixing $\lambda_0 = 1$, say; the number of separation constants will then be k, or one less than the number of variables.

2.8   Problems with boundary conditions

In their role as separation constants, the parameters λ_s in §2.6 are unrestricted. We now consider how far they may be chosen so as to make the solutions y_r of (2.6.5), or their product y, satisfy supplementary conditions. We are concerned with boundary conditions of the standard type, imposing homogeneous linear conditions on the values of y_r and its derivatives at the end-points of the x_r-interval, the number of such conditions being equal to the order of the differential equation concerned.

The effect of suitable boundary conditions for one of the equations (2.6.5) will be that a non-trivial solution $y_r(x_r)$ exists only if $(\lambda_0, \ldots, \lambda_k)$ lies in some proper subset of the collection of non-zero $(k+1)$-tuples considered projectively. This suggests that if we impose such boundary conditions on all $(k+1)$ equations, we are liable to obtain a problem in which one of the y_r must vanish identically, whatever the choice of the λ_s.

This may be seen more clearly in the case in which, for each r, L_{r0} is a differential operator of higher order than any of $L_{r1}, \ldots, L_{rk}$. In this case we take $\lambda_0 = 1$, and the $\lambda_1, \ldots, \lambda_k$, for which (2.6.5) has a non-trivial solution satisfying suitable boundary conditions, will be the zeros of some entire function of $\lambda_1, \ldots, \lambda_k$. If we imposed boundary conditions for all of (2.6.5), we should be looking for sets of which these were common zeros of $k+1$ entire functions of $\lambda_1, \ldots, \lambda_k$. Again, this is liable to yield the empty set.

This suggests that we consider the problem in which boundary conditions are imposed in connection with k of the $k+1$ equations (2.6.5), say, those for $r = 1, \ldots, k$. The boundary value problems formed by (2.6.5) for $r = 1, \ldots, k$, together with the boundary conditions, will serve to determine some subset of admissible sets $(\lambda_1, \ldots, \lambda_k)$, not all zero and considered projectively. The equation (2.6.5) with $r = 0$ may then be set aside, as playing no role in the determination of these constants.

This formulation arises naturally in physical problems in which boundary conditions are imposed in respect of the spatial variables, but not in respect of time, and which the time-dependence of normal modes is determined in terms of eigenvalues of spatial differential equations.

In what follows, we study the situation of k equations with associated boundary conditions, each involving k parameters, with the parameters considered inhomogeneously, with λ_0 taken to be 1.

2.9   Associated partial differential equations

Since in the problem (2.1.4)–(2.1.6) the number of boundary-value problems is equal to the number of parameters $\lambda_1, \ldots, \lambda_k$, we cannot expect to eliminate all the parameters. The problems provide in general just enough information to determine a spectrum. The problems (2.1.4)–(2.1.6) may indeed arise from a partial differential equation with boundary conditions, neither of which involves any parameter. However, in such cases, we arrive at (2.1.4)–(2.1.6) by dispensing with one of the separated equations, commonly that associated with the time variable, as explained in §2.8.

Nevertheless, elimination procedures are important to the theory of (2.1.4)–(2.1.6). What we can do is to eliminate all but one of the parameters. More precisely, a solution of (2.1.4)–(2.1.6) yields a solution of a boundary-value problem for a partial differential equation, involving one of the parameters, at our choice. We thus get an eigenvalue problem of the conventional one-parameter type.

We first pass to what are in effect the induced equations in a tensor product space; as previously, we prefer to rely on explicit constructions rather than on the theory of tensor products. We write

$$y = y(x_1, \ldots, x_k) = \prod_{r=1}^{k} y_r(x_r), \tag{2.9.1}$$

where the $y_r(x_r)$ are solutions of (2.1.4)–(2.1.6). We then have

$$\{\partial^2/\partial x_r^2 - q_r(x_r)\}y + \sum_{s=1}^{k} p_{rs}(x_r)\lambda_s y = 0, \quad r = 1, \ldots, k, \tag{2.9.2}$$

and the boundary conditions

$$y \cos \alpha_r = (\partial y/\partial x_r) \sin \alpha_r, \quad x_r = a_r, \ r = 1, \ldots, k, \tag{2.9.3}$$

$$y \cos \beta_r = (\partial y/\partial x_r) \sin \beta_r, \quad x_r = b_r, \ r = 1, \ldots, k. \tag{2.9.4}$$

These conditions apply on the boundary of the k-dimensional interval

$$I = \{(x_1, \ldots, x_k)| \ a_r \le x_r \le b_r, \ r = 1, \ldots, k\}. \tag{2.9.5}$$

We next apply determinantal procedures, in a similar manner to "Cramer's rule." On I we define the k-th order determinant function

$$P(x_1, \ldots, x_k) = \det \ p_{rs}(x_r), \tag{2.9.6}$$

where the suffixes extend over $1 \leq r, s \leq k$. We write

$$P_{rs} = P_{rs}(x_1, \ldots, \hat{x}_r, \ldots, x_k)$$

for the co-factor of p_{rs} in P; the notation ($\hat{}$) indicates that P_{rs} does not depend on x_r. From determinant theory, we have then that

$$\sum_{r=1}^{k} p_{rs} P_{rt} = \delta_{st} P, \qquad (2.9.7)$$

where $\delta_{st} = 0$, $s \neq t$, $\delta_{st} = 1$, $s = t$. Thus, if we multiply (2.9.2), respectively, by P_{rt}, $r = 1, \ldots, k$, and add, we get

$$\sum_{r=1}^{k} P_{rt}\{\partial^2/\partial x_r^2 - q_r(x_r)\}y + \lambda_t Py = 0, \quad t = 1, \ldots, k. \qquad (2.9.8)$$

Choosing some t-value, we take one of (2.9.8) together with the boundary conditions (2.9.3)–(2.9.4). This gives a boundary-value problem for a partial differential equation, involving one parameter only. In the spectrum of k-tuples $(\lambda_1, \ldots, \lambda_k)$ for (2.1.4)–(2.1.6), the values of λ_t which occur must be among the eigenvalues of the problem given by (2.9.3)–(2.9.4) and the t-th equation in (2.9.8).

Sometimes it is desirable to use a linear combination of the equations (2.9.8). For any $\rho_1, \ldots, \rho_k$, we have

$$\sum_{t=1}^{k} \sum_{r=1}^{k} \rho_t P_{rt}\{\partial^2/\partial x_r^2 - q_r(x_r)\}y + \sum_{t=1}^{k} \{\rho_t \lambda_t\} Py = 0. \qquad (2.9.9)$$

Here the expression $\sum \rho_t \lambda_t$ functions as a parameter. The occasion for this device is that it is desirable in the theory of (2.9.8) that the coefficient P of y, or the coefficients P_{rt} of $\partial^2 y/\partial x_r^2$, or both, be positive. When this fails, it may be possible to choose the ρ_t so that the requisite positivity holds for (2.9.9).

2.10   Generalizations and variations

We shall deal mainly with direct extensions to the multiparameter situation of the classical Sturm-Liouville situation; this we do both for the importance of these cases and for their amenability. Thus, we give little attention to a very wide field of generalisation in which the differential operators may be of higher order. Again, we are not considering cases in which some of the linked eigenvalue problems are of Sturm-Liouville type, and others of a different nature, perhaps involving matrices. We also confine attention to scalar differential equations.

However, even within these limitations there is some scope for variation. For example, it will be noted that in (2.1.4) we have taken the second-order terms in the form $y_r''(x_r)$, rather than in the more general self-adjoint form

$$(d/dx_r)\{f_r(x_r)y_r'(x_r)\}, \tag{2.10.1}$$

where $f_r(x_r)$ is some positive continuous function. We do this to conserve notation, and note that it does not entail any loss of generality. As discussed in Chapter 1, the well-known change of variable

$$d\xi_r = dx_r/f_r(x_r) \tag{2.10.2}$$

$z_r(\xi_r) = y_r(x_r)$, enables us to rewrite (2.10.1) as

$$\{f_r(x_r)\}^{-1} d^2/d\xi_r^2 \{z_r(\xi_r)\} \tag{2.10.3}$$

and so brings us back effectively to the above special case.

We can also relax the usual continuity requirements on the coefficients in (2.1.4). If they are Lebesgue-integrable, the solution $y_r(x_r)$ will have to be such that $y_r'(x_r)$ exists and is absolutely continuous, with the differential equation being satisfied almost everywhere.

We comment next on the boundary conditions (2.1.5)–(2.1.6), which we have taken to be all in the one-point or separated form. Of course, other types of boundary conditions are important, notably periodic boundary conditions. There is no need to take all the boundary conditions of the same type; the choice will, of course, affect such matters as the oscillatory characterization of eigenvalues.

For other types of boundary-value problems, a model is given by the Legendre polynomials, which are finite as $x \to \pm 1$, even though the Legendre differential equation has singularities there. More generally, for an analytic differential equation with a regular singularity, we may impose a boundary condition involving the roots of the indicial equation, limiting the behavior of the solution near the singularity. The relevant situation is given by the Lamé differential equation and generalization, in which we seek to adjust several parameters so that solutions should have suitably limited behavior at several singularities. With the last remark, we verge on another point, namely, as to whether the independent variables x_r in (2.1.4) are to be considered as distinct. If the intervals do not overlap, then it is indeed possible to consider the differential equations (2.1.4) as governing a single independent variable over various disjoint intervals on a single real axis. In some cases, this is quite appropriate. We cite the case of a differential equation

$$z''(x) - q(x)z(x) + \sum_{s=1}^{k} \lambda_s p_s(x)z(x) = 0 \qquad (2.10.4)$$

over some interval $[a, b]$, in which we ask for there to exist a non-trivial solution that vanishes at a, at b, and also at $k - 1$ assigned points in (a, b); the k-tuples $(\lambda_1, \ldots, \lambda_k)$ for which this is possible constitute the spectrum. This problem can of course be re-phrased in terms of (2.1.4)–(2.1.6), with separate independent variables.

For the purposes of the general theory, the formulation in terms of separate independent variables is needed in order to have available the partial differential equations of §2.9.

Certain singular cases, in particular those involving semi-infinite intervals, will be considered later. These, of course, cannot be re-formulated in terms of a single independent variable.

2.11   The half-linear case

We recall that a linear homogeneous equation has the property that the set of solutions is closed under the basic operations of (i) addition and (ii) multiplication by scalars. The term "half-linear" is used for an equation with the latter property only; for such equations, the sum of two solutions need not be a solution. The definition is dependent on the choice of the field of scalars; here this will be the real field. Such equations have been considered mainly in the real domain.

The theory has been developed in the ODE context by Bihari and by Elbert and in the PDE context in connection with the "p-Laplacian," defined by

$$\Delta_p u = \text{div}\{|\text{grad } u|^{p-2} \text{grad} \cdot u\} \qquad (2.11.1)$$

for which there is an extensive literature. It is the former that is most relevant here, particularly the possibility of using suitably adapted Prüfer transformations. A typical single equation would be

$$y''|y'|^{n-1} + p(t)y|y|^{n-1} = 0, \qquad (2.11.2)$$

where $n > 0$. The case $n = 1$ is of course the standard one.

The Prüfer transformation involves a certain generalization of the sine-function. We denote by $S(\phi)$ the solution of

$$S''|S'|^{n-1} + S|S|^{n-1} = 0, \quad S(0) = 0, S'(0) = 1, \quad (2.11.3)$$

and for a non-trivial solution of (2.11.2) introduce amplitude and phase variables $\rho(t) > 0$, $\phi(t)$ by

$$y(t) = \rho(t)S(\phi(t)), \quad y'(t) = \rho(t)S'(\phi(t)). \quad (2.11.4)$$

One finds that

$$\phi'(t) = |S'(\phi)|^{n+1} + p(t)|S(\phi)|^{n+1} \quad (2.11.5)$$

and

$$\rho'(t)/\rho(t) = S'(\phi)S(\phi)|S(\phi)|^{n-1}\{1 - p(t)\}. \quad (2.11.6)$$

See Elbert (1979) for more details and references to this extensive generalization of Sturmian theory.

The multiparameter version of such problems might be

$$y_r''(x_r)|y'(x_r)|^{n_r-1} + \left\{\sum_{s=1}^{k}\lambda_s p_{rs}(x_r) - q_r(x_r)\right\}y_r(x_r)|y(x_r)|^{n_r-1} = 0,$$
$$(2.11.7)$$

$$a_r \le x_r \le b_r, \quad r = 1, \ldots, k,$$

together with the same boundary conditions, (2.1.5)–(2.1.6). Here the n_r are all to be positive.

2.12 A mixed problem

A simple yet non-trivial case is given by the linking of a Sturm-Liouville problem with a matrix equation. An example might be

$$-y'' + q(x)y = \{\lambda_1 p_1(x) + \lambda_2 p_2(x)\}y, \quad a \le x \le b, \quad (2.12.1)$$

with boundary conditions

$$y(a)\cos\alpha = y'(a)\sin\alpha, \quad y(b)\cos\beta = y'(b)\sin\beta, \quad (2.12.2)$$

together with

$$\{\lambda_1 A_1 + \lambda_2 A_2 + B\}u = 0, \quad (2.12.3)$$

where A_1, A_2, and B are constant n-by-n matrices. An eigenvalue is a pair λ_1, λ_2 for which (2.12.1)–(2.12.2) and (2.12.3) both have non-trivial solutions, while an eigenfunction would then be the vector-function $y(x)u$.

Notes for Chapter 2

For a proof of the orthogonality (2.4.7), see Atkinson (1964a, Chapter 6.9). The approach in this section is also used by Faierman (1969). For basic material regarding § 2.5, in particular, operator analogs of (2.5.1)–(2.5.3) in the case of rectangular arrays or when $r = 1, \ldots, k$ and $s = 0, \ldots, k$, see Atkinson (1972, Chapter 6).

The notion of separability defined in §2.7 appears to be new. Similar notions have been defined for specific classes of differential operators (e.g., for first-order differential operators commuting with themselves and with the Dirac operator, as in Miller [1988]), but not in the generality outlined here.

To this day, multiparameter problems associated with a half-linear differential equation (cf. § 2.11) appear not to have been studied. The linked problem as formulated in § 2.12 also appears to lack a follow-up, the only work related to it dating from early work of Ma (1972) in the two parameter case.

2.13   Research problems and open questions

1. Multiparameter problems associated with a half-linear differential equation (cf. § 2.11) appear not to have been studied at all. Thus, for example, the two-parameter case of a half-linear equation should require a detailed study. Some work was undertaken by Bihari (1976) and Eberhard and Elbert (2000), and the references therein, in the one-parameter case.

2. The linked problem as formulated in § 2.12 should be investigated in the case of more than two parameters, starting perhaps with the case of the Dirichlet problem.

3. The case of a single Sturm-Liouville equation in two parameters is of much interest because of a variety of applications. In the singular case where $I = [0, \infty)$ (or I$=(-\infty, \infty)$), the existence of positive solutions of an equation of the form

$$y'' + \{\lambda_1 p_1(x) + \lambda_2 p_2(x)\}y = 0, \quad x \in I \tag{2.13.1}$$

is intimately related with the set of parameter values $(\lambda_1, \lambda_2) \in \mathbb{R}^2$ for which the differential equation (2.13.1) is disconjugate (basically this means that every non-trivial solution of 2.13.1 has at most one zero in I). Call this set $\mathcal{D}$. Such a study was initiated in Mingarelli and Halvorsen (1988) in their study of "disconjugacy domains" of equations of the form (2.13.1). For example, it is known that $\mathcal{D}$ is a closed convex set. Applications include Mathieu's equation and the spectral theory of problems with indefinite weight functions, for example, Allegretto and Mingarelli (1989).

Now, a consequence of Sturm-Liouville theory is that an equation (2.13.1) is either non-oscillatory (i.e., every non-trivial solution as a finite number of zeros) or oscillatory (in the sense that every non-trivial solution has an infinite number of zeros in I). The set of all parameter values $(\lambda_1, \lambda_2) \in \mathbb{R}^2$ such that the differential equation (2.13.1) is non-oscillatory is denoted by $\mathcal{N}$. There are classes of potentials $p_i(x)$, $i = 1, 2$ for which $\mathcal{D} = \mathcal{N}$, for example, if the $p_i(x)$ are both almost periodic functions (in the sense of Bohr), cf. Mingarelli and Halvorsen (1988), §2.7, although periodicity in any sense or generalization is not necessary for this equality is also shown there.

The question is: *Can $\mathcal{D} = \mathcal{N}$ be characterized in terms of properties of the $p_i(x)$? Also, what is the diameter of $\mathcal{D}$ in terms of the given quantities?*

Some consequences are, for example: If for a pair $p_i(x)$ we know that $\mathcal{D} = \mathcal{N}$ on I, then the spectral problem associated with a given set of boundary conditions will have the property that for an eigenpair $(\lambda_1, \lambda_2) \in \mathbb{R}^2$ the resulting eigenfunction will either have at most one zero or will be oscillatory at infinity. This in turn can provide us with information on the discrete spectrum of the boundary problem.

4. Consider the triangular case of § 2.2 once again. For regular boundary value problems with separated boundary conditions much is now known about the spectrum of problems of the form (2.2.3), including cases where p_{11} may change sign in $[a_1, b_1]$ and q is arbitrary. These so-called *non-definite* cases have attracted some attention in the past 30 years. Assuming that $\det p_{rs}(x)$ changes sign in the given interval we take it that $p_{11}(x)$ itself changes sign in $[a_1, b_1]$. We know that (2.2.3), (2.1.4)–(2.1.6) (with $r = 1$) admits an infinite number of both positive and negative eigenvalues. So, working our way *down* the diagonal of the matrix p_{rs} (cf., [2.2.2]) we can repeat the argument at the end of § 2.2 to find an infinite sequence of admissible pairs of eigenvalues and so can ask the following questions:

What is the asymptotic behavior of such sequences of eigenvalues?

For background material regarding the one parameter non-definite case, see the survey paper by Mingarelli (1986), and the references therein and, for example, subsequent papers by Atkinson and Mingarelli (1987), Allegretto and Mingarelli (1989), Chapter 6 in the book by Faierman (1991), Eberhard and Freiling (1992), Binding *et al.* (2002b).

5. In the framework of Question 4 above: What if λ_1 in (2.2.3) is not real

(we know that a finite number of non-real eigenvalues may exist even for real coefficients, see Mingarelli [1986] for references). The resulting Sturm-Liouville equations (2.2.3), (2.2.4), and so on, now have complex coefficients and a complex parameter.

What can be said about the asymptotic behavior of the non-real spectrum of this problem?

Of specific interest here is the paper by Hilb (1911) where an existence theorem for the eigenvalues of complex Sturm-Liouville equations is derived, their asymptotic behavior, and the asymptotic behavior of the Green function (for the Dirichlet problem).

Chapter 3

Definiteness Conditions and the Spectrum

3.1 Introduction

In this chapter, we investigate, in a general way, conditions for the eigenvalues to be real, and to form a discrete set; both the latter features need to be specially interpreted. The treatment is general in that it is not restricted to Sturm-Liouville systems. In Chapter 5, for Sturm-Liouville systems, we give oscillation methods, which can be applied for the same purposes.

The methods of this chapter rely in the first place on the theory of arrays of quadratic forms; the arguments are, however, for the most part self-contained. When specializing in the Sturm-Liouville direction, we will need the results on determinants of functions, developed in the next chapter.

We are concerned with a set of k complex linear spaces, on each of which are defined $k + 1$ quadratic forms. The spaces will now be of functions

$$y_r(x_r), \quad a_r \le x_r \le b_r, \quad r = 1, \ldots, k, \tag{3.1.1}$$

which are to be suitably differentiable, and are to satisfy boundary conditions. We write these boundary conditions as

$$B_r(y_r) = 0, \quad r = 1, \ldots, k, \tag{3.1.2}$$

where $B_r(y_r)$ denotes some set of linear combinations of the values of

$$y_r(a_r), \ y_r(b_r), \ y_r{}'(a_r), \ y_r{}'(b_r), \ldots.$$

The quadratic, or rather sesquilinear, forms are specified by means of a array of ordinary differential operators

$$L_{rs}, \quad 1 \le r \le k, \ 0 \le s \le k, \tag{3.1.3}$$

where the L_{rs}, $s = 0, \ldots, k$, act on $y_r(x_r)$, $a_r \le x_r \le b_r$.

For example, L_{rs} may have the form (1.3.2) though, as noted there, variations on this form are possible. "Suitably differentiable" will mean that $y_r(x_r)$ is such that the derivatives appearing in $L_{rs}(y_r)$ and in $B_r(y_r)$ exist and are in suitable function-classes; some variation is possible in the latter point also.

In the case of main interest, we shall have

$$L_{r0} = d^2/dx_r^2 - q_r(x_r), \ r = 1, \ldots, k, \tag{3.1.4}$$

$$L_{rs} = p_{rs}(x_r), \quad 1 \le r, s \le k, \tag{3.1.5}$$

where $q_r(x_r)$ denote the operations of pointwise multiplication by these functions. For this Sturm-Liouville case, we have a choice between separated and unseparated boundary condition. Emphasizing the former, we take

$$B_r(y_r) = \{y_r(a_r)\cos\alpha_r - y_r{}'(a_r)\sin\alpha_r, \ y_r(b_r)\cos\beta_r - y_r{}'(b_r)\sin\beta_r\}. \tag{3.1.6}$$

The unseparated case may be exemplified by that of periodic boundary conditions, when we should take

$$B_r(y_r) = \{y_r(a_r) - y_r(b_r), y_r{}'(a_r) - y_r{}'(b_r)\}. \tag{3.1.7}$$

If in (3.1.4)–(3.1.5) $p_{rs}(x_r)$, $q_r(x_r)$ are continuous we should ask that $y_r(x_r)$ be continuously twice differentiable; if $p_{rs}(x_r)$, $q_r(x_r)$ are only required to be in $L(a_r, b_r)$, then we should ask that $y_r{}'(x_r)$ exist and be absolutely continuous over the interval in question.

In the general case, without restriction to (3.1.4)–(3.1.5), it is desirable to consider eigenvalues homogeneously. We then say that the $(k+1)$-tuple $(\lambda_0, \ldots, \lambda_k)$ of scalars, not all zero, is an "eigenvalue" if there exists a set of k functions (3.1.1), suitably differentiable and in no case identically zero, satisfying the ordinary differential equations

$$\sum_0^k \lambda_s L_{rs}(y_r) = 0, \quad a_r \le x_r \le b_r, \ r = 1, \ldots, k, \tag{3.1.8}$$

and also the boundary conditions (3.1.2). Clearly, if $(\lambda_0, \ldots, \lambda_k)$ is an eigenvalue, then so is the $(k+1)$-tuple $(c\lambda_0, \ldots, c\lambda_k)$, for any $c \ne 0$. Two eigenvalues related in this way are not considered distinct. In other words, an eigenvalue is an equivalence class of $(k+1)$-tuples $(\lambda_0, \ldots, \lambda_k)$ of which at least one, say, λ_t is not zero, and for which the ratios λ_s/λ_t are all fixed.

This concept of an eigenvalue is open to various modifications and refinements. It may be convenient to normalize the eigenvalue so as to yield a single $(k+1)$-tuple rather than an equivalence class of them. Notably, if it is known for the problem in question, that for some fixed t, $\lambda_t \neq 0$ for every eigenvalue, we may without loss of generality specify that $\lambda_t = 1$, and go over to a notion of an eigenvalue as a k-tuple of scalars, considered inhomogeneously and not as an equivalence class. It is also possible to restrict the signs of the λ_s, yielding the notion of a "signed eigenvalue." We postpone this to §3.4. In another variation, we may work in terms of associated partial differential equations.

As noted in Section 2.6, the ordinary differential operators L_{rs} and boundary operators B_r give rise to partial differential operators $L_{rs}{}^\dagger$, $B_r{}^\dagger$, acting on suitably differentiable functions $y(x_1, \ldots, x_k)$; these operators are derived simply by replacing d/dx_r by $\partial/\partial x_r$. If the functions (3.1.1) satisfy (3.1.2) and also (3.1.8), then their product

$$y(x_1, \ldots, x_k) = \prod_1^k y_r(x_r), \qquad (3.1.9)$$

will satisfy the partial differential equations

$$\sum_{s=0}^k \lambda_s L_{rs}{}^\dagger(y) = 0, \quad r = 1, \ldots, k, \qquad (3.1.10)$$

on the k-dimensional interval

$$I = \{(x_1, \ldots, x_k) \mid a_r \le x_r \le b_r, \ r = 1, \ldots, k\} \qquad (3.1.11)$$

and on the boundary of I the boundary conditions in partial derivatives given by

$$B_r{}^\dagger(y) = 0, \quad r = 1, \ldots, k. \qquad (3.1.12)$$

This gives an alternative, formally broader notion of an eigenvalue, as a non-zero $(k+1)$-tuple $(\lambda_0, \ldots, \lambda_k)$ for which there is a non-zero y satisfying (3.1.10), (3.1.12). In cases of interest to us, the two notions of an eigenvalue turn out to be equivalent.

3.2   Eigenfunctions and multiplicity

Let $(\lambda_0, \ldots, \lambda_k)$ be an eigenvalue, and so a set of scalars, not all zero, such that (3.1.8), (3.1.2) have non-trivial solutions. The term "eigenfunction" may be applied in any of the following senses:

(i) a product (3.1.9) of solutions of (3.1.8), (3.1.2),

(ii) a linear combination of such products,

(iii) a solution of (3.1.10), (3.1.12).

It is convenient here to admit in (ii), (iii) the case of an identically zero function. The definition (ii) is formally broader than (i), and (iii) than (ii).

In the definitions (ii), (iii), the eigenfunctions form a linear space; this is not in general the case for (i). In the Sturm-Liouville case (3.1.4)–(3.1.6) with separated boundary conditions, however, the definitions are equivalent.

For each r-value, the set of solutions of (3.1.8), (3.1.2) will form a linear space, of dimension ν_r, say; if for every $\lambda_0, \lambda_1, \ldots, \lambda_k$ not all zero the operator (3.1.8) is a differential operator that does not vanish identically on any subinterval, then ν_r will not exceed the order of the differential operator $\sum_s \lambda_s L_{rs}$.

In the Sturm-Liouville case (3.1.4)–(3.1.6), if $\lambda_0 \neq 0$, we shall have $\nu_r = 1$ for each r; if we keep to (3.1.4)–(3.1.5), but replace (3.1.6) by unseparated conditions, such as the periodic conditions $y_r(a_r) = y_r(b_r)$, $y_r{}'(a_r) = y_r{}'(b_r)$, we can have $\nu_r = 1$ or 2.

We define the multiplicity of the eigenvalue $(\lambda_0, \ldots, \lambda_k)$ to be the product

$$\prod_1^k \nu_r.$$

It is thus the dimension of the linear space of functions (ii); it is also the dimension of the space of functions (iii), though we shall not prove this.

The collections of functions (i), (ii) will be identical if there is at most one r-value for which $\nu_r > 1$. For example, in the Sturm-Liouville case (3.1.4)–(3.1.5), for one r-value we may replace the separated boundary conditions (3.1.6) by unseparated conditions, such as the periodic conditions, and can still treat (i) as a general notion of an eigenfunction, the collection (i) being a linear space.

Eigenfunctions of the type (i) may be termed "decomposable," following tensor product usage.

3.3 Formal self-adjointness

In this and the next section, we introduce two requirements for boundary-value problems (3.1.8), (3.1.2), which ensure desirable properties, such as the reality of the spectrum. Basically, these requirements are applied to the boundary conditions and differential equations, considered as a single entity; it may happen that the requirements are satisfied if we impose separately restrictions on the boundary conditions and on the differential equations.

To be specific, we introduce the sesquilinear forms

$$\Psi_{rs}(y_r, z_r) = \int_{a_r}^{b_r} L_{rs} y_r(x_r) \overline{z_r(x_r)} \, dx_r, \quad r = 1, \ldots, k, \ s = 0, \ldots, k, \quad (3.3.1)$$

where the bar indicates complex conjugation, and y_r, z_r are such that the right is well defined. We define also the associated quadratic forms

$$\Phi_{rs}(y_r) = \Psi_{rs}(y_r, y_r), \quad r = 1, \ldots, k, \ s = 0, \ldots, k. \quad (3.3.2)$$

We say that the problem represented by (3.1.8), (3.1.2) is "formally self - adjoint" if the $\Phi_{rs}(y_r)$ take real values only, whenever the y_r are suitably differentiable functions satisfying the boundary conditions (3.1.2).

It follows from the theory of quadratic forms that (3.3.1) have the property of hermitian symmetry, or that

$$\Psi_{rs}(z_r, y_r) = \overline{\Psi_{rs}(y_r, z_r)} \quad (3.3.3)$$

if y_r, z_r are suitably differentiable and both satisfy (3.1.2). For since $\Psi_{rs}(y_r + z_r, y_r + z_r)$ is real-valued, along with $\Psi_{rs}(y_r, y_r)$ and $\Psi_{rs}(z_r, z_r)$, we have that $\Psi_{rs}(y_r, z_r) + \Psi_{rs}(z_r, y_r)$ is real. A similar argument shows that

$$\Psi_{rs}(y_r, iz_r) + \Psi_{rs}(iz_r, y_r)$$

is real, and so also

$$i\{\Psi_{rs}(z_r, y_r) - \Psi_{rs}(y_r, z_r)\}.$$

These two facts imply (3.3.3)

In the special case (3.1.4)–(3.1.5), we have

$$\Phi_{rs}(y_r) = \int_{a_r}^{b_r} p_{rs}(x_r)\,|y_r(x_r)|^2\,dx_r, \quad r, s = 1, \ldots, k, \qquad (3.3.4)$$

and these are real if the $p_{rs}(x_r)$ are real, regardless of the choice of boundary conditions. In the case $s = 0$ we have

$$
\begin{aligned}
\Phi_{r0}(y_r) &= \int_{a_r}^{b_r} (y_r'' - q_r y_r)\,\overline{y_r}\,dx_r \\
&= [y_r \overline{y_r}']_{a_r}^{b_r} - \int_{a_r}^{b_r} |y_r'|^2\,dx_r - \int_{a_r}^{b_r} q_r |y_r|^2 dx_r. \qquad (3.3.5)
\end{aligned}
$$

Here the last two terms are real, provided that $q_r(x_r)$ is real, while the integrated term will be real if $y_r(x_r)$ satisfies suitable boundary conditions, such as those of the usual (separated) Sturm-Liouville type.

3.4   Definiteness

We are now ready to investigate additional hypotheses that ensure that the eigenvalues are real. First, we clarify the latter term. An eigenvalue has been defined in §3.1 as an equivalence class of $(k + 1)$-tuples $(c\lambda_0, \ldots, c\lambda_k)$, for arbitrary (real) $c \neq 0$, and where at least one $\lambda_s \neq 0$. It will be said to be "real" if for some non-zero c all the $c\lambda_s$ are real.

It is not sufficient for this purpose that (3.1.2), (3.1.8) be formally self-adjoint; however, we shall assume the latter as a minimum. We must consider the rank of the k-by-$(k+1)$ matrix

$$\Phi_{rs}(y_r), \quad r = 1, \ldots, k, \ s = 0, \ldots, k, \qquad (3.4.1)$$

where $y_1, \ldots, y_k$ are any set of non-trivial solutions of the boundary-value problem (3.1.8), (3.1.2). We say that the problem is "definite" if (3.4.1) always has its maximal possible rank k.

It will of course be sufficient if the matrix (3.4.1) has rank k for a wider class of sets of functions $y_1, \ldots, y_k$. It may, for example, have rank k for all sets of suitably smooth functions $y_1, \ldots, y_k$, not necessarily satisfying differential equations or boundary conditions. Again, in special cases, we can ensure the rank condition by specifying that some particular k-th order minor of the array (3.4.1) is not zero.

As already indicated, we have

Theorem 3.4.1. *If the system* (3.1.8), (3.1.2) *is formally self-adjoint and definite, then all eigenvalues are real.*

For the proof let $(\lambda_0, \ldots, \lambda_k)$ be an eigenvalue, and let the y_r be associated non-trivial solutions of (3.1.8), (3.1.2). Multiplying (3.1.8) by $\overline{y_r}$ and integrating over (a_r, b_r), we get

$$\sum_0^k \lambda_j \Phi_{rj}(y_r) = 0, \quad r = 1, \ldots, k. \tag{3.4.2}$$

The coefficients (3.4.1) in this set of equations are all real, by the assumption of formal self-adjointness, and form a matrix of rank k, by the assumption of definiteness. Hence $(\lambda_0, \ldots, \lambda_k)$ must be a multiple of some real $(k+1)$-tuple, as asserted.

Under these conditions, the notion of an eigenvalue may be refined to that of an equivalence class of real $(k+1)$-tuples, not all zero, for which (3.1.8) are non-trivially possible. It is possible to make a further refinement, in which the signs of the λ_s are determinate.

To see this, we write for $t = 0, \ldots, k$,

$$\Omega_t(y_1, \ldots, y_k) = (-1)^t \det \Phi_{rs}(y_r), \tag{3.4.3}$$

where the determinant is taken over

$$r = 1, \ldots, k, \quad s = 0, \ldots, k, \ s \neq t. \tag{3.4.4}$$

Thus Ω_t will be the determinant obtained by deleting the $(t+1)$-th column of (3.4.1), and multiplying by $(-1)^t$. The definiteness assumption requires that

$$\Omega_t(y_1, \ldots, y_k) \neq 0 \tag{3.4.5}$$

for at least some t, if the y_r are non-trivial solutions of (3.1.8), (3.1.2); it is sufficient that this be so for any set of suitably smooth non-zero functions.

We have from (3.4.2) that an eigenvalue is given by

$$\lambda_j = c\Omega_j(y_1, \ldots, y_k), \quad j = 0, \ldots, k, \tag{3.4.6}$$

for some $c \neq 0$, where the y_r are non-trivial solutions of (3.1.8), (3.1.2). We can take as a "signed eigenvalue" the equivalence class given by (3.4.6) with arbitrary $c > 0$. It should be mentioned that, in the case of a multiple eigenvalue, the numbers $\Omega_j(y_1, \ldots, y_k)$, $j = 0, \ldots, k$, have fixed signs, or are zero, independently of the choice of the non-trivial solutions y_r.

3.5 Orthogonalities between eigenfunctions

As in many other contexts, the hypotheses of formal self-adjointness and definiteness imply here that eigenfunctions associated with distinct eigenvalues are mutually orthogonal. Two special features of the multiparameter case deserve comment. One is that the eigenfunctions are orthogonal with respect to $k+1$ scalar products. Another is that the orthogonality is established in the first place for decomposable eigenfunctions, and may then be extended by linearity to linear combinations of such eigenfunctions.

To formulate these orthogonalities, we must extend the notation (3.4.3); we write

$$\chi_t(y_1,\dots,y_k;z_1,\dots,z_k) = (-1)^t \det \Psi_{rs}(y_r,z_r), \qquad (3.5.1)$$

$r = 1,\dots,k$, $s = 0,\dots,k$, $s \neq t$. Thus, as in (3.4.3), χ_t is the determinant formed by deleting the t-th column of the array (3.3.1), multiplied by a sign-factor $(-1)^t$. The functions forming the arguments of χ_t must be suitably differentiable, so that (3.3.1) should be defined. We have then

Theorem 3.5.1. *Let the problem* (3.1.8)–(3.1.2) *be formally self-adjoint and definite. Let* $y_1,\dots,y_k$ *be solutions of* (3.1.8)–(3.1.2) *for the eigenvalue* $\lambda_0,\dots,\lambda_k$, *and let* $z_1,\dots,z_k$ *be solutions of the same equations, with the* λ_s *in (3.1.7) replaced by* μ_s, *where* $\mu_0,\dots,\mu_k$ *is a distinct eigenvalue. Then*

$$\chi_t(y_1,\dots,y_k;z_1,\dots,z_k) = 0, \quad t = 0,\dots,k. \qquad (3.5.2)$$

We first multiply (3.1.8) by $\bar{z}_r$, respectively, and integrate over (a_r,b_r). This gives

$$\sum_{0}^{k} \lambda_s \Psi_{rs}(y_r,z_r) = 0, \quad r = 1,\dots,k. \qquad (3.5.3)$$

Likewise,

$$\sum_{0}^{k} \mu_s \Psi_{rs}(z_r,y_r) = 0, \quad r = 1,\dots,k. \qquad (3.5.4)$$

Since the problem is formally self-adjoint and definite, we may take it that the λ_s, μ_s are real. Taking complex conjugates in (3.5.4), and using, (3.3.3), we have

$$\sum_{0}^{k} \mu_s \Psi_{rs}(y_r,z_r) = 0, \quad r = 1,\dots,k. \qquad (3.5.5)$$

We compare this with (3.5.3), and note that the vectors $(\lambda_0, \ldots, \lambda_k)$, $(\mu_0, \ldots, \mu_k)$ are linearly independent; this is what we mean by specifying that the eigenvalues are distinct. We deduce that the k-by-$(k+1)$ matrix

$$\Psi_{rs}(y_r, z_r), \quad r = 1, \ldots, k, \ s = 0, \ldots, k \qquad (3.5.6)$$

must have rank less than k. This is equivalent to the conclusion of Theorem 3.5.1.

We next express (3.5.1) as (not necessarily positive definite) scalar products of the functions

$$y(x_1, \ldots, x_k) = \prod_1^k y_r(x_r), \quad z(x_1, \ldots, x_k) = \prod_1^k z_r(x_r). \qquad (3.5.7)$$

For this purpose, we introduce the operators

$$\Delta_t = (-1)^t \det L_{rs}{}^\dagger, \quad r = 1, \ldots, k, \ s = 0, \ldots, k, \quad s \neq t. \qquad (3.5.8)$$

Here the $L_{rs}{}^\dagger$ are the partial differential operators associated with the L_{rs}, and the "determinants" are evaluated in the usual way, the result being also a partial differential operator. With the notation (3.5.7), we then have

$$
\begin{aligned}
\chi_t(y_1, \ldots, y_k; z_1, \ldots, z_k) &= \int_{a_1}^{b_1} \cdots \int_{a_k}^{b_k} (\Delta_t y) \, \bar{z} \, dx_1 \ldots dx_k, \\
&= (\Delta_t y, z), \qquad (3.5.9)
\end{aligned}
$$

say, where $(\ ,\ )$ denotes the obvious scalar product in $L^2(I)$, I being the k-dimensional interval, which is the cartesian product of the (a_r, b_r). The orthogonality of eigenfunctions, associated with distinct eigenvalues, takes the form

$$(\Delta_t y, z) = 0, \quad t = 0, \ldots, k. \qquad (3.5.10)$$

These scalar products retain sense for suitably differentiable functions y, z, not necessarily decomposable. In the case that (3.1.5) hold, the operators L_{rs} and $L_{rs}{}^\dagger$ are simply given by pointwise multiplication by $p_{rs}(x_r)$, while Δ_0 is given by multiplication by

$$P(x_1, \ldots, x_k) = \det \ p_{rs}(x_r).$$

The orthogonality relation (3.5.10) for $t = 0$ then takes the form

$$(Py, z) = 0. \qquad (3.5.11)$$

If (3.1.4) also holds, the remaining orthogonality relations are given by

$$\left(\sum_{r=1}^k P_{rt} L_{r0} y, \ z \right) = 0, \quad t = 1, \ldots, k, \qquad (3.5.12)$$

where, as previously, P_{rt} is the co-factor of p_{rt} in the determinant P.

3.6   Discreteness properties of the spectrum

In Sturm-Liouville theory, the fact that the eigenvalues have no finite limit-point may be seen in at least three ways; we have for this purpose oscillatory arguments, the use of orthogonality properties, and the fact that the eigenvalues are the zeros of an entire function. The first two of these arguments extend easily to the multiparameter case. Of these, the oscillatory argument will be developed in the next chapter. Here we take up the argument based on orthogonality.

The argument is rather generally applicable, and is not confined to the second-order case. We work in terms of the concept of an eigenvalue as a non-zero $(k+1)$-tuple, and topologize the set of such $(k+1)$-tuples by component-wise convergence. We give sufficient conditions that such a $(k+1)$-tuple not be a limit-point of eigenvalues. We take the latter in real terms, as we confine attention to formally self-adjoint definite cases. We have

Theorem 3.6.1. *Let L_{rs} have the form (2.6.2) with continuous coefficients, and let the problem (3.1.8), (3.1.2) be formally self-adjoint and definite. Let $(\mu_0, \dots, \mu_k)$ be such that the differential equations (3.1.8) are of fixed order, with no singular points, for all sets $(\lambda_0, \dots, \lambda_k)$ in some neighborhood of $(\mu_0, \dots, \mu_k)$. Then $(\mu_0, \dots, \mu_k)$ is not a limit-point of the spectrum.*

For the proof, we assume the contrary, that there is a sequence of eigenvalues

$$\lambda_{0n}, \dots, \lambda_{kn}, \quad n = 1, 2, \dots \tag{3.6.1}$$

such that

$$\lambda_{sn} \to \mu_s, \quad s = 0, \dots, k, \tag{3.6.2}$$

as $n \to \infty$. We may assume that the sets (3.6.1) are real, non-zero, and mutually distinct.

Corresponding to (3.6.1) we denote by

$$y_{rn}(x_r), \quad r = 1, \dots, k, \quad n = 1, 2, \dots \tag{3.6.3}$$

non-trivial solutions of (3.1.8), (3.1.2). We normalize them, for example by,

$$\sum_{m=0}^{m_r-1} |(d/dx_r)^m y_r(a_r)| = 1, \tag{3.6.4}$$

where m_r denotes the maximal order of any of the differential operators L_{rs}, and denote the corresponding eigenfunction by

$$y^{(n)} = \prod_1^k y_{rn}, \quad n = 1, 2, \ldots . \tag{3.6.5}$$

We now note that any two members of (3.6.5) are mutually orthogonal, in the sense (3.5.10). Writing

$$N = \prod_1^k m_r + 1, \tag{3.6.6}$$

we take successive batches of N of (3.6.5), that is to say,

$$y^{(jN+1)}, \ldots, y^{(jN+N)}, j = 1, 2, \ldots . \tag{3.6.7}$$

The members of each batch are mutually orthogonal. We then pick out a subsequence such that, for each $r = 1, \ldots, k$, and each $t = 1, \ldots, N$, the $y_{r,jN+t}(x_r)$ converges uniformly, as j runs through this sequence. This may be achieved by bringing about convergence in the initial data appearing in (3.6.4). This gives, in the limit, N eigenfunctions, associated with the eigenvalue $(\mu_0, \ldots, \mu_k)$; these are mutually orthogonal in the sense (3.5.10), and so are linearly independent. However, the multiplicity of an eigenvalue cannot exceed the product of the orders of the differential equations (3.1.8), and so we have a contradiction.

In the Sturm-Liouville case (3.1.4)–(3.1.5), with homogeneous eigenvalues $(\lambda_{0n}, \ldots, \lambda_{kn})$, we cannot have (3.6.2) if $\mu_0 \neq 0$. In other words, with inhomogeneous eigenvalues obtained by taking $\lambda_0 = 1$, the eigenvalues have no finite limit-point.

3.7 A first definiteness condition, or "right-definiteness"

There are two special ways in which the definiteness condition of §3.4 can be realized. The essence of this condition was that a certain matrix (3.4.1) should have its maximal possible rank for all sets of suitably restricted functions $y_1, \ldots, y_k$. This means that of the $(k+1)$ minors of order k, which can be formed from the rectangular matrix (3.4.1) by deleting one column, at least one must not be zero. It was not specified which one of these minors should not be zero, and it is conceivable that the choice of minor might have to vary with the choice of the functions $y_r(x_r)$. Indeed, there might be no fixed linear combination of these minors that did not vanish for some admissible choice of the functions. An analogous situation was noted in the finite-dimensional case when $k \geq 3$ (see Chapters 9, 10 of Atkinson [1972]).

In the notation (3.5.7)–(3.5.9), the definiteness condition was that

$$(\Delta_t y, y) \neq 0, \tag{3.7.1}$$

for at least one t, $0 \leq t \leq k$, whatever the choice of the $y_r(x_r)$, provided that they satisfy the differential equations for some set of λ_s, and satisfy the boundary conditions; in general, t is allowed to depend on y. We say that the problem (3.1.8), (3.1.2), assumed formally self-adjoint, satisfies the "first definiteness condition," or is "right-definite" if (3.7.1) holds with $t = 0$, for y restricted as above. It will, of course, be sufficient for this purpose that (3.7.1) holds, with $t = 0$, for a wider class of functions, for example, not necessarily satisfying the boundary conditions.

The prime example of this situation is given by the Sturm-Liouville case (3.1.4)–(3.1.5), if the $p_{rs}(x_r)$ are real and continuous, and

$$P = \det p_{rs}(x_r) \geq 0, \tag{3.7.2}$$

with inequality for at least one set of $x_1, \ldots, x_k$. We then have, with $y = y_1, \ldots, y_k$,

$$(\Delta_0 y, y) = \int_{a_1}^{b_1} \cdots \int_{a_k}^{b_k} P|y|^2 dx_1 \cdots dx_k. \tag{3.7.3}$$

The right will be positive if the y_r satisfy the differential equations (3.1.4), and do not vanish identically; here the boundary conditions play no role. The above remarks do not depend on the operators L_{r0}, (3.1.4) having the form there given, or in particular being of second order.

We see from (3.4.5) that if this first definiteness condition holds, then $\lambda_0 \neq 0$ for any eigenvalue. On this basis we can go over to an inhomogeneous formulation, as, for example, in (3.1.4), by taking $\lambda_0 = 1$. It follows easily from Theorem 3.6.1 that in the case (3.1.4)–(3.1.5), assumed formally self-adjoint and satisfying the first definiteness condition, a $(k+1)$-tuple $(\mu_0, \ldots, \mu_k)$ for which $\mu_0 \neq 0$ cannot be a limit-point of a sequence of (homogeneous) eigenvalues. This means that in the inhomogeneous form, with $\lambda_0 = 1$, the eigenvalues, now k-tuples $(\lambda_1, \ldots, \lambda_k)$ as in (3.1.4), have no finite limit. Again, this argument is not restricted to the second-order case.

3.8 A second definiteness condition, or "left-definiteness"

The definiteness condition of the previous section and that which is now to be discussed are both included in the situation that there is a fixed set of real numbers $\rho_0, \ldots, \rho_k$ such that the determinant

$$
\begin{vmatrix}
\rho_0 & \cdots & \rho_k \\
\Phi_{10}(y_1) & \cdots & \Phi_{1k}(y_1) \\
\cdots & \cdots & \cdots \\
\Phi_{k0}(y_k) & \cdots & \Phi_{kk}(y_k)
\end{vmatrix} \neq 0,
\tag{3.8.1}
$$

for any set of non-trivial solutions of (3.1.8), (3.1.2). In the first of these conditions, we have the case that this is so with the choice

$$
\rho_0 = 1, \rho_1 = \cdots = \rho_k = 0.
\tag{3.8.2}
$$

In the case (3.1.4)–(3.1.5), and in the Sturm-Liouville case (3.1.6), this is ensured by "condition A" on the $p_{rs}(x_r)$, without the y_r having to satisfy (3.1.2) or (3.1.8) (see § 4.6 below for the definition of *condition A*).

We now take up the complementary situation that (3.8.1) holds with some fixed set of ρ_s with

$$
\sum_1^k |\rho_s| > 0.
\tag{3.8.3}
$$

With primary reference to the case (3.1.4)–(3.1.5), we shall term this the "second definiteness condition."

We now consider special ways in which this condition may be satisfied. We suppose first that $\rho_0 = 0$. Specializing the sign of the left of (3.8.1), we see that it will be sufficient for this condition that

$$
\Phi_{10}
\begin{vmatrix}
\rho_1 & \cdots & \rho_k \\
\Phi_{21} & \cdots & \Phi_{2k} \\
\cdots & \cdots & \cdots \\
\Phi_{k1} & \cdots & \Phi_{kk}
\end{vmatrix}
+ \cdots + \Phi_{k0}
\begin{vmatrix}
\Phi_{11} & \cdots & \Phi_{1k} \\
\Phi_{21} & \cdots & \Phi_{2k} \\
\cdots & \cdots & \cdots \\
\rho_1 & \cdots & \rho_k
\end{vmatrix} < 0.
\tag{3.8.4}
$$

This in turn will hold if

$$
\Phi_{r0}(y_r) < 0, \quad r = 1, \ldots, k,
\tag{3.8.5}
$$

and if the k determinants appearing in (3.8.4) are all positive, that is, if

$$\begin{vmatrix} \rho_1 & \cdots & \rho_k \\ \Phi_{21} & \cdots & \Phi_{2k} \\ \cdots & \cdots & \cdots \\ \Phi_{k1} & \cdots & \Phi_{kk} \end{vmatrix} > 0, \dots, \quad \begin{vmatrix} \Phi_{11} & \cdots & \Phi_{1k} \\ \Phi_{21} & \cdots & \Phi_{2k} \\ \cdots & \cdots & \cdots \\ \rho_1 & \cdots & \rho_k \end{vmatrix} > 0. \tag{3.8.6}$$

Here the row $\rho_1, \dots, \rho_k$ replaces each row in turn of the array $\Phi_{rs}(y_r)$, $1 \le r$, $s \le k$. The definiteness condition is then that (3.8.5)–(3.8.6) hold for every set of non-trivial solutions of (3.1.8), (3.1.2), for some fixed set $\rho_1, \dots, \rho_k$.

For the case (3.1.4)–(3.1.5), this may be simplified. We postulate that (3.8.5) hold for suitably differentiable $y_r(x_r)$, not identically zero, and satisfying the boundary conditions; reference to the boundary conditions can not now be omitted. In the case of (3.8.6), we postulate that these hold for continuous, not identically zero $y_r(x_r)$. This is the role of what will be termed "condition B" in Chapter 4. If the determinants (3.6.1) are all positive-valued, then (3.8.6) is ensured, in the same way as "condition A" ensures the positivity of (3.7.3).

The condition (3.8.5) will be ensured if the one-parameter Sturm-Liouville problems given by

$$L_{r0}y_r + \lambda y_r = 0, \quad B_r(y_r) = 0, \tag{3.8.7}$$

has only positive eigenvalues.

We consider now the case that $\rho_0 \neq 0$, and that (3.8.3) is still in force. This can be reduced to the previous case. We choose any real numbers τ_s such that

$$\sum_1^k \tau_s \rho_s = \rho_0, \tag{3.8.8}$$

and from the first column in the left of (3.8.1) subtract $\tau_1, \dots, \tau_k$ times the remaining columns. This replaces ρ_0 in (3.8.1) by 0, as in the previous case. In this way, we see that the second definiteness condition will be satisfied if (3.8.6) holds, in virtue of condition B, and if for some $\tau_1, \dots, \tau_k$ we have

$$\Phi_{r0}(y_r) - \sum_1^k \tau_s \Phi_{rs}(y_r) < 0, \quad r = 1, \dots, k, \tag{3.8.9}$$

for suitably differentiable y_r, not identically zero, and satisfying the boundary conditions. As for (3.8.7), this can be replaced by the hypotheses that the Sturm-Liouville problems

$$L_{r0}y_r - \sum_1^k \tau_s p_{rs} y_r + \lambda y_r = 0, \quad B_r(y_r) = 0 \tag{3.8.10}$$

all have only positive eigenvalues.

Notes for Chapter 3

For basic material regarding definiteness conditions, see Atkinson (1972, Chapter 7); in particular, §7.2–7.4 therein. The notion of "local definiteness" was introduced in Atkinson (1972, Chapter 10) and then developed by Volkmer (1988, p. 39). Local definiteness was also used in Binding and Volkmer (1986) to prove existence and uniqueness of eigenvalues of abstract multiparameter problems in Hilbert space.

This chapter introduces the notions of "right-definiteness" (resp. "left definiteness"), which subsequently lead to *Condition A* (resp. *Condition B*) or *the first (resp. second) definiteness condition* in later chapters. It is of historical interest to note that the first use of quadratic forms in proving that the eigenvalues of various Sturm-Liouville problems are real is due to Sturm (1829). He did this for what is currently called a right-definite problem. For a review of Sturm's classical 1836 work, see Hinton (2005) while for a general review of his work in differential equations see Lützen and Mingarelli (2008).

3.9 Research problems and open questions

1. A topic of current interest has been the study of multiparameter problems of the form (cf., [2.1.4])

$$(p_r y_r')'(x_r) - q_r(x_r)y_r(x_r) + \sum_1^k \lambda_s p_{rs}(x_r)\, y_r(x_r) = 0,$$

$a_r \leq x_r \leq b_r$, $r = 1 \ldots k$, which satisfy *eigendependent* boundary conditions of the form

$$(a_{r0}\lambda_r + b_{r0})\, y_r(a_r) = (c_{r0}\lambda_r + d_{r0})\, (p_r y_r')(a_r),$$

$$(a_{r1}\lambda_r + b_{r1})\, y_r(b_r) = (c_{r1}\lambda_r + d_{r1})\, (p_r y_r')(b_r),$$

where for simplicity, and without loss of generality, it may be assumed that $[a_r, b_r] = [0, 1]$ and $p_r \equiv 1$ throughout, $r = 1, \ldots, k$ (see Chapter 1).

These eigendependent boundary problems have been studied by various authors Bhattacharyya *et al.* (2002), (2001), Binding *et al.* (1994), Fulton (1977), Walter (1973), under either one of the definiteness conditions mentioned here. As a result, for example, various oscillation-type theorems have been formulated that include the possibility of "repeated oscillation counts," and this even in these definite cases and, in particular, the one parameter Sturm-Liouville case.

Assuming neither left- nor right-definiteness here (also called non-definite) is there a spectral theory that parallels the Haupt-Richardson theory of non-definite Sturm-Liouville problems? For example, can one infer the countability of the spectrum? For reference purposes, the Haupt-Richardson theory is described in Mingarelli (1986).

2. Let $a_i(x) \in C[a,b]$, $i = 1, 2, \ldots, m$ and consider the linear differential equation of order m

$$\sum_{i=1}^{m} a_i(x)\, y^{(m)}(x) = 0, \quad x \in [a, b].$$

It is easy to see that if there exists a subinterval $I \subset [a, b]$ in which $a_i(x) \equiv 0$, for each $i = 1, 2, \ldots, m$, then the dimension of the solution space of this equation (denoted by ν_r in § 3.2) must exceed the order, m. Is this condition necessary for $\nu_r > m$?

3. Find an abstract version of the results in § 3.2, where the L_{rs} are more generally, linear operators in Hilbert space.

Chapter 4

Determinants of Functions

4.1 Introduction

We recall that in the standard Sturm-Liouville problem

$$y''(x) - q(x)y(x) + \lambda p(x)y(x) = 0, \quad a \le x \le b, \qquad (4.1.1)$$

where $[a, b]$ is a finite real interval, in which $p(x), q(x)$ are real and, say, continuous; we impose boundary conditions of the form

$$y(a)\cos\alpha = y'(a)\sin\alpha, \quad y(b)\cos\beta = y'(b)\sin\beta. \qquad (4.1.2)$$

There are basically three situations in regard to "definiteness," namely,

(i) the usual "right-definite," or classically "orthogonal" case, in which $p(x)$ is positive in $[a, b]$,

(ii) the "left-definite," or classically "polar" case, in which $p(x)$ changes sign, but in which the differential operator has a certain sign-definiteness; there is to be a number μ such that the quadratic form

$$\int_a^b \{y'' - q(x)y - \mu p(x)y\}\, y\, dx \qquad (4.1.3)$$

is negative-definite on the space of real continuously twice differentiable functions satisfying (4.1.2),

53

(iii) the "indefinite," or classically "non-definite" case, in which $p(x)$ changes sign, and the form (4.1.3), subject to (4.1.2) is indefinite for every μ.

This classification is not quite exhaustive; for example, $p(x)$ might have zeros, without changing sign. Likewise, the form (4.1.3) might be semi-definite for some μ. While such borderline cases are often of special interest and importance, the above represent the main possibilities.

Both cases (i) and (ii) have the desirable feature that the eigenvalues are all real and have no finite limit-point. However, these cases differ in regard to the oscillatory characterization and limiting behavior of the eigenvalues. For any eigenvalue λ, we use the term "oscillation number" for the number of zeros in the open interval (a, b) of a solution of (4.1.1)–(4.1.2). As is well known, in case (i), the usual one, there is precisely one eigenvalue for which the oscillation number is any assigned non-negative integer; furthermore, if the eigenvalues are indexed according to ascending oscillation numbers, they form an ascending sequence, with $+\infty$ as its only limit. In case (ii), the polar case, there are precisely two eigenvalues for each oscillation number; these form an ascending and a descending sequence, tending to $+\infty$ and also to $-\infty$. In case (iii), we have the ascending and descending sequence of eigenvalues, for sufficiently large oscillation numbers, but may also have a finite number of complex eigenvalues. In degenerate cases, every complex number may be an eigenvalue.

This classification for Sturm-Liouville operators separates into two components, the sign-behavior of the scalar function $p(x)$ and the definiteness of a quadratic form. This situation extends to the standard single-column type multiparameter Sturm-Liouville problem, in which the differential operators appear in the first column only of the operator-array. We have now to consider, for the analogue of (i), the sign-constancy or otherwise of the determinant $\det p_{rs}(x_r)$ in (2.1.4) and for the analogue of (ii) certain subdeterminants. This topic turns out to be richer than might appear at first sight and will be the subject of this chapter.

Accordingly, we study here, with only incidental reference to differential operators, determinants of functions, in which those in each row are functions of the same variable, a different one for each row. The term "Stäckel determinant" is sometimes used for such determinants.

4.2 Multilinear property

We first note some simple identities. The first expresses the fact that a determinant is a linear function of each of its rows. In each of the intervals $[a_r, b_r]$, we select m_r points, and attach to them weights. We write $M = m_1 + \cdots + m_k$. We have then

Theorem 4.2.1. *For any set of* k^2 *functions*

$$p_{rs}(x_r), \quad a_r \le x_r \le b_r, \ 1 \le r, s \le k, \tag{4.2.1}$$

any set of M *points*

$$x_{ru} \in [a_r, b_r], \quad u = 1, \ldots, m_r, \ r = 1, \ldots, k, \tag{4.2.2}$$

and any M *scalars*

$$\sigma_{ru}, \quad u = 1, \ldots, m_r, \ r = 1, \ldots, k, \tag{4.2.3}$$

we have

$$\det \sum_{u=1}^{m_r} \sigma_{ru} p_{rs}(x_{ru}) = \sum_{u_1=1}^{m_1} \cdots \sum_{u_k=1}^{m_k} \left(\prod_{1}^{k} \sigma_{ru_r} \right) \det p_{rs}(x_{ru_r}). \tag{4.2.4}$$

In the above, the determinants are extended over $1 \le r, s \le k$. This is a case of a general identity for multilinear functions. As a corollary we note

Theorem 4.2.2. *If the determinant* $\det p_{rs}(x_r)$ *is strictly positive, for all sets of arguments* $x_1, \ldots, x_k$, *and the* σ_{ru_r} *are all positive, then the determinant on the left of* (4.2.4) *is also positive.*

In connection with the orthogonality of eigenfunctions, we need an analogous result, in which sums are replaced by integrals. For this purpose, the functions $p_{rs}(x_r)$ must be restricted in some way.

Theorem 4.2.3. *Let the functions* (4.2.1) *be Lebesgue-integrable over the respective intervals* (a_r, b_r), *and let the functions*

$$f_r(x_r), \quad a_r \le x_r \le b_r, \ r = 1, \ldots, k, \tag{4.2.5}$$

be continuous. Then

$$\det \int_{a_r}^{b_r} p_{rs}(x_r) f_r(x_r) \, dx_r = \int_{a_1}^{b_1} \cdots \int_{a_k}^{b_k} \{\det p_{rs}(x_r)\} \left\{ \prod_{1}^{k} f_r(x_r) \, dx_r \right\} \tag{4.2.6}$$

In this case, we give an explicit proof. Using the expansion of a determinant as a sum of signed products, we see that left of (4.2.6) equals

$$\sum \varepsilon_\sigma \prod_1^k \int_{a_r}^{b_r} p_{r\sigma(r)}(x_r) f_r(x_r) dx_r,$$

where ε_σ is 1 or -1, according to whether the permutation $\sigma(1), \ldots, \sigma(k)$ of $1, \ldots, k$, is even or odd. We thus derive

$$\int_{a_1}^{b_1} \cdots \int_{a_k}^{b_k} \sum \varepsilon_\sigma \left\{ \prod_1^k p_{r\sigma(r)}(x_r) \right\} \left\{ \prod_1^k f_r(x_r) \, dx_r \right\}$$

which is the same as the right of (4.2.6).

4.3   Sign-properties of linear combinations

In this and the next section, we derive two distinct criteria for

$$\det p_{rs}(x_r) \tag{4.3.1}$$

to have fixed sign, for all sets of arguments

$$a_r \le x_r \le b_r, \quad r = 1, \ldots, k. \tag{4.3.2}$$

Here, to begin with, "fixed sign" is meant strictly, and excludes that (4.3.1) should vanish. Both the criteria are in terms of the sign-properties of linear combinations of the p_{rs} of the form

$$\sum \mu_s p_{rs}(x_r), \quad r = 1, \ldots, k, \tag{4.3.3}$$

for real sets $\mu_1, \ldots, \mu_k$, not all zero. The results of this section are concerned with arbitrary sets μ_s, not all zero; in the next section, we are concerned to choose the μ_s in special ways.

In the basic result of the first kind, we do not need that the $p_{rs}(x_r)$ be continuous.

Theorem 4.3.1. *In order that* (4.3.1) *have fixed sign, it is necessary and sufficient that for every real set $\mu_1, \ldots, \mu_k$ not all zero, at least one of the functions* (4.3.3) *should have fixed sign.*

Again, "fixed sign" excludes that the function in question may vanish.

We prove first the sufficiency. Since this can be proved very simply when the $p_{rs}(x_r)$ are continuous, we deal with this case first. Assuming that for every set of μ_s, not all zero, at least one of (4.3.3) does not vanish, we have that $\det p_{rs}(x_r)$ does not vanish; since it is continuous, and since its domain, the k-dimensional interval I (2.6.3) is connected, we deduce that $\det p_{rs}(x_r)$ has fixed sign.

We now prove the sufficiency in the general case. As before, we have that

$$P(x_1, \ldots, x_k) = \det p_{rs}(x_r)$$

does not vanish, where the determinant extends over $1 \le r, s \le k$; we need a different argument to show that it has fixed sign, since it need not be continuous. For some set of arguments

$$x_{r0} \in [a_r, b_r], \quad r = 2, \ldots, k, \tag{4.3.4}$$

we choose $\mu_s = P_{1s}(x_{20}, \ldots, x_{k0})$, $s = 1, \ldots, k$, that is to say the co-factors of $p_{1s}(x_1)$, $s = 1, \ldots, k$, in the determinant $P(x_1, x_{20}, \ldots, x_{k0})$; here we use the notation above, and pass over the trivial case $k = 1$. With this choice for the μ_s, we have

$$\sum \mu_s p_{1s}(x_1) = P(x_1, x_{20}, \ldots, x_{k0}), \tag{4.3.5}$$

$$\sum \mu_s p_{rs}(x_{r0}) = 0, \quad r = 2, \ldots, k. \tag{4.3.6}$$

Since $\det p_{rs}(x_r) = P(x_1, \ldots, x_k)$ does not vanish, we see that the μ_s are not all zero, so that at least one of the functions (4.3.3) must have fixed sign. In view of (4.3.6), this must be the left of (4.3.5). This shows that $P(x_1, x_2, \ldots, x_k)$ has fixed sign as a function of x_1, for any set of fixed $x_2, \ldots, x_k$.

The same reasoning may be applied to the other variables, and we deduce that the sign of $P(x_1, \ldots, x_k)$ remains unchanged if any one of its arguments is varied, the others remaining fixed. Since we may pass from any point of I to any other point by changing one coordinate at a time, we deduce the result.

We now prove the necessity, and so assume that $\det p_{rs}(x_r)$ has fixed sign. Arguing by contradiction, we suppose that for some set of μ_s, not all zero, none of the functions (4.3.3) has fixed sign, so that they either take both signs or have a zero. Thus, for each r, $1 \le r \le k$, we can choose a pair $x_{r1}, x_{r2} \in [a_r, b_r]$ such that the values

$$\sum \mu_s p_{rs}(x_{rt}), \quad t = 1, 2,$$

either have opposite signs or are both zero, with $x_{r1} = x_{r2}$. Thus, in either case, we can find positive numbers σ_{r1}, σ_{r2}, such that

$$\sum_{t=1}^{2} \sigma_{rt} \sum_{1}^{k} \mu_s p_{rs}(x_{rt}) = 0, \quad r = 1, \ldots, k; \tag{4.3.7}$$

Since the μ_s are not all zero, it follows from (4.3.7) by elimination that

$$\det \sum_{t=1}^{2} \sigma_{rt} p_{rs}(x_{rt}) = 0. \qquad (4.3.8)$$

Here the left can be expanded as a linear combination of expressions of the form $\det p_{rs}(x_r)$, coefficients that are positive, since the σ_{rt} are all positive. This contradicts Theorem 4.2.2, and so completes the proof of Theorem 4.3.1.

We need also a uniform version of this result.

Theorem 4.3.2. *Let the $p_{rs}(x_r)$, $a_r \leq x_r \leq b_r$, $1 \leq r$, $s \leq k$, be continuous, and let $\det p_{rs}(x_r)$ have fixed sign. Then there is a $\delta > 0$ such that the inequality*

$$\left| \sum \mu_s p_{rs}(x_r) \right| \geq \delta \sum |\mu_s|, \quad r = 1, \ldots, k, \qquad (4.3.9)$$

is valid for at least one r; here r, but not δ, may depend on the choice of the μ_s.

It is clearly sufficient to prove this on the supposition that

$$\sum |\mu_s| = 1. \qquad (4.3.10)$$

If the result were untrue, there would be sequences

$$\mu_{sn}, x_{rn}, \delta_n, \quad n = 1, 2, \ldots; \quad r, s = 1, \ldots, k \qquad (4.3.11)$$

such that

$$\sum |\mu_{sn}| = 1, \quad x_{rn} \in [a_r, b_r], \quad \delta_n \to 0, \qquad (4.3.12)$$

and such that

$$\left| \sum \mu_{sn} p_{rs}(x_{rn}) \right| < \delta_n, \quad r = 1, \ldots, k. \qquad (4.3.13)$$

We can then choose an n-sequence such that all the μ_{sn}, x_{rn} converge; we write

$$\mu_s^* = \lim \mu_{sn}, \quad x_r* = \lim x_{rn},$$

as $n \to \infty$ through this sequence. Passing to the limit in (4.3.13) we get

$$\sum \mu_s^* p_{rs}(x_r*) = 0, \quad \sum |\mu_s^*| = 1,$$

and this contradicts the hypothesis that $\det p_{rs}(x_r) \neq 0$.

The continuity assumptions on $p_{rs}(x_r)$ can be relaxed. It will often be convenient to consider the set of points

$$C_r = \{(p_{r1}(x_r), \ldots, p_{rk}(x_r)) |\ a_r \leq x_r \leq b_r\}. \qquad (4.3.14)$$

It is easily seen that Theorem 4.3.2 remains in force if we assume that $\det p_{rs}(x_r)$ does not vanish, and that the sets $C_1, \ldots, C_k$ are compact. One may also relax the hypothesis that the domains of the $p_{rs}(x_r)$ are finite closed intervals.

4.4 The interpolatory conditions

We now pass to a second type of equivalence between the positivity of the determinant (4.3.1) and the sign-behavior of linear combinations (4.3.3). These conditions deal not with the behavior of arbitrary linear combinations, but with the existence of those with special sign-behavior.

Theorem 4.4.1. *Let the sets of points C_r (see (4.3.14)) be compact. Then in order that $\det p_{rs}(x_r)$ be of fixed sign it is necessary and sufficient that for any set of sign-factors $(\varepsilon_1, \ldots, \varepsilon_k)$, all equal to 1 or -1, there should exist a set $(\mu_1, \ldots, \mu_k)$ such that*

$$\varepsilon_r \sum \mu_s p_{rs}(x_r) > 0, \quad a_r \le x_r \le b_r, \quad r = 1, \ldots, k. \qquad (4.4.1)$$

We start with the sufficiency. Let the $p_{rs}(x_r)$, $a_r \le x_r \le b_r$, $1 \le r$, $s \le k$, be continuous.[1] We suppose the condition regarding (4.4.1) fulfilled, but that $P(x_1, \ldots, x_k)$, defined in §3.5, satisfies $P(x_{10}, \ldots, x_{k0}) = 0$ for some set of x_{r0}, and derive a contradiction. Since $P(x_{10}, \ldots, x_{k0}) = 0$, there must exist a set of real $(\beta_1, \ldots, \beta_k)$, not all zero, such that

$$\sum \beta_r p_{rs}(x_{r0}) = 0, \quad s = 1, \ldots, k. \qquad (4.4.2)$$

Thus, for arbitrary $\mu_1, \ldots, \mu_k$, we have

$$\sum \beta_r \sum \mu_s p_{rs}(x_{r0}) = 0. \qquad (4.4.3)$$

We now suppose the ε_r in (4.4.1) chosen so that $\varepsilon_r = +1$ if $\beta_r \ge 0$, and $\varepsilon_r = -1$ if $\beta_r < 0$, so that $\varepsilon_r \beta_r \ge 0$ for all r, with inequality in at least one case. If the μ_s are chosen so that (4.4.1) holds, then the terms in the r-sum on the left of (4.4.3) will be non-negative, with at least one being positive. This gives a contradiction, and proves the sufficiency.

We note that the compactness of the C_r has not been used in this part of the proof. For an argument that does not use the continuity of the p_{rs}, see the Notes at the end.

We pass to the proof of the necessity, which depends on a topological argument. We assume that $\det p_{rs}(x_r)$ has fixed sign, and forms a certain map from real k-dimensional space $\mathbf{R}^k$ to itself, which we write

$$(\mu_1, \ldots, \mu_k) \mapsto (\nu_1, \ldots, \nu_k). \qquad (4.4.4)$$

---

[1] A slightly more general argument here due to Prof. H. Volkmer may be found in the Notes.

Here we define

$$\nu_r = \min \sum \mu_s p_{rs}(x_r), \tag{4.4.5}$$

if the minimum over $[a_r, b_r]$ is positive,

$$\nu_r = \max \sum \mu_s p_{rs}(x_r), \tag{4.4.6}$$

if this maximum is negative, and otherwise take $\nu_r = 0$. Since the C_r are compact, these extrema are attained, so that we may legitimately write "min" end "max," rather than "inf" and "sup."

We need three properties of the map (4.4.4). Firstly, if the μ_s are not all zero, then the ν_r are not all zero. This follows from Theorem 4.3.1; if the μ_s are not all zero, then at least one of (4.3.3) is bounded away from zero. Secondly, if the μ_s are reversed in sign, then so are the ν_r; (4.4.4) implies a correspondence

$$(-\mu_1, \ldots, -\mu_k) \mapsto (-\nu_1, \ldots, -\nu_k). \tag{4.4.7}$$

Thirdly, the map (4.4.4) is continuous; for each r, ν_r is a continuous function of $(\mu_1, \ldots, \mu_k)$.

We discuss this last point in more detail, and for this purpose express the definitions (4.4.5)–(4.4.6) in another way. Let $f_r(\mu_1, \ldots, \mu_k)$ denote the right of (4.4.5) and $g_r(\mu_1, \ldots, \mu_k)$ the right of (4.4.6). These are continuous functions; this follows from the compactness of the C_r. Our definition of ν_r may be re-formulated as

$$\nu_r = \max\{f_r(\mu_1, \ldots, \mu_k), 0\} + \min\{g_r(\mu_1, \ldots, \mu_k), 0\}.$$

Here the first summand on the right is continuous, since it is the maximum of two continuous functions; similarly, the last summand is continuous, and so ν_r must be a continuous function of the μ_s.

We derive from (4.4.4) a normalized map, involving unit vectors. We write for this purpose

$$\|\mu\| := \sqrt{\sum \mu_s^2},$$

and restrict the domain of (4.4.4) to the unit sphere $\|\mu\| = 1$. With each such point $(\mu_1, \ldots, \mu_k)$, we associate the point $(\omega_1, \ldots, \omega_k)$, where

$$\omega_r = \nu_r / \|\nu_k\|, \quad r = 1, \ldots, k.$$

We thus derive a map

$$(\mu_1, \ldots, \mu_k) \mapsto (\omega_1, \ldots, \omega_k),$$

which takes the unit sphere into itself, is continuous, and takes a pair of antipodal points into another such pair, as in (4.4.4), (4.4.7). According to the well-known theorem of Borsuk, such a map must be onto. In particular, it must be possible to choose the μ_s so that the ω_r, and so also the ν_r, have any prescribed collection of signs. This is what we had to prove for the "necessity" in Theorem 4.4.1.

4.5   Geometrical interpretation

We suppose the $p_{rs}(x_r)$ to be continuous, so that the point-sets C_r (4.3.14) will be continuous arcs in $\mathbf{R}^k$. Theorem 4.4.1 may be re-formulated by saying that in order that no $(k-1)$-dimensional subspace of $\mathbf{R}^k$ intersect all of the C_r, it is necessary and sufficient that there exist subspaces separating the C_r in any manner. For example, there must exist a subspace with all of the C_r on the same side of it, another subspace with C_1 on one side and $C_2, \dots, C_k$ on the other side, and so on.

In the case $k = 2$, we have two continuous arcs, C_1 and C_2, in the real plane. The statement is that in order that no line through the origin meet both curves, there must exist two lines through the origin, one of which separates the curves, the other having both curves on the same side of it; neither of these two lines should meet the curves.

We can re-formulate this simple proposition in polar-coordinate terms, as

Theorem 4.5.1. *Let the non-zero vector-functions*

$$\{p_{r1}(x_r), p_{r2}(x_r)\}, \quad a_r \le x_r \le b_r, \ r = 1, 2, \tag{4.5.1}$$

be continuous. Then in order that $\det p_{rs}(x_r)$ *be positive, for all pairs* x_1, x_2, *it is necessary and sufficient that there exist* θ_1, θ_2, *such that*

$$\theta_1 < \theta_2 < \theta_1 + \pi, \tag{4.5.2}$$
$$\theta_1 < \arg\{p_{11}(x_1) + ip_{12}(x_1)\} < \theta_2, \quad a_1 \le x_1 \le b_1, \tag{4.5.3}$$
$$\theta_2 < \arg\{p_{21}(x_2) + ip_{22}(x_2)\} < \theta_1 + \pi, \quad a_2 \le x_2 \le b_2, \tag{4.5.4}$$

for suitable determinations of the "arg" function.

4.6   An alternative restriction

So far we have been concerned with the condition that $\det p_{rs}(x_r)$ have fixed sign; this forms a natural extension of the classical condition on (4.1.1) that $p(x)$ be positive. There is a second type of restriction of scarcely less interest, on which we rely in the "polar" case in which $\det p_{rs}(x_r)$ changes sign; this second condition is non-trivial only if $k > 1$.

The condition that $\det p_{rs}(x_r)$ have fixed sign will be termed "condition (A)." We are now concerned with a different "condition (B)," which may also be imposed on our k-by-k array of functions

$$p_{rs}(x_r), \quad a_r \leq x_r \leq b_r, \ r, s = 1, \ldots, k.$$

This is that there should exist a set of real numbers $\rho_1, \ldots, \rho_k$, such that the k determinants, of order k, given by substituting this set for each row in turn of the array $p_{rs}(x_r)$, should all be positive. Thus there should exist $\rho_1, \ldots, \rho_k$, such that the k determinants

$$
\begin{vmatrix} \rho_1 & \cdots & \rho_k \\ p_{21} & \cdots & p_{2k} \\ \cdots & \cdots & \cdots \\ p_{k1} & \cdots & p_{kk} \end{vmatrix}, \ldots,
\begin{vmatrix} p_{11} & \cdots & p_{1k} \\ p_{21} & \cdots & p_{2k} \\ \cdots & \cdots & \cdots \\ \rho_1 & \cdots & \rho_k \end{vmatrix}
\tag{4.6.1}
$$

take positive values only, for all values of the x_r.

Equivalently, in condition (B) we ask that there exist $\rho_1, \ldots, \rho_k$, such that, for all sets $x_1, \ldots, x_k$,

$$\sum \rho_s P_{rs} > 0, \quad r = 1, \ldots, k, \tag{4.6.2}$$

where

$$P_{rs} = P_{rs}(x_1, \ldots, \hat{x}_r, \ldots, x_k)$$

is the co-factor of p_{rs} in $P = \det p_{rs}(x_r)$; the notation indicates that P_{rs} does not depend on x_r. Here the r-th expression in (4.6.1)–(4.6.2) is a function of all the $x_1, \ldots, x_k$, except x_r.

In particular, it is sufficient for condition (B), but not necessary, that the co-factors of the elements in any one column be positive, for all sets $x_1, \ldots, x_k$.

It is possible for condition (B) to hold, with or without condition (A). The following result is useful in the latter case.

Theorem 4.6.1. *Let the sets C_r, $r = 1, \ldots, k$, of (4.3.14) be compact, and let condition (B) hold. Then precisely one of the following eventualities holds:*

(i) *for every real set $\mu_1, \ldots, \mu_k$, not all zero, the inequality*

$$\sum \mu_s p_{rs}(x_r) < 0 \tag{4.6.3}$$

is valid for at least one r, $1 \leq r \leq k$, and at least one $x_r \in [a_r, b_r]$

(ii) *for some $\tau = \pm 1$ we have*

$$\tau \det p_{rs}(x_r) \geq 0. \tag{4.6.4}$$

with inequality for at least one set of x_r.

(iii) for some set $\mu_1, \ldots, \mu_k$, not all zero,

$$\sum \mu_s p_{rs}(x_r) = 0, \quad a_r \leq x_r \leq b_r, \ r = 1, \ldots, k. \tag{4.6.5}$$

We remark that if (i) holds, then the reversed inequality must also hold, again for some r and some associated x_r; this follows on applying (i) with the signs of the μ_s reversed.

We show first that if (i), (iii) do not hold, then (ii) must hold; this will show that at least one of the cases (i), (ii), (iii) must hold. We complete the proof by showing that the three cases are mutually exclusive.

Supposing that (i) does not hold, we have that there exists a set $\mu_1, \ldots, \mu_k$, not all zero, such that

$$\sum \mu_s p_{rs}(x_r) \leq 0, \quad a_r \leq x_r \leq b_r, \ r = 1, \ldots, k, \tag{4.6.6}$$

and since (iii) does not hold, we must have inequality in (4.6.6) for at least one pair r, x_r. We suppose, in virtue of condition (B), that $\rho_1, \ldots, \rho_k$ are chosen so that (4.6.1)–(4.6.2) take only positive values. We denote the r-th expression in (4.6.1) or (4.6.2) by D_r. By the standard determinantal identity

$$\sum_{r=1}^{k} p_{rs} P_{rt} = \begin{cases} 0, & s \neq t \\ P, & s = t \end{cases},$$

we have

$$\sum D_r \sum \mu_s p_{rs}(x_r) = \left\{ \sum \mu_t \rho_t \right\} \det p_{rs}(x_r). \tag{4.6.7}$$

Here the left is non-negative, but does not vanish identically; more precisely, there is a certain r, and a certain x_r, such that the left of (4.6.6) is positive, and so also the left of (4.6.7). Thus

$$\sum \mu_t \rho_t \neq 0 \tag{4.6.8}$$

and the determinant in (4.6.7) cannot change sign. This shows that (ii) holds.

It remains to show that no two of (i), (ii), and (iii) can hold together. Since (iii) implies that $\det p_{rs}(x_r)$ vanishes identically, we cannot have (i) and (iii) together, since (4.6.3), (4.6.5) are incompatible. Similarly, we cannot have (ii) and (iii) together.

The case of (i) and (ii) is more difficult. Suppose first that we have (ii) with strict inequality in (4.6.4). By Theorem 4.4.1, we can then choose the μ_s so that the left of (4.6.3) is positive for all r and x_r, in contradiction to (i). Consider next the case that (ii) holds as stated. We consider a perturbed matrix given by

$$p_{rs}(x_r, \varepsilon) = p_{rs}(x_r) + \varepsilon\tau\rho_s, \quad 1 \le r, s \le k. \tag{4.6.9}$$

For this, we have, for small ε

$$\det p_{rs}(x_r, \varepsilon) = \det p_{rs}(x_r) + \varepsilon\tau \sum D_r + \mathrm{O}(\varepsilon^2) \tag{4.6.10}$$

Thus, for small $\varepsilon > 0$, the left is strictly positive, or strictly negative, for all sets x_r, according to the sign of τ in (4.6.4). The perturbation (4.6.9) does not affect the compactness of the sets C_r, and so we can apply Theorem 4.4.1 to the perturbed matrix. Hence for some set $\mu_s(\varepsilon)$, we have

$$\sum \mu_s(\varepsilon) p_{rs}(x_r, \varepsilon) > 0, \quad a_r \le x_r \le b_r, \ r = 1, \ldots, k. \tag{4.6.11}$$

We normalize the $\mu_s(\varepsilon)$ by $\sum |\mu_s(\varepsilon)| = 1$, make $\varepsilon \to 0$ through some positive sequence, and select a subsequence such that the $\mu_s(\varepsilon)$ all converge. Proceeding to the limit in (4.6.11) we obtain a contradiction to (4.6.3). This completes the proof of Theorem 4.6.1.

The situation (i) of the last theorem may equivalently be expressed in a uniform version.

Theorem 4.6.2. *Let the functions $p_{rs}(x_r)$ be bounded, and for every set $\mu_1, \ldots,$ μ_k, not all zero, suppose that there exists an r and an x_r satisfying (4.6.3). Then there is a fixed $\delta > 0$ such that the inequality*

$$\sum \mu_s p_{rs}(x_r) < -\delta \sum |\mu_s| \tag{4.6.12}$$

holds for some r, and some x_r. If, more specially, the $p_{rs}(x_r)$ are continuous, there exist $\delta > 0$, $\eta > 0$, such that (4.6.12) holds, for some r, in an x_r-interval of length η.

Like Theorem 4.3.2, this can be proved by contradiction, using the Bolzano-Weierstrass theorem. As is commonly the case, the Heine-Borel theorem can also be used. It is sufficient to consider sets μ_s such that

$$\sum |\mu_s| = 1. \tag{4.6.13}$$

For every point $\mu_{10}, \ldots, \mu_{k0}$ on this unit sphere, there will be a $\zeta > 0$ such that

$$\sum \mu_{s0} p_{rs}(x_r) < -\zeta$$

for some r and some x_r, and so a $\xi > 0$ such that

$$\sum \mu_s p_{rs}(x_r) < -\zeta/2$$

if

$$\sum |\mu_s - \mu_{so}| < \xi;$$

here we use the boundedness of the $p_{rs}(x_r)$. By the Heine-Borel theorem, we can cover the compact set described by (4.6.13) by means of a finite number of such neighbourhoods; we then have (4.6.12), where δ is the least of the numbers $\zeta/2$ occurring.

The second part of the theorem follows from general properties of continuous functions on compact intervals; finiteness of the intervals was not actually required for the first part of the theorem.

4.7 A separation property

For the property (condition (A)) that $\det p_{rs}(x_r)$ had fixed sign, we found in Section 4.3 an equivalent property for the sign-behavior of arbitrary non-trivial linear combinations $\sum \mu_s p_{rs}(x_r)$. We now find a rather similar requirement in connection with condition (B), defined in the last section, with the ρ_s taken as known. Passing over the trivial case $k = 1$, we have

Theorem 4.7.1. *In order that (4.6.1) all have the same fixed sign, for some given set $\rho_1, \ldots, \rho_k$, and all sets of values of the x_r, it is necessary and sufficient that for every set of μ_s, not all zero and such that*

$$\sum \mu_s \rho_s = 0 \tag{4.7.1}$$

the k functions

$$\sum \mu_s p_{rs}(x_r), \quad a_r \leq x_r \leq b_r, \ r = 1, \ldots, k, \tag{4.7.2}$$

must contain one taking only positive values, and one taking only negative values.

We start with the necessity; changing the signs of the ρ_s if necessary, we assume that all of (4.6.1) (or (4.6.2)) are positive-valued. Denoting them as before by $D_1, \ldots, D_k$ we have by (4.6.7), (4.7.1) that

$$\sum D_r \sum \mu_s p_{rs}(x_r) = 0. \tag{4.7.3}$$

Since the D_r are all positive, we conclude that for any set $x_1, \ldots, x_k$, either the functions (4.7.2) are all zero, or else the values of these functions contain at least one positive and at least one negative. The first case may be rejected, since it would then follow, in view of (4.7.1) that all of the determinants (4.6.1), were zero. Thus, for each set $x_1, \ldots, x_k$, the k numbers

$$\sum \mu_s p_{rs}(x_r), \quad r = 1, \ldots, k, \tag{4.7.4}$$

must contain a positive and a negative number. Thus, if the μ_s are not all zero and satisfy (4.7.1) one of the functions (4.7.2) takes positive values only, and one negative values only.

We now prove the sufficiency, and so assume that for any non-zero set $\mu_1, \ldots, \mu_k$ satisfying (4.7.1), at least one of (4.7.2) is positive-valued, and at least one negative-valued. We show that the first of (4.6.1), which we now write as $D_1(x_2, \ldots, x_k)$, has the same sign as $D_n = D_n(x_1, \ldots, x_{n-1}, x_{n+1}, \ldots, x_k)$, for any $n \geq 2$.

We choose $\mu_1, \ldots, \mu_k$ to be the co-factors of $p_{n1}(x_n), \ldots, p_{nk}(x_n)$ in D_1, where $n \neq 1$. Then $\sum \mu_s \rho_s$ will be the expansion of a determinant in which the row $\rho_1, \ldots, \rho_k$ is repeated, and so (4.7.1) will hold. For similar reasons, we have

$$\sum \mu_s p_{rs}(x_r) = 0, \quad r \neq 1, n, \tag{4.7.5}$$

and

$$\sum \mu_s p_{ns}(x_n) = D_1(x_2, \ldots, x_k), \tag{4.7.6}$$

$$\sum \mu_s p_{1s}(x_1) = -D_n(x_1, \ldots, x_{n-1}, x_{n+1}, \ldots, x_k). \tag{4.7.7}$$

Here (4.7.6) is immediate, while the left of (4.7.7) is the expansion of a determinant derived from D_1 by replacing the $p_{ns}(x_n)$ by the $p_{1s}(x_1)$; this determinant may be derived from D_n by interchanging two rows, and this gives the minus sign in (4.7.7).

Since (4.7.5)–(4.7.7) must contain on the left a positive and a negative number, this being our hypothesis, and since those in (4.7.5) are all zero, we have that (4.6.6)–(4.6.7) are not zero and have opposite signs. Thus

$$D_1(x_2, \ldots x_k), \quad D_n(x_1, \ldots, x_{n-1}, x_{n+1}, \ldots, x_k)$$

have the same sign. Since the latter does not depend on x_n we have that the sign of $D_1(x_2, \ldots, x_k)$ remains unchanged if we vary any one of the arguments $x_2, \ldots, x_k$. Since we can change any set of values of these arguments into any other set of values by varying one argument at a time, we have that $D_1(x_2, \ldots, x_k)$ has fixed sign; likewise, $D_n(x_1, \ldots, x_{n-1}, x_{n+1}, \ldots, x_k)$ must have fixed sign for each $n = 2, \ldots, k$, the same sign as that of $D_1(x_2, \ldots, x_k)$. This proves Theorem 4.7.1.

It is useful, particularly in the construction of examples, to express the result in the geometrical language of Section 4.5. We have that in order that the determinants (4.6.1) have the same fixed sign, it is necessary and sufficient that every $(k-1)$-dimensional subspace containing the vector $(\rho_1, \ldots, \rho_k)$ should separate two of the sets $C_1, \ldots, C_k$.

We have the following uniform variant.

Theorem 4.7.2. *Let the $p_{rs}(x_r)$, $a_r \leq x_r \leq b_r$, be continuous, and let the determinants (4.6.1) be all positive, for some fixed set $\rho_1, \ldots, \rho_k$ and all $x_1, \ldots, x_k$. Then there is a $\delta > 0$ such that for any set $\mu_1, \ldots, \mu_k$ satisfying (4.7.1) and such that*

$$\sum |\mu_s| = 1 \tag{4.7.8}$$

we have

$$\sum \mu_s p_{us}(x_u) \geq \delta, \quad a_u \leq x_u \leq b_u, \tag{4.7.9}$$

$$\sum \mu_s p_{vs}(x_v) \leq -\delta, \quad a_v \leq x_v \leq b_v, \tag{4.7.10}$$

for some u, $v \in \{1, \ldots, k\}$.

It follows from Theorem 4.7.1 and our continuity assumptions that, for every set $\mu_1, \ldots, \mu_k$ satisfying (4.7.1), (4.7.8), there is a $\delta > 0$ and some u, v such that (4.7.9)–(4.7.10) hold. Thus, for some $\delta > 0$, we shall have, throughout the respective intervals,

$$\sum \mu_s{}^* p_{us}(x_u) \geq \delta/2, \quad \sum \mu_s{}^* p_{vs}(x_v) \leq -\delta/2, \tag{4.7.11}$$

for sets of values of $\mu_s{}^*$ such that

$$\sum |\mu_s{}^* - \mu_s| < \varepsilon. \tag{4.7.12}$$

By the Heine-Borel theorem, the set of $\mu_1, \ldots, \mu_k$ satisfying (4.7.1), (4.7.8) can be covered by a finite number of sets of the form (4.7.12). We then get the result on replacing δ by the least of the lower bounds appearing in results of the form (4.7.11).

4.8 Relation between the two main conditions

We now compare the two situations with which we were concerned above in regard to the array of functions $p_{rs}(x_r)$, $1 \leq r$, $s \leq k$, $a_r \leq x_r \leq b_r$. We refer to these conditions as

(A) the k-th order determinant $\det p_{rs}(x_r)$ has fixed sign, for all sets of arguments $x_1, \ldots, x_k$,

(B) there exists a real set $\rho_1, \ldots, \rho_k$ such that the determinants (4.6.1) have the same fixed sign, for all sets of arguments $x_1, \ldots, x_k$.

We have

Theorem 4.8.1. *If the $p_{rs}(x_r)$ are continuous, then condition (A) is more special than condition (B) if $k = 1, 2$, while if $k \geq 3$, either condition may hold with or without the other.*

If $k = 1$, condition (A) means that a single function $p_{11}(x_1)$ has fixed sign, as in the standard case for boundary-value problems of Sturm-Liouville type. If $k = 1$, condition (B) means that in the first-order determinant formed by $p_{11}(x_1)$ we may replace the only row by a constant so as to get a determinant of fixed sign; thus condition (B) holds trivially if $k = 1$, while condition A may or may not do so.

We discuss the case $k = 2$ in the geometrical language of Section 4.5. Here condition (A) means that the arcs C_1, C_2 can be separated by a line L_1 through the origin, while a second such line L_2 has C_1, C_2 both strictly on the same side of it. It is easily seen that condition (B) means that there is such a line L_1 separating C_1 and C_2, without requiring the existence of L_2, so that (B) is more general than (A). It is obvious that pairs of arcs C_1, C_2 can be drawn that can be separated by a line through the origin, but which do not both lie on the same side of any such line; these would yield examples of (B) without (A).

Example We now show that conditions (A), (B) are independent when $k = 3$. We show first that they can hold together. Let us take the matrix $p_{rs}(x_r)$ to be constant and equal to the unit matrix. Then this has constant positive determinant, so that condition (A) holds. We get a set of ρ_s such that (4.6.1) are all positive on taking $\rho_1 = \rho_2 = \rho_3 = 1$.

A similar example will show that (B) can hold without (A). We take the matrix or array $p_{rs}(x_r)$ to be constant and equal to

$$\begin{vmatrix} 2 & -1 & -1 \\ -1 & 2 & -1 \\ -1 & -1 & 2 \end{vmatrix}. \tag{4.8.1}$$

This has vanishing determinant, so that condition (A) does not hold. If, however, we replace any one row by the set $(1, 1, 1)$, we get a positive determinant, so that condition (B) does hold.

Example To show that (A) can hold without (B) when $k = 3$, we need a more elaborate example. We specify this in terms of the sets C_r defined as in (4.5.1), for the case $k = 2$. We take

$$C_r = \left\{ (\xi_1, \xi_2, \xi_3) \mid \xi_r = 1, \sum_{s \neq r} |\xi_s| = \delta \right\}, \tag{4.8.2}$$

where $\delta \in [1/2, 1)$; it is clear that these sets can be parametrized in the form (4.5.1), but the details would play no part in the argument and will be omitted.

We first show that condition (A) holds for (4.8.2), or that any set of three vectors chosen one from each of these sets is linearly independent. Suppose if possible that for same set of μ_s, not all zero, the equation

$$\sum_{1}^{3} \mu_s \xi_s = 0 \tag{4.8.3}$$

has a solution in each of the C_r. Let us determine necessary and sufficient conditions for the plane (4.8.3) to meet C_1. We must have in this event

$$\mu_1 = -\sum_{2}^{3} \mu_s \xi_s, \quad \sum_{2}^{3} |\xi_s| = \delta, \tag{4.8.4}$$

and so

$$|\mu_1| \leq \delta \max\{|\mu_2|, |\mu_3|\}. \tag{4.8.5}$$

Conversely, (4.8.5) is sufficient in order that (4.8.3) meet C_1. For the set of pairs ξ_2, ξ_3 such that $|\xi_2| + |\xi_3| = \delta$ is connected, and includes the pairs $\pm\delta$, 0, and 0, $\pm\delta$. Thus the range of values of $\mu_2\xi_2 + \mu_3\xi_3$ subject to $|\xi_2| + |\xi_3| = \delta$ is also connected, and includes the values $\pm\delta\mu_2$, $\pm\delta\mu_3$; it therefore includes μ_1 if (4.8.5) holds.

Similar arguments apply to the question of whether (4.8.3) meets C_2 or C_3. Thus if (4.8.3) meets all of the C_r, we must have (4.8.5), and also

$$|\mu_2| \leq \delta \max\{|\mu_3|, |\mu_1|\}, \tag{4.8.6}$$

$$|\mu_3| \leq \delta \max\{|\mu_1|, |\mu_2|\}. \tag{4.8.7}$$

However, (4.8.5)–(4.8.7) cannot hold together, if $\delta < 1$ and the μ_s are not all zero. Thus condition (A) must hold for the set (4.8.2) if $\delta \in (0, 1)$.

We must now show that condition (B) does not hold for the set (4.8.2) if $\delta \in [1/2, 1)$. We have to show that for any non-zero set (ρ_1, ρ_2, ρ_3) there is a non-zero set μ_1, μ_2, μ_3 such that

$$\sum_{1}^{3} \mu_s \rho_s = 0 \tag{4.8.8}$$

and such that the plane (4.8.3) fails to separate two of the C_r; in fact, we show that such a plane exists that meets two of the C_r.

By symmetry, it will sufficient to treat the case that

$$|\rho_1| \leq |\rho_2| \leq |\rho_3|. \tag{4.8.9}$$

If $\rho_1 = 0$, we may take any pair μ_2, μ_3 such that $\mu_2\rho_2 + \mu_3\rho_3 = 0$, and then choose μ_1 so large that (4.8.6)–(4.8.7) hold. Then the plane (4.8.3) will meet both C_2 and C_3. Suppose then that $\rho_1 \neq 0$. We now take

$$\mu_1 = -2/\rho_1, \mu_2 = 1/\rho_2, \mu_3 = 1/\rho_3, \tag{4.8.10}$$

so that (4.8.8) will hold. The conditions (4.8.6)–(4.8.7) for (4.8.3) to meet C_2 and C_3 now take the form

$$|\rho_2|^{-1} \leq \delta \max\{|\rho_3|^{-1}, 2|\rho_1|^{-1}\}.$$
$$|\rho_3|^{-1} \leq \delta \max\{2|\rho_1|^{-1}, |\rho_2|^{-1}\}.$$

These are true, in view of (4.8.9), if $\delta \geq 1/2$. This completes the proof that (A) can hold without (B), if $k = 3$.

Finally, we note that the situation for $k = 3$ extends to higher values of k. To adapt an example for the case k to the case $k + 1$, we border the array with zeros and a "1," as in the example

$$\begin{vmatrix} p_{11} & p_{12} & \cdots & p_{1k} & 0 \\ p_{21} & p_{22} & \cdots & p_{2k} & 0 \\ \vdots & \vdots & & \vdots & \vdots \\ p_{k1} & p_{k2} & \cdots & p_{kk} & 0 \\ 0 & 0 & \cdots & 0 & 1 \end{vmatrix}. \tag{4.8.11}$$

Clearly, the new array will also have a determinant of fixed sign, and so satisfy condition (A). Suppose if possible that (4.8.11) satisfies condition (B), so that on replacing the rows (4.8.11) by some set of scalars $(\rho_1, \ldots, \rho_{k+1})$ we get a set of $k+1$ determinants of fixed sign. It is then seen that the determinants (4.6.1) must have fixed sign, so that the original k-by-k array satisfied condition (B), then neither does (4.8.11). This proves the result.

4.9   A third condition

We discuss here a slightly more abstruse condition, which may be satisfied by the matrix or array $p_{rs}(x_r)$, and which is relevant to the distribution of eigenvalues for higher values of k. We term this condition

(C) there exists a pair of vectors $(\rho_1, \ldots, \rho_k)$, $(\sigma_1, \ldots, \sigma_k)$ such that for any non-zero set $\mu_1, \ldots, \mu_k$ satisfying

$$\sum \mu_s \rho_s = 0, \quad \sum \mu_s \sigma_s = 0, \tag{4.9.1}$$

at least one of the functions

$$\sum \mu_s p_{rs}(x_r), \quad a_r \leq x_r \leq b_r, \ r = 1, \ldots, k, \tag{4.9.2}$$

should take only positive values.

By reversing the signs of the μ_s, we see that at least one of the functions (4.9.2) must take negative values only. Thus, in geometrical terms, we are asking that any $(k-1)$-dimensional subspace of $\mathbf{R}^k$ that contains the vectors $(\rho_1, \ldots, \rho_k)$, $(\sigma_1, \ldots, \sigma_k)$ should separate strictly two of the C_r (4.3.14). In particular, such a subspace can meet at most $k-1$ of the C_r.

The condition has little significance in the cases $k = 1, 2$. In these cases, one can choose the ρ_s, σ_s so that there is no non-zero set μ_s satisfying (4.9.1); this means that condition (C) is satisfied trivially.

We consider now the relation between this condition and conditions (A), (B), considered in the foregoing sections. We note first that in the case $k = 2$ it is clear that condition (B) implies condition (C), since condition (C) holds automatically.

Theorem 4.9.1. *For $k \geq 3$, condition (B) implies condition (C).*

We have that by Theorem 4.7.1, condition (B) is equivalent to the requirement that there exist a set $\rho_1, \ldots, \rho_k$ such that if the μ_s satisfy (4.7.1), then the functions (4.9.2) contain a pair with fixed opposite signs; it is sufficient to ask that this set contain a function taking only positive values. The requirement of condition (C) is then satisfied if in (4.9.1) we take $\sigma_s = \rho_s$, $s = 1, \ldots, k$.

It is easily seen that condition (C) is more general than condition (B), in that the former can hold without the latter.

The relation between conditions (A), (C) is more delicate. We have

Theorem 4.9.2. *Let the $p_{rs}(x_r)$ be continuous. Then (A) implies (C) if $k = 3, 4$, but not for $k \geq 5$.*

Again we note that the implication is trivial in the case $k = 2$, since condition (C) then holds in any case. In the case $k = 3$, we put the argument in geometric terms. By Theorem 4.4.1, condition (A) means that we can separate the curves C_1, C_2, C_3 by a two-dimensional subspace in any manner. For condition (C) in this case, it is sufficient that we can separate some two of the C_1, C_2, C_3 by such a subspace; we can then take the vectors (ρ_1, ρ_2, ρ_3), $(\sigma_1, \sigma_2, \sigma_3)$ to be any pair that span this subspace.

We use Theorem 4.4.1 also for the case $k = 4$. There must exist a real set $\varphi_1, \ldots, \varphi_4$ such that the functions

$$f_r(x_r) = \sum \varphi_s p_{rs}(x_r), \quad r = 1, \ldots, 4, \tag{4.9.3}$$

satisfy throughout their respective intervals the inequalities

$$f_1(x_1) > 0, f_2(x_2) > 0, f_3(x_3) < 0, f_4(x_4) < 0, \tag{4.9.4}$$

and a second set ψ_s, $s = 1, \ldots, 4$ such that, if

$$g_r(x_r) = \sum \psi_s p_{rs}(x_r), \quad r = 1, \ldots, 4, \tag{4.9.5}$$

then

$$g_1(x_1) < 0, g_2(x_2) > 0, g_3(x_3) > 0, g_4(x_4) < 0. \tag{4.9.6}$$

We now consider a set of functions of the form

$$h_r(x_r) = \alpha f_r(x_r) + \beta g_r(x_r), \quad r = 1, \ldots, 4. \tag{4.9.7}$$

We claim that for any pair α, β, not both zero, two of the set (4.9.7) have fixed, opposite signs. If $\alpha = 0$, or if $\beta = 0$, this follows at once from (4.9.4) and (4.9.6). Suppose then that neither of α, β is zero. If they have the same sign, it will follow from (4.9.4), (4.9.6) that $h_2(x_2)$, $h_4(x_4)$ have fixed opposite signs. If α, β have opposite signs, the same statement is true of $h_1(x_1)$, $h_3(x_3)$.

Thus if we set

$$\mu_s = \alpha \varphi_s + \beta \psi_s, \quad s = 1, \ldots, 4, \tag{4.9.8}$$

where α, β are not both zero, two of the functions

$$\sum_1^4 \mu_s p_{rs}(x_r), \quad r = 1, \ldots, 4, \tag{4.9.9}$$

will have fixed opposite signs. Now it follows from (4.9.3)–(4.9.6) that the vectors $(\varphi_1, \ldots, \varphi_4)$ and $(\psi_1, \ldots, \psi_4)$ are linearly independent. Thus the set of $(\mu_1, \ldots, \mu_4)$ defined by (4.9.8), for varying real α, β forms a two-dimensional subspace of $\mathbf{R}^4$ and so can be identified with the set of solutions of a pair of equations of the form (4.9.1). Thus there exists a pair of such equations, such that any non-trivial solution $(\mu_1, \ldots, \mu_4)$ has the property that the set of four functions (4.9.9) contains members of fixed signs, one positive and one negative.

This completes the proof that (A) implies (C) if $k = 3, 4$. The case $k = 5$ will be discussed in the next Section.

4.10   Conditions (A), (C) in the case $k = 5$

It is a question of providing an example in which condition (A) holds, but not condition (C). The example will be similar to that given in Section 4.8, to show that (A) can hold without (B). For some $\delta \in (0, 1)$, we define the subset D_r of $\mathbf{R}^5$ by

$$D_r = \left\{ (\xi_1, \ldots, \xi_5) | \xi_r = 1, \sum_{s \neq r} |\xi_s| = \delta \right\}, \quad r = 1, \ldots, 5. \tag{4.10.1}$$

We will take C_r to be a continuous arc, which is a subset of D_r, including all its extremal points, being those for which $|\xi_s| = \delta$ for some $s \neq r$. For example, C_1 might consist of a sequence of straight segments joining the points $(1, \pm\delta, 0, 0, 0)$, $(1, 0, \pm\delta, 0, 0)$, $(1, 0, 0, \pm\delta, 0)$, and $(1, 0, 0, 0, \pm\delta)$ in some order. The convex hull of C_r will then be D_r.

We show first that condition (A) is satisfied; the proof is similar to that for (4.8.2). We note that for any set $\mu_1, \ldots, \mu_5$, not all zero, the subspace defined by

$$\sum_1^5 \mu_s \xi_s = 0 \tag{4.10.2}$$

will meet C_r if and only if

$$|\mu_r| \leq \delta \max_{s \neq r} |\mu_s|. \tag{4.10.3}$$

This cannot be true for all $r = 1, \ldots, 5$, if $0 < \delta < 1$ and the μ_s are not all zero. Thus our example satisfies condition (A).

We next wish to show that condition (C) does not hold in this case, for some $\delta \in (0, 1)$. Again our discussion is similar to that of Section 4.8, though the argument is more lengthy. We have to show that for any choice of ρ_s, σ_s, $s = 1, \ldots, 5$, the pair of equations

$$\sum \mu_s \rho_s = 0, \quad \sum \mu_s \sigma_s = 0 \tag{4.10.4}$$

admits a non-zero solution $\mu_1, \ldots, \mu_5$, such that the set of five functions given by (4.9.2) with $k = 5$ fails to contain a pair with fixed opposite signs. It is sufficient to show that four of the functions have a zero. In our present geometrical language, we wish to show that there is a hyperplane (4.10.2), satisfying (4.10.4), and meeting four of the five sets C_r. Since the condition for such an intersection is given by (4.10.3), we wish to show that any set of equations (4.10.4) has a non-trivial solution, which satisfies (4.10.3) for four values of r with $1 \leq r \leq 5$. If t is the exceptional value of r, which must certainly exist, then $|\mu_t|$ is the greatest of the numbers $|\mu_s|$, or one of the greatest, and our requirement is that

$$|\mu_s| \leq \delta |\mu_t|, \quad s \neq t. \tag{4.10.5}$$

Thus, in order to show that condition (A) does not imply condition (C) in the case $k = 5$, it is sufficient to show that there exists a $\delta \in (0, 1)$ such that every three-dimensional subspace of $\mathbf{R}^5$ contains a non-zero element satisfying (4.10.5), for some t. We deduce this from

Lemma 4.10.1. *Every three-dimensional subspace of $\mathbf{R}^5$ contains an element of which one component has absolute value exceeding the absolute values of the others.*

We prove this by contradiction. If the result is false, there is a three-dimensional subspace T such that for every element of T, the maximum of $|\mu_s|$ is attained for two or more s-values. Writing μ for the vector $(\mu_1, \ldots, \mu_5)$ we denote by $m(\mu)$ the number of s-values, for which $|\mu_s|$ attains its maximum. We have to eliminate the possibility that $m(\mu) \geq 2$ for every non-zero $\mu \in T$.

Suppose first that $\min m(\mu) = 2$, for every $\mu \in T$, $\mu \neq 0$. Then there is a $\mu \in T$ such that two of the $|\mu_s|$ are equal and exceed the others. For definiteness, we suppose that

$$|\mu_1| = |\mu_2| > \max_{s>2} |\mu_s|. \tag{4.10.6}$$

We must have $\mu_1 = \pm \mu_2$; it will be sufficient to discuss the case that

$$\mu_1 = \mu_2. \tag{4.10.7}$$

We now note that for any $\nu = (\nu_1, \ldots, \nu_5) \in T$ and real ε we must have $m(\mu + \varepsilon \nu) \geq 2$. For small ε, this implies that

$$\mu_1 + \varepsilon \nu_1 = \mu_2 + \varepsilon \nu_2,$$

so that $\nu_1 = \nu_2$. Thus (4.10.7) must hold for all elements of T.

Since T is three-dimensional, it is the set of solutions of (4.10.7) and a second equation, which we may take to be the first of (4.10.4). We consider solutions such that $\mu_1 = \mu_2 = 0$. Then μ_3, μ_4, μ_5 may be any solutions of

$$\sum_3^5 \mu_s \rho_s = 0. \tag{4.10.8}$$

If one of ρ_s, $s = 3, 4, 5$, is zero, we may take the corresponding μ_s to have any value, the others being zero, and then have a solution satisfying the assertion of the lemma. The case that none of ρ_3, ρ_4, ρ_5 is zero is exemplified by situation that

$$0 < |\rho_3| \leq |\rho_4| \leq |\rho_5|.$$

As shown in (4.8.8)–(4.8.10), there exists in this case also a solution satisfying the requirements of the lemma. Other cases in which $m(\mu) = 2$ involve only sign-variations on (4.10.7) and permutations of the suffixes; this completes our discussion of this case.

Suppose next that $\min m(\mu) = 3$, for $\mu \in T$. In a sufficiently typical case, there will be an element such that

$$|\mu_1| = |\mu_2| = |\mu_3| > \max(|\mu_4|, |\mu_5|), \qquad (4.10.9)$$

and with

$$\mu_1 = \mu_2 = \mu_3. \qquad (4.10.10)$$

The above perturbation argument then shows that (4.10.10) must hold throughout T. Since T is three-dimensional, this describes T precisely. But then μ_4, μ_5 are arbitrary, and so there exists a set satisfying the requirements of the lemma.

The supposition that $\min m(\mu)$, $\mu \in T$, is 4 would imply that T was contained either in the set described by $\mu_1 = \mu_2 = \mu_3 = \mu_4$, or by a set derived from this by variations in signs or suffixes. Since T is to be three-dimensional, this is impossible. The case that $\min m(\mu) = 5$ may be disposed of similarly. This completes the proof of the lemma.

We pass now to the proof of the main result, that there is a fixed δ such that every three-dimensional subspace contains an element satisfying (4.10.5) for some t. We suppose the contrary, that there is sequence T_1, $T_2, \ldots$, of three-dimensional subspaces and an increasing sequence of positive δ_1, $\delta_2, \ldots$, with $\delta_n \to 1$, such that T_n contains no element satisfying (4.10.5) with $\delta = \delta_n$. More explicitly, let us define for non-zero vectors μ a functional $\chi(\mu)$ as follows. Let $|\mu_s|$, $s = 1, \ldots, 5$, attain its maximum when $s = t$; we then specify that

$$\chi(\mu) = \max_{s \neq t} |\mu_s|/|\mu_t|. \qquad (4.10.11)$$

The hypothesis to be disposed of is that

$$\chi(\mu) \geq \delta_n, \mu \in T_n, \mu \neq 0. \qquad (4.10.12)$$

We note that $\chi(\mu)$ is continuous on the set of non-zero elements of $\mathbf{R}^5$.

With the usual inner product, we may suppose that T_n is generated by an orthonormal set u_n, v_n, w_n. By transition to a subsequence, if necessary, we may suppose that these converge, respectively, to an orthonormal set u, v, w, which span a subspace T, say. By the lemma, T will contain a non-zero element, say

$$\mu = \alpha u + \beta v + \gamma w,$$

such that $\chi(\mu) < 1$. However, by (4.10.12), we have $\chi(\alpha u_n + \beta v_n + \gamma w_n) \geq \delta_n$, and so, making $n \to \infty$, we get $\chi(\mu) \geq 1$. This gives a contradiction, and completes our discussion of the case $k = 5$.

The extension of the fact, that condition (A) does not imply condition (C), to higher k-values may be accomplished as at the end of Section 4.8, for the case of conditions (A), (B). For the array (4.8.11), and for the array formed by deleting the last row and column, condition (A) will hold in both cases or in neither. If, however, condition (C) holds for (4.8.11), it will also hold for the k-by-k array formed by these deletions. In fact, if ρ_s, σ_s, $s = 1, \ldots, k+1$, satisfy the requirements of condition (C) for the (4.8.11), then ρ_s, σ_s, $s = 1, \ldots, k$ will satisfy these requirements for the array formed by the $p_{rs}(x_r)$, $1 \leq r$, $s \leq k$. Thus if condition (C) fails to hold for the latter, it will also not hold for (4.8.11). This completes the proof of Theorem 4.9.1.

4.11   Standard forms

If the k-by-k matrix of functions $p_{rs}(x_r)$ is multiplied on the right by a constant matrix, we obtain a new matrix of a similar form, in which the elements of the r-th row are again all functions of x_r. This process corresponds to a linear transformation of the spectral parameters λ_s, so that the boundary-value problem is not essentially altered. It is sometimes convenient to suppose such a transformation carried out, so that the resulting matrix of functions should have some special sign-properties.

We take first the case $k = 2$, that of a set of functions

$$p_{rs}(x_r), \quad a_r \leq x_r \leq b_r, \ r, s = 1, 2, \tag{4.11.1}$$

which we suppose continuous. For some set γ_{st}, $1 \leq s, t \leq 2$, with non-zero determinant, we form new functions

$$p_{rs}{}^\dagger(x_r) = \sum_{t=1}^{2} p_{rt}(x_r)\gamma_{ts}, \quad r, s = 1, 2. \tag{4.11.2}$$

By Theorem 4.4.1, or the discussion in §4.5, we have that if (4.11.1) satisfy condition (A), that of having a determinant of fixed sign, then we can choose the γ_{st} so that

$$p_{11}{}^\dagger(x_1) > 0, p_{12}{}^\dagger(x_1) < 0, \quad a_1 \leq x_1 \leq b_1, \tag{4.11.3}$$

$$p_{21}{}^\dagger(x_2) > 0, p_{22}{}^\dagger(x_2) > 0, \quad a_2 \leq x_2 \leq b_2. \tag{4.11.4}$$

Conversely, if (4.11.3)–(4.11.4) hold, the new matrix satisfies condition (A).

Suppose next that condition (B) holds in the case $k = 2$; as we know, this is a weaker hypothesis than condition (A) in this case. Again assuming continuity, we can choose a transformation so that

$$p_{12}{}^\dagger(x_1) > 0, \quad a_1 \leq x_1 \leq b_1, \tag{4.11.5}$$

$$p_{22}{}^\dagger(x_2) < 0, \quad a_2 \leq x_2 \leq b_2. \tag{4.11.6}$$

This is rather immediate consequence of condition (B). Conversely, if (4.11.5)–(4.11.6) holds, then so does condition (B).

We now give some results concerning normal forms under conditions (A) or (B) in the general case.

Theorem 4.11.1. *If condition (A) holds, there is a linear transformation such that the k m-th order determinants*

$$\det_{1 < r,s \leq m} p_{rs}{}^\dagger(x_r), \quad m = 1, \ldots, k, \tag{4.11.7}$$

take only positive values.

For some set x_{r0}, $r = 1, \ldots, k$ of values of the x_r, we define the matrix γ_{st}, $1 \leq s, t \leq k$, by

$$(\gamma_{st}) = (p_{rs}(x_{r0}))^{-1}; \tag{4.11.8}$$

condition (A) ensures that the matrix inverse on the right exists. Defining $p_{rs}{}^\dagger(x_r)$ as in (4.11.2), we then have that

$$p_{rs}{}^\dagger(x_{r0}) = \begin{cases} 0, & r \neq s, \\ 1, & r = s. \end{cases} \tag{4.11.9}$$

The new matrix $p_{rs}{}^\dagger(x_r)$ will also satisfy condition (A), and so is positive for all sets of the x_r. We then derive the result of the theorem, in view of (4.11.9), on taking $x_r = x_{r0}$, $m < r \leq k$. The result can be used in cases where we seek to determine in turn the possible values of $\lambda_k, \lambda_{k-1}, \ldots$, which can appear in an eigenvalue of (2.1.4)–(2.1.6). In the case of condition (B), we can transform to a case in which condition (B) is satisfied with

$$\rho_1 = 1, \rho_2 = \cdots = \rho_k = 0. \tag{4.11.10}$$

In other words, the co-factors of the entries of the first column must take positive values only.

Theorem 4.11.2. *Let the k-by-k matrix $p_{rs}(x_r)$ satisfy condition (B), where $k \geq 2$. Then there is a linear transformation such that the new matrix $p_{rs}{}^\dagger(x_r)$ satisfies condition (B) with the choice (4.11.10).*

For some matrix γ_{st}, we define

$$\rho_t{}^\dagger = \sum_{s=1}^{k} \rho_s \gamma_{st}, \quad p_{rt}{}^\dagger(x_r) = \sum_{s=1}^{k} p_{rs}(x_r)\gamma_{st}$$

where the ρ_t are such as to render (4.6.1) positive-valued. We need to choose the γ_{st} such that $\rho_1{}^\dagger = 1$, $\rho_s{}^\dagger = 0$ for $2 \leq s \leq k$, and such that $\det \gamma_{st} > 0$. Evidently, this can be done.

4.12   Borderline cases

In the foregoing, we emphasized situations in which various determinants of functions, or linear combinations of functions, had fixed sign, without vanishing, over finite closed domains. While this makes for simplicity and definiteness, applications have a way of involving slight departures from such hypotheses. We give here some modifications of earlier results, in which functions may vanish without changing sign, instead of having fixed sign, or in which the domains concerned are not closed.

The results of Section 4.3 were formulated for the case that the x_r-domains were compact intervals. In fact, we have

Theorem 4.12.1. *The result of Theorem 4.3.1 holds good if the x_r-domains are arbitrary intervals.*

The intervals may be open, closed, or neither, and may be bounded or unbounded; the proof is unaffected.

In the next result, we weaken the "fixed sign" situation.

Theorem 4.12.2. *We retain the continuity of the p_{rs}. Let*

$$\det p_{rs}(x_r) \geq 0, \quad x_r \in I_r, \ r = 1, \ldots, k, \tag{4.12.1}$$

where $I_1, \ldots, I_k$ are any intervals, and let inequality in (4.12.1) hold for at least one set of arguments. Then for any set $\mu_1, \ldots, \mu_k$, not all zero, at least one of the functions

$$\sum \mu_s p_{rs}(x_r), \quad x_r \in I_r, \ r = 1, \ldots, k, \tag{4.12.2}$$

does not change sign, and does not vanish identically.

Let inequality hold in (4.12.1) when $x_r = x_{r0} \in I_r, r = 1, \ldots, k$. We consider the perturbed set

$$p_{rs}(x_r, \eta) = p_{rs}(x_r) + \eta p_{rs}(x_{r0}). \qquad (4.12.3)$$

This has the property that, for all $\eta > 0$,

$$\det p_{rs}(x_r, \eta) > 0, \quad x_r \in I_r, \ r = 1, \ldots, k.$$

We see this by expanding the determinant on the left in powers of η, and noting that all terms are non-negative, the coefficient of η^k being positive. The last theorem shows that for any set of μ_s, not all zero, at least one of

$$\sum \mu_s p_{rs}(x_r, \eta) = \sum \mu_s p_{rs}(x_r) + \eta \sum \mu_s p_{rs}(x_{r0}), \qquad (4.12.4)$$

$r = 1, \ldots, k$, has fixed sign in I_r, without vanishing. In the rest of this section, all sums are over $s = 1, \ldots, k$.

We now make $\eta \to 0$ through some positive sequence $\eta_1, \eta_2, \ldots$. By transition to a subsequence if necessary, we may assume that for some fixed r, (4.12.4) has fixed sign for all $\eta = \eta_n$. Taking this sign to be positive, say, we have

$$\sum \mu_s p_{rs}(x_r) + \eta_n \sum \mu_s p_{rs}(x_{r0}) > 0, \quad x_r \in I_r. \qquad (4.12.5)$$

Making $n \to \infty$, we get

$$\sum \mu_s p_{rs}(x_r) \geq 0, x_r \in I_r.$$

Here we cannot have equality when $x_r = x_{r0}$, in view of (4.12.5). This completes the proof.

We prove next a more delicate result along these lines, with a slightly stronger hypothesis than (4.12.1) and an element of uniformity in the conclusion.

Theorem 4.12.3. *Let the $p_{rs}(x_r)$ be continuous on the compact intervals $[a_r, b_r]$, and let $\det p_{rs}(x_r)$ be positive for all sets $x_{rn} \in [a_r, b_r]$ such that, for at least one r, $x_r \in (a_r, b_r)$. Then for any ε, with $0 < \varepsilon < \min(b_r - a_r)/2$, there is a $\delta > 0$ such that, for any set $\mu_1, \ldots, \mu_k$ with $\sum |\mu_s| = 1$, we have, for some r*

$$\sum \mu_s p_{rs}(x_r) \neq 0, \quad a_r < x_r < b_r, \qquad (4.12.6)$$

$$\left| \sum \mu_s p_{rs}(x_r) \right| \geq \delta, \quad a_r + \varepsilon \leq x_r \leq b_r - \varepsilon. \qquad (4.12.7)$$

For fixed ε, we can define a positive-valued function $\delta(\mu_1, \ldots, \mu_k)$ as the greatest of the numbers

$$\min \left| \sum \mu_s p_{rs}(x_r) \right|, \quad a_r + \varepsilon \leq x_r \leq b_r - \varepsilon, \qquad (4.12.8)$$

taken over the set of r satisfying (4.12.6); this set is non-empty by Theorem 4.3.1. We have to show that this function is bounded from 0, for fixed ε and over the set of k-tuples $(\mu_1, \ldots, \mu_k)$ such that $\sum |\mu_s| = 1$.

We suppose the contrary, that there is a sequence of k-tuples

$$(\mu_{1n}, \ldots, \mu_{kn}), \ n = 1, 2, \ldots, \text{with} \sum |\mu_{sn}| = 1,$$

such that $\delta(\mu_{1n}, \ldots, \mu_{kn}) \to 0$. By choosing a subsequence, we may take it that the μ_{sn} converge as $n \to \infty$, and write $\mu_s = \lim \mu_{sn}$. We choose a number x_{rn} realizing the definition of δ in (4.12.8), so that

$$\delta(\mu_{1n}, \ldots, \mu_{kn}) = |\sum \mu_{sn} p_{rs}(x_{rn})|; \tag{4.12.9}$$

we also have $x_{rn} \in [a_r + \varepsilon, b_r - \varepsilon]$. Without loss of generality, we assume that r here is independent of n, and also that x_{rn} tends to a limit, say x_{r0} as $n \to \infty$. We then have, in view of the continuity of the $p_{rs}(x_r)$,

$$\sum \mu_s p_{rs}(x_{r0}) = 0, \tag{4.12.10}$$

and also $x_{r0} \in [a_r + \varepsilon, b_r - \varepsilon]$. We observe next that there must be a $t \neq r$ such that

$$\sum \mu_s p_{ts}(x_t) = 0, \quad a_t \leq x_t \leq b_t. \tag{4.12.11}$$

For if this were not so, $\det p_{ts}(x_t)$ would have a zero with one of its arguments, namely, x_{r0}, in the interior of the corresponding interval, contrary to hypothesis. Since $\mu_{sn} \to \mu_s$ we have, for all large n,

$$\sum \mu_{sn} p_{ts}(x_t) = 0, \quad a_t \leq x_t \leq b_t,$$

and so for large n,

$$\delta(\mu_{1n}, \ldots, \mu_{kn}) \geq \min_{x_t \in [a_t + \varepsilon, b_t - \varepsilon]} |\sum \mu_{sn} p_{ts}(x_t)|.$$

In view of (4.12.11) this minimum does not tend to 0 as $n \to \infty$, and so we have a contradiction. This proves Theorem 4.12.3.

4.13   Metric variants on condition (A)

In the foregoing, our various conditions, requiring the positivity of certain determinants, were usually reinforced with continuity hypotheses, so as to yield some degree of uniformity in the results. In an alternative procedure, we make more specific the positivity requirements for the determinants in question.

Reviewing our notation, we denote by P the determinant $\det p_{rs}(x_r)$, by P_{rs} the co-factor of p_{rs}, and now write $P_r = P_r(x_r)$ for the r-th row formed by $p_{rs}(x_r)$, $s = 1, \ldots, k$. As a norm of P_r we take

$$\|P_r\| = \sum_s |p_{rs}(x_r)|. \tag{4.13.1}$$

We have trivially that

$$|P| \leq \prod \|P_r\| \tag{4.13.2}$$

and similarly

$$|P_{rs}| \leq \prod_{t \neq r} \|P_t\|, \tag{4.13.3}$$

The idea is then to impose as a hypothesis an inequality in the opposite sense to (4.13.2).

We assume that

$$P(x_1, \ldots, x_k) \geq \left\{ \sum_r f_r(x_r) \right\} \prod_1^k \|P_m(x_m)\|, \tag{4.13.4}$$

where the $f_r(x_r)$ are in any case non-negative, and satisfy positivity conditions at our disposal. For definiteness, we will assume that the $f_r \in C[a_r, b_r]$, and that

$$f_r(x_r) > 0, \quad a_r < x_r < b_r, \quad r = 1, \ldots, k. \tag{4.13.5}$$

We refer to this as "condition AM."

For some real set $\mu_1, \ldots, \mu_k$ with $\max |\mu_s| = 1$, we write

$$u_r(x_r) = \sum \mu_s p_{rs}(x_r), \quad r = 1, \ldots, k, \tag{4.13.6}$$

and claim

Theorem 4.13.1. *Let the p_{rs} be continuous. Then for some $r \in \{1, \ldots, k\}$ we have either*

$$u_r(x_r) \geq f_r(x_r)\|P_r(x_r)\|, \quad x_r \in [a_r, b_r] \tag{4.13.7}$$

or else

$$u_r(x_r) \leq -f_r(x_r)\|P_r(x_r)\|, \quad x_r \in [a_r, b_r]. \tag{4.13.8}$$

We first establish the combined situation of (4.13.7)–(4.13.8), namely,

$$|u_r(x_r)| \geq f_r(x_r)\|P_r(x_r)\|, \quad x_r \in [a_r, b_r]. \tag{4.13.9}$$

Supposing the contrary, we would have for each $r = 1, \ldots, k$ an $x'_r \in [a_r, b_r]$ such that

$$|u_r(x'_r)| < f_r(x'_r)\|P_r(x'_r)\|. \tag{4.13.10}$$

We show that this leads to a contradiction.

Let P' denote $P(x_1', \ldots, x_k')$, and likewise write P_{rs}' for the co-factor of $p_{rs}(x'_r)$, $u_r(x'_r) = \sum \mu_s p_{rs}(x'_r)$. For any $t \in \{1, \ldots, k\}$, we multiply the latter by P_{rt}' and sum over r, to get

$$\sum P_{rt}' u_r(x'_r) = \mu_t P'. \tag{4.13.11}$$

On the left, we use (4.13.10), and (4.13.3) as applied to the set $x_1', \ldots, x_k'$, to get

$$|P_{rt}' u_r(x'_r)| < f_r(x'_r) \prod_1^k \|P_m(x_m')\|.$$

The inequality is strict since, by (4.13.10), the factors on the right are positive. We deduce that

$$\left| \sum P_{rt}' u_r(x'_r) \right| < \left\{ \sum f_r(x'_r) \right\} \prod_1^k \|P_m(x_m')\|.$$

Comparing this with (4.13.11), and using the basic hypothesis (4.13.4), we deduce that $|\mu_t| < 1$, whereas we assumed that $\max |\mu_s| = 1$. This completes the proof of (4.13.9).

In the continuous case, we can deduce from (4.13.8) that just one of (4.13.7)–(4.13.8) must be valid.

As an L^1-variant, we have

Theorem 4.13.2. *Let the p_{rs}, $f_r \in L(a_r, b_r)$ satisfy (4.13.4) almost everywhere in the region $\prod[a_r, b_r]$. Then for some r, (4.13.7) holds a.e. in $[a_r, b_r]$.*

The proof is similar to that of Theorem 4.3.1 in this case, using the argument of (4.3.7)–(4.3.8).

Notes for Chapter 4

The proof of Theorem 4.4.1 is similar to that of Theorem 9.4.1 in Atkinson (1972). The continuity assumption in the sufficiency part of the proof of Theorem 4.4.1 may be relaxed by introducing the convex hull of the sets C_r. "Then the determinant of the matrix $c_1, c_2, \ldots, c_k$ is still of fixed sign whenever c_r is in the convex hull of C_r. Since the set of values of this determinant is an interval, this completes the proof of the sufficiency" (H. Volkmer).

Theorem 4.2.3 is contained in the doctoral dissertation of M. Faierman (1966); see also §1.2 therein.

Conditions (A) and (B) become part of conditions of right- and left-definiteness, when a column of Sturm-Liouville operators is placed to the right of the p_{rs}. The term "ellipticity condition" is also used in the (B) case. See Sleeman (1978b), where the relation between the conditions is discussed.

The term "Hilbert type" was introduced in Farenick (1986) for the case of condition (B), where the co-factors of elements in the last column are positive. The general case, reducible to this by a transformation, was termed "conformable" for such a transformation. For the use of this condition in connection with the completeness problem for eigenfunctions, see Faierman (1966), (1969), (1991a), Sleeman (1978b), and references there given.

The determinant $\det p_{rs}(x_r)$, acting multiplicatively on functions $f(x_1, \ldots, x_k)$ forms an example of a determinantal operator, or operator-valued determinant. Such operators appear in almost all multiparameter work; see Atkinson (1968), Volkmer (1986), (1988), Farenick (1986), Källström-Sleeman (1975b), and so forth.

For a recent review of the one-parameter case under all definiteness conditions, one may consult Zettl (2005). The non-definite two-parameter case is treated in Faierman (1991a, Chapter 6).

4.14 Research problems and open questions

1. As noted in this chapter, the study of determinantal operators is of marked importance when dealing with the spectral theory of multiparameter problems. In Aslanov (2004), the main problem is the proof of the following result in the case where $n = 3$: If the 3×3 tensor determinant is positive (non-negative) on decomposable tensors, then the operator is positive (non-negative) on the whole Hilbert tensor product space. For $n = 2$, the positivity question was solved in Almamedov et al. (1985) (English translation, vol. 32 (1), 225–227) and Almamedov et al. (1987). In the finite dimensional case, this positivity problem is solved by Atkinson (1972; see Theorems 7.8.2 and 9.4.1), while in the strongly positive case by Binding and Browne (1978a).

The question here is[2] to *determine whether Aslanov's theorem on the positivity of the determinantal operator extends to the case $n > 3$. Is the non-negativity result also true?*

2. The natural question here is: *Assuming continuity of the p_{rs}, what are the corresponding results for Sections 4.3 and so forth when (4.3.1) takes on both signs?* For basic work regarding this question in the two-parameter case, see Faierman (1991).

---

[2]We are indebted to Professor A. A. Aslanov for these comments.

Chapter 5

Oscillation Theorems

5.1   Introduction

We now specialize the discussion to the second-order Sturm-Liouville case with several parameters. We wish to extend to this case two classical oscillation theorems for the one-parameter case of Sturm-Liouville, dealing with the "orthogonal" and the "polar" cases.

We have referred to this distinction already in Section 4.1. The "orthogonal" case is the standard one, in which the coefficient of the spectral parameter is positive, or at least does not change sign; in this case, there is just one eigenvalue for each "oscillation number," the number of zeros of the eigenfunction in the interior of the basic interval. In the polar case, the coefficient of λ changes sign, but there is a compensating sign-definiteness associated with the differential operator and the boundary conditions; in this case, there are two eigenvalues for each oscillation number.

It turns out that these two situations extend to the multiparameter case. That corresponding to the orthogonal case, in which there is one eigenvalue (k-tuple) to each set of oscillation numbers, is the subject of what is known as the "Klein oscillation theorem." Although this does not seem to have been formulated by F. Klein in its full generality, it seems clear that he would have expected it. The other type of result, in which there are two eigenvalues for each set of oscillation numbers, we term the "Richardson oscillation theorem," since for certain cases it was stated by him.

As in the standard one-parameter case, it is possible to use these oscillation theorems for the approximate or asymptotic location of eigenvalues.

5.2   Oscillation numbers and eigenvalues

We are concerned with the following boundary-value problem: we seek values of $\lambda_1,\ldots,\lambda_k$ such that each of the k equations

$$y_r''(x_r) + \left\{\sum_{s=1}^{k}\lambda_s\,p_{rs}(x_r) - q_r(x_r)\right\}y_r(x_r) = 0, \quad r = 1,\ldots,k, \qquad (5.2.1)$$

has a solution in its respective finite interval $[a_r, b_r]$, not identically zero, and satisfying the boundary conditions

$$y_r(a_r)\cos\alpha_r = y_r'(a_r)\sin\alpha_r,$$
$$y_r(b_r)\cos\beta_r = y_r'(b_r)\sin\beta_r, \quad r = 1,\ldots,k. \qquad (5.2.2)$$

Here, α_r, β_r will be assumed real, and without loss of generality subject to

$$0 \le \alpha_r < \pi, \quad 0 < \beta_r \le \pi, \quad r = 1,\ldots,k. \qquad (5.2.3)$$

The $p_{rs}(x_r)$ will be assumed real-valued and continuous, as also the $q_r(x_r)$; some relaxation of these continuity restrictions is possible, but will not be carried out at this point.

The situation differs in an essential manner according to whether we assume the first, or the second definiteness condition, in the terminology of the last chapter. In the former of these, we impose some positivity requirement on $\det p_{rs}(x_r)$; we investigate this situation first. As a minimal requirement, we ask that

$$\det p_{rs}(x_r) \ge 0 \qquad (5.2.4)$$

for all sets of arguments $x_r \in [a_r, b_r]$ in the closed intervals, and

$$\int_{a_1}^{b_1}\cdots\int_{a_k}^{b_k}\det p_{rs}(x_r)\,dx_1\cdots dx_k > 0. \qquad (5.2.5)$$

Later on, we shall have to make further restrictions.

However, (5.2.4)–(5.2.5) certainly ensure that all eigenvalues, if any exist are real. We can indeed go further, and claim that there is at most one eigenvalue corresponding to any given set of oscillation numbers. This statement forms part of the Klein oscillation theorem. We have

Theorem 5.2.1. *Let* (5.2.4)–(5.2.5) *hold. Then to each set of non-negative integers* $n_1, \ldots, n_k$, *there is at most one eigenvalue* $(\lambda_1, \ldots, \lambda_k)$ *for which these are the oscillation numbers.*

By the latter we mean that the non-trivial solutions $y_r(x_r)$ of the problem (5.2.1)–(5.2.2) have, respectively, n_r zeros in the open intervals (a, b).

For the proof, we suppose on the contrary that there are two such eigenvalues $(\lambda_1, \ldots, \lambda_k)$, and $(\lambda_1^\dagger, \ldots, \lambda_k^\dagger)$, distinct in that they differ in at least one entry. By Theorem 4.12.2, the k functions

$$\sum_{s=1}^{k} (\lambda_s - \lambda_s^\dagger) p_{rs}(x_r), \quad r = 1, \ldots, k \tag{5.2.6}$$

contain one that does not change sign in its interval (a_r, b_r), and which does not vanish identically. From classical Sturmian principles, it follows that the solutions of

$$y_r'' + \left\{ \sum_{s=1}^{k} \lambda_s p_{rs} - q_r \right\} y_r = 0, \quad y_r'' + \left\{ \sum_{s=1}^{k} \lambda_s^\dagger p_{rs} - q_r \right\} y_r = 0$$

satisfying the boundary conditions (5.2.2) for the r-value in question, cannot have the same number of zeros in (a_r, b_r), and so cannot have the same oscillation numbers. This gives a contradiction, and so proves the uniqueness of the eigenvalue.

We have incidentally a fresh proof of the discreteness of the spectrum. For in any convergent sequence of distinct real eigenvalues, the sets of oscillation numbers would have to be all different, and so could not be uniformly bounded. However, it follows from the Sturm comparison theorem that the number of zeros of any non-trivial solution of any of (5.2.1) must be uniformly bounded in any bounded $(\lambda_1, \ldots, \lambda_k)$-region.

This last line of argument can be much developed, so as to provide quantitative information on the distribution of the eigenvalues.

5.3 The generalized Prüfer transformation

It was shown by H. Prüfer that a certain polar-coordinate transformation provides an effective method of establishing the standard Sturmian oscillation

properties in the one-parameter case. This approach can be usefully extended
to the multiparameter case.

For all $\lambda_1, \ldots, \lambda_k$, we determine solutions

$$y_r(x_r) = y_r(x_r; \lambda_1, \ldots, \lambda_k), \quad r = 1, \ldots, k, \qquad (5.3.1)$$

of (5.2.1) by the initial conditions

$$y_r(a_r) = \sin \alpha_r, \quad y_r'(a_r) = \cos \alpha_r; \qquad (5.3.2)$$

the first of the boundary conditions (5.2.2) will then be satisfied automatically. We then determine continuous functions of the x_r, and also of the real
parameters $\lambda_1, \ldots, \lambda_k$,

$$\theta_r(x_r) = \theta_r(x_r; \lambda_1, \ldots, \lambda_k), \quad r = 1, \ldots, k, \qquad (5.3.3)$$

by

$$y_r(x_r) \cos \theta_r(x_r) = y_r'(x_r) \sin \theta_r(x_r), \qquad (5.3.4)$$

and fixed initial values

$$\theta_r(a_r) = \alpha_r, \quad r = 1, \ldots, k. \qquad (5.3.5)$$

As in the ordinary case, the requirement (5.3.4) fixes $\theta_r(x_r)$ to within an arbitrary additive multiple of π; this is to be determined so as to ensure that θ_r is
continuous.

We need the differential equations satisfied by the θ_r both as functions of the
x_r, respectively, and also as functions of the λ_s. The calculations are essentially
as in the one-parameter case.

Differentiating (5.3.4) and multiplying by $\sin \theta_r(x_r)$ we get, writing (') for
d/dx_r,

$$y_r' \sin \theta_r \cos \theta_r - y_r \theta_r' \sin^2 \theta_r = y_r'' \sin^2 \theta_r + y_r' \theta_r' \sin \theta_r \cos \theta_r.$$

Using (5.3.4), we obtain

$$y_r \cos^2 \theta_r - y_r \theta_r' \sin^2 \theta_r = y_r'' \sin^2 \theta_r + y_r \theta_r' \cos^2 \theta_r.$$

After rearrangement and use of (5.2.1), we get

$$\theta_r' = \cos^2 \theta_r + \left\{ \sum_{s=1}^{k} \lambda_s p_{rs} - q_r \right\} \sin^2 \theta_r. \qquad (5.3.6)$$

As in the standard case, we have

Theorem 5.3.1. *For fixed* $\lambda_1, \ldots, \lambda_k$, $\theta_r(x_r)$ *is an increasing function of* x_r
when $\theta_r(x_r)$ *is a multiple of* π.

For the right of (5.3.6) is then positive; these x_r-values are those at which $y_r(x_r) = 0$. Thus if $\theta_r(x_r)$ reaches multiples of π as x_r increases in (a_r, b_r) it will reach them in ascending order, reaching each once only, and reaching only positive multiples of π; the last remark follows from (5.2.3) and (5.3.5). Hence, we have

Theorem 5.3.2. *For the r-th equation in (5.2.1) to have a non-trivial solution satisfying (5.2.2), with precisely n_r zeros in (a_r, b_r), it is necessary and sufficient that*

$$\theta_r(b_r; \lambda_1, \ldots, \lambda_k) = \beta_r + n_r \pi. \qquad (5.3.7)$$

Thus an eigenvalue of (5.2.1)–(5.2.2), with oscillation numbers $n_1, \ldots, n_k$ is a set $(\lambda_1, \ldots, \lambda_k)$ satisfying (5.3.7) with $r = 1, \ldots, k$. Our approach will be to study the solution of these equations with the aid of the implicit function theorem.

5.4   A Jacobian property

We turn next to the behavior of the functions (5.3.3) as functions of the λ_s. We have to calculate the partial derivatives $\partial\theta/\partial\lambda_s$. Differentiating (5.3.4) with respect to λ_s we get

$$(\partial y_r/\partial\lambda_s)\cos\theta_r - y_r(\partial\theta_r/\partial\lambda_s)\sin\theta_r =$$

$$= (\partial y_r'/\partial\lambda_s)\sin\theta_r + y_r'(\partial\theta_r/\partial\lambda_s)\cos\theta_r. \qquad (5.4.1)$$

Multiplying by $y_r \sin\theta_r$ and using (5.3.4) we get after rearrangement that

$$(y_r'\partial y_r/\partial\lambda_s - y_r\partial y_r'/\partial\lambda_s)\sin^2\theta = y_r^2\partial\theta_r/\partial\lambda_s. \qquad (5.4.2)$$

We now complete the polar-coordinate transformation by introducing positive-valued functions

$$R_r(x_r; \lambda_1, \ldots, \lambda_k), \quad r = 1, \ldots, k, \qquad (5.4.3)$$

by

$$y_r = R_r \sin\theta_r, \quad y_r{}' = R_r \cos\theta_r, \qquad (5.4.4)$$

in accordance with (5.3.4). Then (5.4.2) yields

$$\partial\theta_r/\partial\lambda_s = R_r^{-2}\{y_r'\partial y_r/\partial\lambda_s - y_r\partial y_r'/\partial\lambda_s\}. \qquad (5.4.5)$$

We must now evaluate the expression in the braces in (5.4.5). We differentiate (5.2.1) with respect to λ_s, getting

$$\partial y_r{''}/\partial\lambda_s + \left\{\sum_{s=1}^{k}\lambda_s p_{rs} - q_r\right\}\partial y_r/\partial\lambda_s + p_{rs}y_r = 0.$$

Taking account of (5.2.1) itself, we derive

$$y_r{''}\partial y_r/\partial\lambda_s - y_r\partial y_r{''}/\partial\lambda_s = p_{rs}y_r^2. \tag{5.4.6}$$

The left is the x_r-derivative of the expression in the braces in (5.4.5), which vanishes when $x_r = a_r$. Thus an integration gives, for $a_r \le x_r \le b_r$,

$$y_r' \,\partial y_r/\partial\lambda_s - y_r\partial y_r'/\partial\lambda_s = \int_{a_r}^{x_r} p_{rs}(u_r)\{y_r(u_r)\}^2 du_r.$$

Hence we may re-write (5.4.5) as

$$\partial\theta_r/\partial\lambda_s = R_r^{-2}\int_{a_r}^{x_r} p_{rs}(u_r)\{y_r(u_r)\}^2 du_r. \tag{5.4.7}$$

From this, we draw the main conclusion of this argument.

Theorem 5.4.1. *For* $a_r \le x_r \le b_r$, $r = 1,\ldots,k$,

$$\partial(\theta_1,\ldots,\theta_k)/\partial(\lambda_1,\ldots,\lambda_k)$$

$$= \left\{\prod_{r=1}^{k}R_r(x_r)\right\}^{-2}\int_{a_1}^{x_1}\cdots\int_{a_k}^{x_k}\det p_{rs}(u_r)\left\{\prod_{r=1}^{k}(y_r(u_r))^2\,du_r\right\}. \tag{5.4.8}$$

For the Jacobian on the left is the determinant of the expressions (5.4.7) over r, $s = 1,\ldots,k$, and this, by Theorem 4.2.3, equals the right of (5.4.8).

5.5   The Klein oscillation theorem

For this, we need to make a slightly stronger hypothesis, than that used in Section 5.2. We prove

Theorem 5.5.1. *Let the* $p_{rs}(x_r)$, $a_r \le x_r \le b_r$, r, $s = 1,\ldots,k$, *be continuous, and let*

$$\det p_{rs}(x_r) > 0, \tag{5.5.1}$$

for all sets of $x_1, \ldots, x_k$ *such that at least one of the* x_r *lies in the interior of* (a_r, b_r). *Then to each set of non-negative integers* $n_1, \ldots, n_k$ *there is precisely one eigenvalue* $\lambda_1, \ldots, \lambda_k$ *such that the* n_r *are the oscillation numbers of the associated non-trivial solutions* $y_r(x_r)$ *of* (5.2.1)–(5.2.2).

We showed in Section 5.2 that there was at most one such eigenvalue, subject to a slightly weaker hypothesis on the $p_{rs}(x_r)$. We have now to prove the existence. It is a question of showing that the equations (5.3.7), with $r = 1$, ..., k, have at least one solution $\lambda_1, \ldots, \lambda_k$ or, in other words, that the map

$$(\lambda_1, \ldots, \lambda_k) \to (\theta_1(b_1; \lambda_1, \ldots, \lambda_k), \ldots, \theta_k(b_k; \lambda_1, \ldots, \lambda_k)), \qquad (5.5.2)$$

acting on $\mathbf{R}^k$ into itself has as its image a set S, say, which includes the "octant"

$$0 < \theta_r < \infty, \ r = 1, \ldots, k. \qquad (5.5.3)$$

Actually, S is included in this octant. It follows from Theorem 5.3.1, and the first of (5.2.3), that $\theta_r(b_r; \lambda_1, .., \lambda_k) > 0$ for all r and all $\lambda_1, \ldots, \lambda_k$. Thus we wish to show that S coincides with this octant.

The map (5.5.2) is continuous, and (5.5.1) ensures that the Jacobian is positive, by (5.4.8). The mapping is therefore open. In order to prove the result, it is sufficient to show that the boundary ∂S of S is contained in the boundary of (5.5.3). Suppose then that P_1, P_2 are points in the octant, of which P_1 is in S and P_2 is not in S. The segment $P_1 P_2$ would then contain a point of the boundary of S, and since the segment does not meet the boundary of the octant, we would have a contradiction.

It will be sufficient to show that for any point of ∂S, at least one of θ_r vanishes. Let then

$$(\lambda_{1n}, \ldots, \lambda_{kn}), \quad n = 1, 2, \ldots, \qquad (5.5.4)$$

be a sequence of points whose images under (5.5.2) tend to a point of ∂S. This sequence cannot have a convergent subsequence, since the image of the limit of such a subsequence would be in S and not in ∂S; the two sets are disjoint since the mapping is open. We suppose therefore that

$$\sum |\lambda_{sn}| \to \infty, \quad n \to \infty. \qquad (5.5.5)$$

We introduce associated normalized sets

$$(\mu_{1n}, \ldots, \mu_{kn}), \quad n = 1, 2, \ldots \qquad (5.5.6)$$

by

$$\sum |\mu_{sn}| = 1, \quad \lambda_{sn} = \chi_n \mu_{sn}, \quad \chi_n > 0, \qquad (5.5.7)$$

so that

$$\chi_n \to \infty. \qquad (5.5.8)$$

We appeal now to Theorem 4.12.3, according to which for any ε, with

$$0 < \varepsilon < (b_r - a_r)/2, \quad r = 1, \ldots, k,$$

there is a $\delta > 0$ such that for some r we have

$$\sum \mu_{sn} p_{rs}(x_r) \neq 0, \quad a_r < x_r < b_r, \tag{5.5.9}$$

$$\left| \sum \mu_{sn} p_{rs}(x_r) \right| \geq \delta, \quad a_r + \varepsilon \leq x_r \leq b_r - \varepsilon. \tag{5.5.10}$$

We keep ε independent of n, and then δ is also independent of n. By selection of a subsequence, if necessary, we can arrange that r in (5.5.9)–(5.5.10) is independent of n; in the same way, we can arrange that the sign of $\sum \mu_{sn} p_{rs}(x_r)$ in (a_r, b_r) is independent of n also.

Thus for any ε, $0 < \varepsilon < \min_{1 \leq r \leq k}(b_r - a_r)/2$, there is a $\delta > 0$, an r, $1 \leq r \leq k$, and an infinite sequence of n-values such that either

$$\sum \mu_{sn} p_{rs}(x_r) > 0, \quad a_r < x_r < b_r, \tag{5.5.11}$$

$$\sum \mu_{sn} p_{rs}(x_r) \geq \delta, \quad a_r + \varepsilon \leq x_r \leq b_r - \varepsilon, \tag{5.5.12}$$

or else

$$\sum \mu_{sn} p_{rs}(x_r) < 0, \quad a_r < x_r < b_r, \tag{5.5.13}$$

$$\sum \mu_{sn} p_{rs}(x_r) \leq -\delta, \quad a_r + \varepsilon \leq x_r \leq b_r - \varepsilon. \tag{5.5.14}$$

The first alternative can be eliminated. For (5.5.12) would imply that

$$\sum \lambda_{sn} p_{rs}(x_r) \geq \chi_n \delta, \quad a_r + \varepsilon \leq x_r \leq b_r - \varepsilon,$$

so that the left would tend uniformly to ∞ as $n \to \infty$. By the Sturm comparison theorem, this would imply that the number of zeros of a non-trivial solution of

$$y_r'' + \left\{ \sum \lambda_{sn} p_{rs} - q_r \right\} y_r = 0$$

increases without limit as $n \to \infty$ and this, by Theorem 5.3.1 implies that

$$\theta_r(b_r; \lambda_{1n}, \ldots, \lambda_{kn}) \to +\infty,$$

as $n \to \infty$ through the appropriate subsequence. Thus the sequence of images of (5.5.4) would not converge.

It thus follows that for any ε as above, there is a $\delta > 0$, r and an infinite n-sequence such that (5.5.13)–(5.5.14) hold. We apply this principle with a sequence ε_1, ε_2, ... tending to zero. By selection if necessary, we can arrange that the same r, $1 \leq r \leq k$, can be used in each case; we denote the associated δ-values by δ_1, δ_2, ...

Then for some fixed r, a sequence ε_1, ε_2, ... of positive numbers tending to zero, and associated positive numbers δ_1, δ_2, ..., we have (5.5.13) and

$$\sum \mu_{sn} p_{rs}(x_r) \leq -\delta_m, \quad a_r + \varepsilon_m \leq x_r \leq b_r - \varepsilon_m. \qquad (5.5.15)$$

for some infinite n-sequence.

From (5.5.13) it follows that

$$\sum \lambda_{sn} p_{rs}(x_r) - q_r(x_r) \leq -q_r(x_r), \quad a_r \leq x_r \leq b_r \qquad (5.5.16)$$

so that the left is uniformly bounded above, for the same n-sequence. From (5.5.15), we have

$$\sum \lambda_{sn} p_{rs}(x_r) \leq -\chi_n \delta_m, \quad a_r + \varepsilon_m \leq x_r \leq b_r - \varepsilon_m. \qquad (5.5.17)$$

For any δ_m, we can choose n from the appropriate sequence so large that $\chi_n \delta_m$ is as large as we please, for example, so that $\chi_n \delta_m \geq m$. If n_m is the n-value in question, we have then that the sequence of functions

$$\sum \lambda_{sn_m} p_{rs}(x_r) - q_r(x_r), \quad n = 1, 2, \ldots$$

is uniformly bounded above, and tends to $-\infty$, uniformly on closed subsets of (a_r, b_r). Thus, by Theorem A.6.1, we have

$$\theta_r(b_r; \lambda_{1n}, \ldots, \lambda_{kn}) \to 0.$$

This shows, as required, that at any point of ∂S, at least one of the θ_r vanishes.

The argument can be simplified if we assume that (5.5.1) holds for all sets $x_1, \ldots, x_k$ without restriction. It then follows from Theorem 4.3.2 that one of the functions on the left of (5.5.11) is either bounded from below by δ, or bounded from above by $-\delta$, so that for some n-sequence the left of (5.5.16) either tends uniformly to $+\infty$, or uniformly to $-\infty$. These two eventualities are then discussed as above.

Example Observe that condition (5.5.1) is necessary that is, Klein's oscillation theorem, Theorem 5.5.1, may fail if, say, the determinant appearing in (5.5.1) vanishes identically on some subinterval, J_r, say of $[a_r, b_r]$ for some r.

To see this, consider the (de-coupled) case $k = 2$ with $p_{12} = p_{21} = 0$ and $[a_r, b_r] = [-1, 1]$, $r = 1, 2$ with zero boundary conditions at the ends (i.e., the Dirichlet problem). Let $p_{11} = 0$ on $[-1, 0]$ and $p_{11} \geq 0$ on $[0, 1]$ be defined so that it is continuous on $[-1, 1]$. Let $M > 0$ be a given positive integer. Choose q_1 to be a constant function that is so large and negative to ensure that any solution of the first of (5.2.1) has at least M zeros on $[-1, 0]$ (this is always possible by Sturm's oscillation theorem). Then regardless of how we define p_{22}, q_2, and so forth, it is simple to see that for any value of λ_1 it will be impossible to find a solution of the first of (5.2.1) with anything less than M zeros (since such a solution must satisfy an equation that is independent of λ_1 on $[-1, 0]$). So, in this case it follows that if $n_1 < M$ in Theorem 5.5.1 there are no eigenvalues λ_1, λ_2 of (5.2.1)–(5.2.2) such that n_1 is the oscillation number of an associated non-trivial solution of (5.2.1)–(5.2.2).

5.6   Oscillations under condition (B), without condition (A)

In subsequent sections, we shall pass to a counterpart of the Klein oscillation theorem, in which there are two eigenvalues to each set of oscillation numbers. For this purpose, we need a preliminary result.

Theorem 5.6.1. *Let the $p_{rs}(x_r)$, $a_r \leq x_r \leq b_r, r,s = 1, \ldots, k$ be continuous, and let $\det p_{rs}(x_r)$ change sign. Let there be a set $\rho_1, \ldots, \rho_k$ such that the determinants (4.6.1) are all strictly positive, for all sets of arguments $x_1, \ldots, x_k$. Let $N(\lambda_1, \ldots, \lambda_k)$ denote the total number of zeros of some set of non-trivial solutions $z_r(x_r)$, $r = 1, \ldots, k$, of*

$$z_r'' + \left\{ \sum \lambda_s p_{rs} - q_r \right\} z_r = 0, \quad r = 1, \ldots, k, \tag{5.6.1}$$

in the respective intervals (a_r, b_r). Then

$$N(\lambda_1, \ldots, \lambda_k) \to \infty. \tag{5.6.2}$$

as

$$\sum |\lambda_s| \to \infty. \tag{5.6.3}$$

In view of the Sturm separation theorem, the number of such zeros can vary by at most k under a change from one set of solutions of (5.6.1) to another. The result is thus insensitive to the choice of solutions of (5.6.1).

The requirement that $\det p_{rs}(x_r)$ change sign cannot be dropped. If, for example, $\det p_{rs}(x_r)$ were positive throughout, we could in view of Theorem 4.4.1 make (5.6.3) take place subject to

$$\sum \lambda_s p_{rs}(x_r) < 0, \quad r = 1, \ldots, k, \tag{5.6.4}$$

and then the number of zeros of each z_r would be bounded, in view of the Sturm comparison theorem.

We pass to the proof of the theorem. For any sequence (5.5.4) satisfying (5.5.5), we take a representation in the form (5.5.7)–(5.5.8). By selection of a subsequence, we may take it that the μ_{sn} all converge. We write $\mu_s = \lim \mu_{sn}$.

We appeal now to Theorem 4.6.1. The requirement that $\det p_{rs}(x_r)$ change sign excludes cases (ii) and (iii) of the theorem, and so we have that at least one of the functions

$$\sum \mu_s p_{rs}(x_r), \quad r = 1, \ldots, k, \tag{5.6.5}$$

takes a positive value somewhere in $[a_r, b_r]$. Thus, by continuity, there is a sub-interval $[x_{r1}, x_{r2}]$, of positive length, and a $\delta > 0$ such that for this r,

$$\sum \mu_s p_{rs}(x_r) \geq \delta, \quad x_{r1} \leq x_r \leq x_{r2}. \tag{5.6.6}$$

We note now that

$$\sum \lambda_{sn} p_{rs}(x_r) = \chi_n \sum \mu_s p_{rs}(x_r) \geq \chi_n \sum \mu_{sn} p_{rs}(x_r) - K\chi_n \sum |\mu_{sn} - \mu_s|,$$

where $K = \max |p_{rs}(x_r)|$. In view of (5.6.6), this gives

$$\sum \lambda_{sn} p_{rs}(x_r) \geq \chi_n \delta - K\chi_n \sum |\mu_{sn} - \mu_s|, \quad x_{r1} \leq x_r \leq x_{r2}$$

so that, for large n,

$$\sum \lambda_{sn} p_{rs}(x_r) \geq \chi_n \delta/2, \quad x_{r1} \leq x_r \leq x_{r2}.$$

It follows that

$$\sum \lambda_{sn} p_{rs}(x_r) - q_r(x_r) \to +\infty,$$

uniformly in $[x_{r1}, x_{r2}]$. The result (5.6.2) now follows from the Sturm comparison theorem.

5.7 The Richardson oscillation theorem

As stated in Section 5.1, we give this name to an oscillation theorem in which there are two eigenvalues for each set of oscillation numbers. We are concerned with what, in Section 3.8, we termed the "second definiteness condition"; since this can overlap with the first condition, for which we have the Klein oscillation theorem, we suppose that the second of these conditions holds but that the first of them is violated. We take the second definiteness condition in standardized form, that is to say with $\rho_1 = 1$, $\rho_2 = 0, \ldots, \rho_k = 0$ in (3.8.1); unlike our treatment of the Klein theorem, we shall not attempt to cover "borderline cases," in which determinants maintain in general the same sign but can have zeros. We have

Theorem 5.7.1. *Let the* $p_{rs}(x_r)$, $q_r(x_r)$ *be continuous in* $[a_r, b_r]$, r, $s = 1, \ldots, k$. *Let* $\det p_{rs}(x_r)$ *take both positive and negative values. Let the cofactors of the elements in the first column of the array* $p_{rs}(x_r)$ *all be positive, for all sets of arguments* $x_1, \ldots, x_k$. *Let the hermitian forms*

$$\Phi_{r0}(y_r) = [\overline{y_r} y_r']_{a_r}^{b_r} - \int_{a_r}^{b_r} \{|y_r'|^2 + q_r |y_r|^2\} \, dx_r, \quad r = 1, \ldots, k, \qquad (5.7.1)$$

all be negative-definite, on the spaces of continuously differentiable functions satisfying the boundary conditions (5.2.2). *Then to each set of non-negative integers* $n_1, \ldots, n_k$ *there correspond precisely two eigenvalues* $\lambda_1, \ldots, \lambda_k$ *for which these are the oscillation numbers of the non-trivial solutions of* (5.2.1)–(5.2.2). *In one of these eigenvalues, we have* $\lambda_1 > 0$, *and in the other* $\lambda_1 < 0$.

Our assumption concerning co-factors means in particular that

$$\begin{vmatrix} p_{22}(x_2) & \cdots & p_{2k}(x_2) \\ p_{32}(x_3) & \cdots & p_{3k}(x_3) \\ \vdots & & \vdots \\ p_{k2}(x_k) & \cdots & p_{kk}(x_k) \end{vmatrix} > 0, \qquad (5.7.2)$$

for all sets $x_r \in [a_r, b_r]$, $r = 2, \ldots, k$. The Klein oscillation theorem then ensures that if λ_1 is given, there is precisely one set of values of $\lambda_2, \ldots, \lambda_k$ such that (5.2.1)–(5.2.2) have non-trivial solutions for $r = 2, \ldots, k$, the solutions having $n_2, \ldots, n_k$ as their oscillation numbers. We may therefore treat $\lambda_2, \ldots, \lambda_k$ as functions of the real variable λ_1. The problem is then to adjust λ_1 so that (5.2.1)–(5.2.2) with $r = 1$ has a non-trivial solution with the oscillation number n_1 .

In terms of the polar-coordinates used earlier in this chapter, we have to solve the equation

$$\theta_1(b_1; \lambda_1, \ldots, \lambda_k) = n_1 \pi + \beta_1, \qquad (5.7.3)$$

in which $\lambda_2, \ldots, \lambda_k$ are treated as functions of λ_1 in the manner just described.

Let us write $\varphi(\lambda_1)$ for the left of (5.7.3), where $\lambda_2, \ldots, \lambda_k$ are treated as functions of λ_1. In addition to $\varphi(\lambda_1)$ being continuous we need the following three properties, in order to obtain the result of Theorem 5.7.1:

(i) $\varphi(\lambda_1) \to +\infty$ as $|\lambda_1| \to \infty$,

(ii) $\varphi(0) < \beta_1$,

(iii) $\varphi(\lambda_1)$ is an increasing (resp. decreasing) function when $\lambda_1 > 0$ (resp. $\lambda_1 < 0$) and $\varphi(\lambda_1) \equiv \beta_1 \mod \pi$.

For (i), we remark that by Theorem 5.6.1, the total number of zeros of any set of solutions of (5.2.1) must tend to ∞ as $|\lambda_1| \to \infty$. If, therefore, we make $\lambda_2, \ldots, \lambda_k$ depend on λ_1 in such a way that the $y_r(x_r)$, $r = 2, \ldots, k$, have fixed oscillation numbers, then the number of zeros of a non-trivial solution of (5.2.1) with $r = 1$ must tend to ∞, as $|\lambda_1| \to \infty$. Thus $\varphi(\lambda_1) \to \infty$ as $|\lambda_1| \to \infty$. We remark that, by general implicit function theorems, $\varphi(\lambda_1)$ is a continuous function of λ_1, and indeed a differentiable one. At this stage, we can therefore say that for sufficiently large n_1 there will be at least one positive and one negative λ_1 satisfying (5.7.3), so that for given $n_2, \ldots, n_k$, and sufficiently large n_1 there are at least two eigenvalues.

To improve this result, we prove next (ii), which will remove the "sufficiently large" proviso on n_1 in the statement just made. We shall establish (ii) if we show that the eigenvalues Λ of the problem

$$z''(x_1) + \{\Lambda + \sum_2^k \lambda_s(0)p_{1s}(x_1) - q_1(x_1)\}z(x_1) = 0, \qquad (5.7.4)$$

$$z(a_1)\cos\alpha_1 = z'(a_1)\sin\alpha_1, \quad z(b_1)\cos\beta_1 = z'(b_1)\sin\beta_1 \quad (5.7.5)$$

are all positive; here $\lambda_s(0)$, $s = 2, \ldots, k$ denote the values of $\lambda_2, \ldots, \lambda_k$ associated with $\lambda_1 = 0$. For suppose that $\theta(x_1; \Lambda)$ is the polar angle function associated with (5.7.4)–(5.7.5) by

$$z(x_1)\cos\theta = z'(x_1)\sin\theta, \quad \theta(a_1; \Lambda) = \alpha_1, \qquad (5.7.6)$$

where $z(x_1)$ is a non-trivial solution of (5.7.4), which satisfies the first of (5.7.5). It is then a standard property that $\theta(b_1; \Lambda)$ is a strictly increasing function of Λ. Thus if (5.7.4), (5.7.5) give a problem with only positive eigenvalues, so that the equations

$$\theta(b_1; \Lambda) = \beta_1 + n_1\pi, \quad n_1 = 0, 1, \ldots$$

have only positive roots, and we have that $\theta(b_1; 0) < 0$, then since

$$\theta(b_1; 0) = \theta_1(b_1; 0, \lambda_2(0), \ldots, \lambda_k(0)) = \varphi(0),$$

this will prove (ii).

We observe next that the eigenvalues of (5.7.4)-(5.7.5) will all be positive if the form

$$\Psi(z) = \Phi_{10}(z) + \sum_{s=2}^k \lambda_s(0)\Phi_{1s}(z), \qquad (5.7.7)$$

is negative-definite on the space of continuously differentiable $z(x_1)$ satisfying (5.7.5). Here we use the notation in Section 3.3, cf., (3.3.4); the $\Phi_{r0}(y_r)$ are displayed explicitly in equivalent form in (5.7.1) for the special differential operators with which we are now concerned. We see this by taking $z(x_1)$ to be a (real) eigenfunction of (5.7.4)–(5.7.5), multiplying (5.7.4) by $z(x_1)$ and integrating the result over (a_1, b_1); the statement can also be viewed as a corollary of the extremal expression for the lowest eigenvalue of a Sturm-Liouville problem.

Thus, to prove (ii), we have to show that (5.7.7) is negative-definite, subject to (5.7.5). To see this, we denote by $y_r(x_r)$, $r = 2, \ldots, k$, non-trivial real solutions of (5.2.1)–(5.2.2) with $\lambda_1 = 0$, $\lambda_s = \lambda_s(0)$, $s = 2, \ldots, k$. Then on multiplying (5.2.1) by the $y_r(x_r)$ and integrating, we get

$$\Phi_{r0}(y_r) + \sum_2^k \lambda_s(0)\Phi_{rs}(y_r) = 0, \quad r = 2, \ldots, k. \tag{5.7.8}$$

Let D denote the determinant of the k-by-k array

$$\begin{vmatrix} \Phi_{10}(z) & \Phi_{12}(z) & \cdots & \Phi_{1k}(z) \\ \Phi_{20}(y_2) & \Phi_{22}(y_2) & \cdots & \Phi_{2k}(y_2) \\ \vdots & \vdots & & \vdots \\ \Phi_{k0}(y_k) & \Phi_{k2}(y_k) & \cdots & \Phi_{kk}(y_k) \end{vmatrix}.$$

It follows from (5.7.7), (5.7.8), and Cramer's rule that

$$D = \Psi(z) \det \Phi_{rs}(y_r). \tag{5.7.9}$$

Here the determinant is now over $2 \leq r, s \leq k$. The determinant on the right is positive, in view of Theorem 4.2.3 and the fact that $\det p_{rs}(x_r)$, $2 \leq r, s \leq k$, is positive. The whole determinant (5.7.9) is negative, since the elements in the first column are negative, while the co-factors are positive, for the reason just mentioned. It thus follows that $\Psi(z)$ is negative, as needed. This proves (ii).

It remains to prove (iii), which we do by calculating $d\varphi/d\lambda_1$. Differentiating (5.3.7) with $r = 2, \ldots, k$, we have

$$\partial\theta_r/\partial\lambda_1 + \sum_2^k (\partial\theta_r/\partial\lambda_s)(d\lambda_s/d\lambda_1) = 0, \quad r = 2, \ldots, k,$$

while

$$\varphi'(\lambda_1) = \partial\theta_1/\partial\lambda_1 + \sum_2^k (\partial\theta_1/\partial\lambda_s)(d\lambda_s/d\lambda_1).$$

Using Cramer's rule again, we deduce that

$$\varphi'(\lambda_1)\partial(\theta_2, \ldots, \theta_k)/\partial(\lambda_2, \ldots, \lambda_k) = \partial(\theta_1, \ldots, \theta_k)/\partial(\lambda_1, \ldots, \lambda_k). \tag{5.7.10}$$

It follows from (5.4.8), and the corresponding result for $2 \leq r, s \leq k$, that these Jacobians have the same sign as, respectively,

$$\det \int_{a_r}^{b_r} p_{rs}(x_r)|y_r(x_r)|^2 \, dx_r, \quad 2 \leq r, s \leq k, \tag{5.7.11}$$

$$\det \int_{a_r}^{b_r} p_{rs}(x_r)|y_r(x_r)|^2\, dx_r, \quad 1 \le r, s \le k, \tag{5.7.12}$$

where the $y_r(x_r)$ are solutions of (5.2.1) fixed by (5.3.2). Here (5.7.11) is certainly positive, in view of (5.7.2) and Theorem 4.2.3. In the case of (5.7.12), we appeal to Theorem 4.6.1, which shows that this has the same sign as λ_1 at an eigenvalue. Thus $d\varphi/d\lambda_1$ has the same sign as λ_1 at an eigenvalue, as was to be proved. This completes the proof of Theorem 5.7.1.

5.8   Unstandardized formulations

It was convenient to prove the Richardson oscillation theorem subject to two specializations, firstly that we assumed the matrices (4.6.1) have positive determinants when $\rho_1 = 1$, $\rho_2 = \cdots = \rho_k = 0$, and secondly that we assumed the forms (5.7.1) negative-definite, rather than the more general forms (3.8.9). We now translate this special form of the result into a version more generally applicable. This procedure not only obviates the need for preliminary transformations, but also yields additional information.

We first remove the specialization on the ρ_s. We prove

Theorem 5.8.1. *Let the $p_{rs}(x_r), q_r(x_r)$, $a_r \le x_r \le b_r$, r, $s = 1,\ldots,k$ be continuous, and let the determinants of (4.6.1) take only positive values. Let the forms (5.7.1) all be negative-definite, on the spaces of continuously differentiable functions satisfying the boundary conditions (5.2.2). Then to each set of non-negative integers $n_1,\ldots,n_k$ there correspond just two eigenvalues $\lambda_1,\ldots,\lambda_k$ of the problem (5.2.1)–(5.2.2); for these two eigenvalues the expression $\sum_1^k \rho_s\lambda_s$ takes opposite signs.*

We make the linear transformations of the ρ_s, λ_s described in the proof of Theorem 4.11.2. We require that

$$\sum_{s=1}^{k} \rho_s\gamma_{st} = \delta_{1t}, \tag{5.8.1}$$

where $\delta_{1t} = 1$, if $t = 1$ and 0 otherwise in order to achieve the situation of Theorem 5.8.1. We then have, if $\lambda_1^\dagger,\ldots,\lambda_k^\dagger$ are the new spectral parameters,

$$\sum_{1}^{k} \rho_s\lambda_s = \sum_{s=1}^{k}\sum_{t=1}^{k} \rho_s\gamma_{st}\lambda_t^\dagger = \lambda_1^\dagger. \tag{5.8.2}$$

By Theorem 5.7.1, the two eigenvalues associated with the given oscillation numbers will have $\lambda_1^\dagger > 0$ in one case, and $\lambda_1^\dagger < 0$ in the other. This proves the result.

We now remove the second specialization.

Theorem 5.8.2. *Let the assumptions of Theorem 5.8.1 hold, except that instead of the negative-definiteness of the forms (5.7.1), we assume that for a certain set of $\tau_1, \ldots, \tau_k$, the same holds for the forms (3.8.9), and this for the same spaces of functions. Then, to each set of non-negative integers $n_1, \ldots, n_k$ there correspond just two eigenvalues of (5.2.1)–(5.2.2), and for these the expression*

$$\sum \rho_s(\lambda_s + \tau_s) \tag{5.8.3}$$

takes opposite signs.

We introduce new spectral parameters

$$\lambda_s{}^* = \lambda_s + \tau_s, \quad s = 1, \ldots, k,$$

so that the differential equations (5.2.1) now read

$$y_r'' + \sum \lambda_s{}^* p_{rs} y_r - q_r{}^* y_r = 0, \quad r = 1, \ldots, k, \tag{5.8.4}$$

where the sum is over $s = 1, \ldots, k$, and

$$q_r{}^* = q_r + \sum \tau_s p_{rs}, \quad r = 1, \ldots, k. \tag{5.8.5}$$

The assumption that the forms (3.8.9) are negative-definite then means that the same properties hold for the forms (5.7.1) with q_r replaced by $q_r{}^*$. Thus, by the last theorem, there are two eigenvalues for each set of oscillation numbers, for which $\sum \lambda_s{}^* \rho_s$ has opposite signs. Thus, in terms of the original spectral parameters, (5.8.3) takes opposite signs on these two eigenvalues, as was to be proved.

5.9   A partial oscillation theorem

Let us recall that for the one-parameter problem, $k = 1$ in (5.2.1)–(5.2.2), in the polar case that $p_{11}(x)$ changes sign, we can assert that there exists an ascending and a descending sequence of eigenvalues, where to each sufficiently large oscillation number there corresponds one eigenvalue in each sequence. In

order to claim that there is one such eigenvalue corresponding to each oscillation number $n = 0, 1, 2, \ldots$, we have to impose a further restriction, such as the
definiteness of a certain quadratic form. A similar situation holds in the multiparameter case. We give here a result concerning the existence of eigenvalues
for sufficiently large oscillation numbers.

Theorem 5.9.1. *Let the $p_{rs}(x_r)$, $q_r(x_r)$ be continuous in $[a_r, b_r]$, let $\det p_{rs}(x_r)$
change sign, and let there exist a set $\rho_1, \ldots, \rho_k$ such that the determinants
(4.6.1) are positive for all sets $x_1, \ldots, x_k$. Then for some $C > 0$, and all sets
of non-negative integers $n_1, \ldots, n_k$ such that*

$$\sum n_r{}^2 > C, \tag{5.9.1}$$

*there are at least two eigenvalues $\lambda_1, \ldots, \lambda_k$, such that the problem (5.2.1)–
(5.2.2) has solutions with the n_r as oscillation numbers.*

By means of a linear transformation, we may take it that condition (B) holds
in the standardized form with $\rho_2 = \cdots = \rho_k = 0$. We use the argument of the
proof of Theorem 5.7.1. We write

$$\varphi(\lambda_1) = \varphi(\lambda_1; n_2, \ldots, n_k) \tag{5.9.2}$$

for the value of $\theta_1(b; \lambda_1, \ldots, \lambda_k)$, where $\lambda_2, \ldots, \lambda_k$ are determined as functions
of λ_1 by the requirement that (5.2.1)–(5.2.2) have non-trivial solutions for $r =
2, \ldots, k$ and the given λ_1. We no longer claim that we have $\varphi(0) < \beta_1$. Instead,
we claim that for some fixed K, we have

$$\varphi(0) = \varphi(0; n_2, \ldots, n_k) < K, \tag{5.9.3}$$

for all sets $n_2, \ldots, n_k$. This implies that if n_1 is sufficiently large then there is
an eigenvalue with oscillation count $(n_1, n_2, \ldots, n_k)$. Let

$$\lambda_{s0} = \lambda_{s0}(n_2, \ldots, n_k), \quad s = 2, \ldots, k \tag{5.9.4}$$

be the values of $\lambda_2, \ldots, \lambda_k$ determined as above, with $\lambda_1 = 0$. The equations

$$y_r'' + \left\{ \sum \lambda_{s0} p_{rs} - q_r \right\} y_r = 0, \quad r = 2, \ldots, k, \tag{5.9.5}$$

then have non-trivial solutions satisfying (5.2.2), having the oscillation numbers
$n_2, \ldots, n_k$. Thus if we make

$$\sum n_s{}^2 \to \infty, \tag{5.9.6}$$

where the $n_2, \ldots, n_k$ run though some sequence of $(k-1)$-tuples of non-negative
integers, we must have

$$\sum |\lambda_{s0}| \to \infty. \tag{5.9.7}$$

For it follows from the Sturm comparison theorem that if the oscillation number of a solution of (5.9.5) for some r becomes unbounded, then the λ_{s0} must become unbounded; for a more exact exploitation of this argument, we refer to Section 5.9.

We write

$$\lambda_{s0} = \chi \mu_{s0}, \quad \chi \geq 0, \quad \sum |\mu_{s0}| = 1, \tag{5.9.8}$$

all these quantities being functions of $n_2, \ldots, n_k$. In view of (5.9.7), we have that (5.9.6) implies that

$$\chi = \chi(n_2, \ldots, n_k) \rightarrow \infty \tag{5.9.9}$$

We now use Theorem 4.7.2, which ensures that for some $k > 0$ we have

$$\sum \mu_{s0} p_{r's}(x_{r'}) \geq \delta, \tag{5.9.10}$$

$$\sum \mu_{s0} p_{r''s}(x_{r''}) \leq -\delta \tag{5.9.11}$$

for some r', r'', as we vary $n_2, \ldots, n_k$ so as to achieve (5.9.6). It then follows from (5.9.11) and (5.9.9) that

$$\sum \lambda_{s0} p_{r''s}(x_{r''}) - q_{r''}(x_{r''}) \rightarrow -\infty \tag{5.9.12}$$

uniformly in $[a_{r''}, b_{r''}]$. This could not occur if r'' were equal to any of $2, \ldots, k$, since then the equation (5.9.5) with $r = r''$ could ultimately not have a solution with fixed oscillation number satisfying the boundary conditions. Thus we must have $r'' = 1$. It now follows from Theorem A.6.1 that $\varphi(0) \rightarrow 0$, so that (5.9.3) is satisfied. We now repeat the argument starting with (5.9.2) by interchanging the roles of λ_1 and λ_s for each fixed s, $1 < s \leq k$. This completes the proof.

Notes for Chapter 5

The positivity of the Jacobian $\partial \theta / \partial \lambda$ is used in Atkinson (1964a, Section 6.10), in considering the discrete analogue, that of orthogonal polynomials. For the differential equation case, see Atkinson (1964a, p. 551); Faierman (1966, Chapter 2). See also Binding and Browne (1984).

Ince (1956) considered the case of several disjoint intervals on the same axis, without singularities. He used induction over k, together with a Sturm-type theorem with nonlinear parameter-dependence.

Partial oscillation theorems, to the effect that eigentuples exist ensuring sufficiently large oscillation numbers, have been discussed by many authors. See in particular Turyn (1980) and Richardson (1918).

Many authors assume that "condition (A)" holds on the closed region $\prod[a_r, b_r]$, though cases occur in which the determinant $\det p_{rs}$ vanishes at points of the boundary.

Theorem 5.9.1 is a multiparameter extension of old results of Richardson (1918) and Haupt (1911) in the one parameter Sturm-Liouville case. Under weaker assumptions on the coefficients (Lebesgue integrability), this result was extended by Everitt *et al.* (1983) in the same setting. In this vein, it is important to note that there may be non-negative integers $n_1, \ldots, n_k$ such that there is *no* eigenvalue for which these are the oscillation numbers. This eventuality, of course, is not precluded by Theorem 5.9.1.

5.10   Research problems and open questions

1.   Give a different proof of Klein's oscillation theorem using the fact that we can vary the p_{rs}, q_r continuously into constants, retaining the validity of condition (A), by some homotopy, as functions of τ in $[0, 1]$. The eigenvalues will then satisfy a differential equation, as functions of τ, and then each will remain bounded through this variation.

2.   The continuity of the q_r in this chapter seems unnecessary, and can be replaced by Lebesgue integrability. However, relaxing the continuity assumption on the p_{rs} to Lebesgue integrability over (a_r, b_r) for all $r, s = 1, 2, \ldots, k$ seems much more challenging. Can this be done?

3.   Consider the one-parameter case of (5.2.1)–(5.2.2). This is of course the Sturm-Liouville equation with a weight;

$$y''(x) + (\lambda p(x) - q(x))\, y(x) = 0, \quad a \le x \le b, \tag{5.10.13}$$

where y satisfies (5.2.2). Assume for simplicity that p, q are both continuous in $[a, b]$ and that

$$\int_a^b |p(x)|\, dx > 0,$$

there being no sign restrictions upon $p(x)$ (the so-called *non-definite* case). The smallest non-negative integer n_h such that there is at least one eigenfunction of (5.10.13)– (5.2.2) is called the Haupt index of the problem. The smallest non-negative integer n_R such that there is exactly one eigenfunction of (5.10.13) satisfying (5.2.2) is called the Richardson index (or Richardson number) of the boundary problem. The existence of these quantities can be found in the quoted papers of Haupt and Richardson (see Mingarelli [1986] for further details concerning these numbers). In addition, it is known that there is always an at most finite number of pairs say, $\lambda_{\mathbb{C}}$, of non-real eigenvalues (see Mingarelli [1983]).

We note that whenever $p(x)$ changes sign there is an infinite sequence of both positive, λ_n^+, and negative, λ_n^-, eigenvalues having no finite limit point. In this case, there exists two such indices, $n_h^\pm$ and $n_R^\pm$ corresponding in turn to the positive, respectively, negative eigenvalues.

Is it the case that

$$\lambda_{\mathbb{C}} \leq \min \{n_R^+, n_R^-\}?$$

4. *What is the multiparameter analogue of the Haupt and Richardson index in the cases $k \geq 3$?* In the two-parameter case, this question is answered by Faierman (1991, Theorem 1.1).

5. Prove or disprove the following general conjecture (cf. Problem 3 above):

$$\lambda_{\mathbb{C}} \leq \min \{n_h^+, n_h^-\}.$$

The only result known in this connection appears to be the special case treated in Allegretto-Mingarelli (1987); that is, whenever (5.10.13)–(5.2.2) has a non-real eigenvalue there is no real eigenfunction whatsoever that is of one sign in (a, b) (at least for the Dirichlet problem). An abstract version of this last result that includes difference and partial differential operators can be found in Mingarelli (1994).

6. *Find the spectral radius of the non-real spectrum of* (5.10.13)–(5.2.2) *in the non-definite case* (cf. Problems 3–5 above). A more tractable though not quite solved problem is to *find a priori estimates for the location of non-real eigenvalues in terms of the known quantities p, r, or their integrals*, and so forth.

Chapter 6

Eigencurves

6.1 Introduction

In this chapter, we present an alternative approach to the study of multiparameter Sturm-Liouville problems in the case of two equations and two parameters. This is based on a more detailed study of the case of a single boundary-value problem with two parameters. We apply Sturmian methods to the boundary-value problem in which we ask whether the equation

$$y''(x) + \{\sum_{j=1}^{2} \lambda_j p_j(x) - q(x)\} y(x) = 0, \quad a \leq x \leq b, \qquad (6.1.1)$$

has a non-trivial solution satisfying

$$y(a) \cos \alpha = y'(a) \sin \alpha, \quad y(b) \cos \beta = y'(b) \sin \beta. \qquad (6.1.2)$$

We take $p_1(x)$, $p_2(x)$ and $q(x)$ to be real and, as a rule, continuous. We take α, β to be real and, without loss of generality, choose them so that

$$0 \leq \alpha < \pi, \quad 0 < \beta \leq \pi. \qquad (6.1.3)$$

Pairs (λ_1, λ_2) such that (6.1.1)–(6.1.2) have a non-trivial solution may be considered as a sort of eigenvalue. When we are dealing with a single boundary-value problem with two parameters, we have no reason to expect that such admissible pairs will have to be real. We shall however confine attention to real pairs, since only they will be relevant to simultaneous problems with real spectra.

Frequently, though not invariably, we consider the case that some linear combination of $p_1(x)$, $p_2(x)$ does not vanish, and so has fixed sign. Since the p_1, p_2 are to be continuous, there will be a second linear combination, linearly independent of the first, which also has fixed sign. By a linear transformation, we can then replace (6.1.1) by a similar equation in which the coefficients of the parameters have any assigned fixed signs.

A degenerate case deserves comment, namely, that in which the $p_1(x)$, $p_2(x)$ are linearly dependent, and not both identically zero. In this case, one of them will be a constant multiple of the other. If $p_1(x) = cp_2(x)$, the differential equation (6.1.1) may be replaced by

$$y'' + \{(c\lambda_1 + \lambda_2)p_2 - q\}y = 0, \quad a \le x \le b.$$

The spectrum of this, together with (6.1.2), will clearly take the form of the set of lines $c\lambda_1 + \lambda_2 = const.$; here the "const." will run through a set of values without finite limit-point, since $p_2(x)$ does not vanish identically.

We conclude this section by collecting some general properties of the spectrum.

Theorem 6.1.1. *Let the $p_j(x)$, $j = 1$, 2, and $q(x)$ be real and continuous, and the $p_j(x)$ linearly independent. Then the real spectrum of (6.1.1)–(6.1.2) is a non-empty closed set that meets any line in a set without finite limit-point.*

Let $y(x; \lambda_1, \lambda_2)$ denotes the solution of (6.1.1) such that

$$y(a; \lambda_1, \lambda_2) = \sin\alpha, \quad y'(a; \lambda_1, \lambda_2) = \cos\alpha. \tag{6.1.4}$$

The points (λ_1, λ_2) of the spectrum are then the zeros of

$$\Delta(\lambda_1, \lambda_2) = y(b; \lambda_1, \lambda_2)\cos\beta - y'(b; \lambda_1, \lambda_2)\sin\beta. \tag{6.1.5}$$

Since $\Delta(\lambda_1, \lambda_2)$ is continuous, indeed entire, the set of its zeros, or of its real zeros, is closed, as asserted.

We prove next that the real spectrum is non-empty. Since p_1, p_2 are linearly independent, at least one of them, say, $p_1(x)$, does not vanish identically. We can then obtain points of the real spectrum by fixing λ_2 at any real value, and treating (6.1.1)–(6.1.2) as an eigenvalue problem for λ_1. The fact that $p_1(x)$ is not identically zero then ensures that there is an infinity of eigenvalues.

Finally, we show that the spectrum meets any real line in a set without finite limit-point; to see this, we take an arbitrary real line in the (λ_1, λ_2)-plane in the parametric form

$$\lambda_j = \sigma\rho_j + \gamma_j, \quad j = 1, 2, \tag{6.1.6}$$

where ρ_1, ρ_2 are not both zero, and σ is a real parameter. The intersections of this line with the real spectrum are then the real eigenvalues σ of the problem formed by

$$y'' + \{\sigma(\rho_1 p_1 + \rho_2 p_2) + \gamma_1 p_1 + \gamma_2 p_2 - q\}y = 0,$$

together with (6.1.2). Since the coefficient of σ in this equation does not vanish identically, there will be an infinity of such σ-values, without finite limit-point. This completes the proof.

6.2   Eigencurves

In what follows, we assume, unless otherwise stated, that $p_1(x)$, $p_2(x)$, $q(x)$ are continuous in the finite interval $[a, b]$, and that at least one of the $p_s(x)$ does not vanish identically. The "spectrum" of (6.1.1)–(6.1.2) is then a non-trivial closed subset of the real (λ_1, λ_2)-plane.

The spectrum may be divided up into connected components, on each of which we shall have

$$\theta(b; \lambda_1, \lambda_2) \equiv \beta \quad \mathrm{mod}\ \pi, \tag{6.2.1}$$

where θ is the Prüfer angle defined as in (5.3.3-5), with the obvious modifications. By continuity, we shall then have on each component

$$\theta(b; \lambda_1, \lambda_2) = \beta + n\pi, \tag{6.2.2}$$

where the integer n is fixed for each component. By Theorem 5.3.1, we have $n \geq 0$; in fact n will be the oscillation number of the corresponding solutions of (6.1.1)–(6.1.2), the number of zeros they have in the open interval (a, b).

It need not be the case that different connected components of the spectrum are associated with different oscillation numbers. However, we shall deal with such a case in what follows; roughly speaking, this will be the case if some linear combination of the $p_s(x)$ has fixed sign. We use the term "eigencurve" for a connected subset of the spectrum; we shall be concerned with a case in which eigencurves are characterized by oscillation numbers.

The case in which some linear combination of the $p_s(x)$ has fixed sign is most conveniently investigated in a standardized form, which may be achieved by a linear transformation. This was discussed in §4.11.

Theorem 6.2.1. *Let either*

$$p_2(x) > 0, \quad a \leq x \leq b \tag{6.2.3}$$

or

$$p_2(x) < 0, \quad a \le x \le b. \tag{6.2.4}$$

Then to each non-negative integer n there is a curve $\Gamma(n)$ defined by

$$\lambda_2 = f(\lambda_1, n), \quad -\infty < \lambda_1 < \infty, \tag{6.2.5}$$

such that $\Gamma(n)$ consists precisely of those λ_1, λ_2 satisfying (6.2.2). The functions (6.2.5) are continuous. If (6.2.3) holds, we have

$$f(\lambda_1, 0) < f(\lambda_1, 1) < \dots \tag{6.2.6}$$

while if (6.2.4) holds, we have (6.2.6) with the inequalities reversed.

Suppose for definiteness that (6.2.3) holds; the case of (6.2.4) can then be handled by an obvious transformation. We consider the eigenvalue problem (6.1.1)–(6.1.2) as a standard Sturm-Liouville eigenvalue problem in which λ_1 is given and λ_2 is the spectral parameter. By the Sturm oscillation theorem (a special case of the Klein oscillation theorem given by $k = 1$), there exists an ascending sequence of values of λ_2 such that the associated solution has n zeros in (a, b). Denoting these as in (6.2.5), we have the result (6.2.6).

It remains to establish the continuity of the functions (6.2.5). We give here a proof based on the Sturm comparison theorem. We have, for any $\lambda_1^\dagger \ne \lambda_1$, that the equations

$$y'' + \{\lambda_1 p_1 + f(\lambda_1, n)p_1 - q\}y = 0, \tag{6.2.7}$$

$$y'' + \{\lambda_1^\dagger p_1 + f(\lambda_1^\dagger, n)p_2 - q\}y = 0, \tag{6.2.8}$$

both have solutions with n zeros in (a, b), satisfying the same boundary conditions (6.2.1)–(6.2.2). Thus the difference between the two coefficients of y in (6.2.7)–(6.2.8), that is to say,

$$(\lambda_1^\dagger - \lambda_1)p_1(x) + (f(\lambda_1^\dagger, n) - f(\lambda_1, n))p_2(x) \tag{6.2.9}$$

cannot have constant sign in (a, b); in fact it must either change sign or else vanish identically. The same is therefore true of

$$\{f(\lambda_1^\dagger, n) - f(\lambda_1, n)\}/\{\lambda_1^\dagger - \lambda_1\} + p_1(x)/p_2(x), \tag{6.2.10}$$

Thus $\{f(\lambda_1^\dagger, n) - f(\lambda_1, n)\}/\{\lambda_1^\dagger - \lambda_1\}$ cannot exceed

$$\max\{-p_1(x)/p_2(x)\}, \tag{6.2.11}$$

nor be less than

$$\min\{-p_1(x)/p_2(x)\}, \tag{6.2.12}$$

where these extrema are taken over $[a, b]$. This shows that the functions $f(\lambda_1, n)$ are not merely continuous but satisfy a uniform Lipschitz condition.

6.3   Slopes of eigencurves

We start by formulating a result that was essentially proved in the last section.

Theorem 6.3.1. *Let $p_2(x)$, $a \leq x \leq b$, be positive. Then if $\lambda_1 \neq \lambda_1{}^\dagger$, the ratio*

$$\{f(\lambda_1{}^\dagger, n) - f(\lambda_1, n)\}/\{\lambda_1{}^\dagger - \lambda_1\}$$

lies in the interior of the interval

$$[\min\{-p_1(x)/p_2(x)\},\ \max\{-p_1(x)/p_2(x)\}], \tag{6.3.1}$$

unless the interval reduces to a point; in this case said ratio equals the constant $-p_1(x)/p_2(x)$.

We have shown that this ratio lies in the closed interval (6.3.1). However, the function (6.2.10) must actually change sign, unless it vanishes identically, and this proves the result.

This result may also be seen by evaluating $df(\lambda_1, n)/d\lambda_1$. Differentiating (6.2.2) with respect to λ_1 we have

$$f'(\lambda_1, n)\partial\theta/\partial\lambda_2 + \partial\theta/\partial\lambda_1 = 0; \tag{6.3.2}$$

this is a special case of the calculation performed in §5.4. Using (5.4.7) with minor changes we get

$$f'(\lambda_1, n) = -\int_a^b p_1\, y^2 dx / \int_a^b p_2\, y^2\, dx, \tag{6.3.3}$$

where $y(x) = y(x; \lambda_1, \lambda_2)$. Thus, if the interval (6.3.1) does not reduce to a point, the slopes of the eigencurves (6.2.5) all lie strictly in the interior of (6.3.1); the same will be true of arbitrary chords.

The result of Theorem 6.3.1 remains true if $p_2(x)$ is negative throughout $[a, b]$.

We pursue this topic in more detail later, but pause to note that the result of this section permits an alternative proof of the Klein oscillation theorem in the case of two parameters. This time we shall not cover the "borderline case" in which $\det p_{rs}(x_r)$ can have zeros.

6.4   The Klein oscillation theorem for $k = 2$

We shall assume the differential equations concerned to have been brought to a certain standard form by a preliminary linear transformation of the spectral parameters. In the system

$$y_r''(x_r) + \left\{\sum \lambda_s p_{rs}(x_r) - q_r(x_r)\right\} y_r(x_r) = 0, \quad a_r \leq x_r \leq b_r, \quad (6.4.1)$$

with continuous $p_{rs}(x_r)$, $q_r(x_r)$, we shall take it that

$$p_{11}(x_1) > 0, p_{12}(x_1) < 0, \quad a_1 \leq x_1 \leq b_1, \quad (6.4.2)$$

$$p_{21}(x_2) > 0, p_{22}(x_2) > 0, \quad a_2 \leq x_2 \leq b_2. \quad (6.4.3)$$

It was noted in Section 4.11 that this can be arranged by a linear transformation of the spectral parameters, if in the original system det $p_{rs}(x_r)$ had fixed sign. We have then, as a rather special case of Theorem 5.5.1,

Theorem 6.4.1. *Let the $p_{rs}(x_r), q_r(x_r)$ be continuous and satisfy (6.4.2)–(6.4.3). Then for any pair n_1, n_2 of non-negative integers, there is precisely one eigenvalue (λ_1, λ_2) of the problem (6.4.1) with boundary conditions (5.2.2) with $k = 2$, such that the $y_r(x_r)$ have the n_r as their oscillation numbers.*

Since the coefficients of λ_2 in (6.4.1) have fixed signs, Theorem 6.2.1 ensures that there exist continuous functions

$$f(\lambda_1, n_r, r), \quad r = 1, 2, \quad (6.4.4)$$

such that if $\lambda_2 = f(\lambda_1, n_r, r)$, then the r-th equation in (6.4.1) has a non-trivial solution satisfying the associated boundary conditions (5.2.2), and having n_r zeros in the interior of the associated basic interval. We denote the graph of $f(\lambda_1, n_r, r)$ by $\Gamma(n_r, r)$. We have to show that there is a unique λ_1 such that

$$f(\lambda_1, n_1, 1) = f(\lambda_1, n_2, 2) \quad (6.4.5)$$

or that the eigencurves $\Gamma(n_r, r), r = 1, 2$, have a unique intersection.

This follows from the results of the last section. These show that the slopes of chords of the $\Gamma(n_r, r)$ lie in the intervals

$$[\min\left(-p_{r1}(x_r)/p_{r2}(x_r)\right), \max\left(-p_{r1}(x_r)/p_{r2}(x_r)\right)], \quad r = 1, 2, \quad (6.4.6)$$

where the extrema are taken over $[a_r, b_r]$, $r = 1, 2$. From (6.4.2)–(6.4.3) it follows that if $r = 1$, the interval contains only positive numbers, while if $r = 2$ it contains only negative numbers. This shows that

$$f(\lambda_1, n_1, 1) \to \pm\infty \text{ as } \lambda_1 \to \pm\infty, \quad (6.4.7)$$

where the signs are to correspond, while

$$f(\lambda_1, n_2, 2) \to \mp\infty \quad \text{as} \quad \lambda_1 \to \pm\infty. \tag{6.4.8}$$

Thus the equation (6.4.5) certainly has a root. Since the left of (6.4.5) is an increasing function, and the right a decreasing function of λ_1, this root must be unique. This completes the proof.

Example The simplest application of this theorem follows: For $k = 2$, let $[a_r, b_r] = [0, 1]$ and $p_{11} = 1$, $p_{22} = 1$, $p_{21} = 1$. Next, let $p_{12} = -1$ and consider the Dirichlet problem on the various intervals. Then (6.4.1)takes the form

$$y_1''(x_1) + (\lambda_1 - \lambda_2)y_1(x_1) = 0, \quad y_2''(x_2) + (\lambda_1 + \lambda_2)y_2(x_2) = 0.$$

A moment's notice shows that the eigenvalues are now given by pairs (λ_1, λ_2), where $\lambda_1 - \lambda_2 = n_1^2\pi^2$ and $\lambda_1 + \lambda_2 = n_2^2\pi^2$ for $n_1, n_2 = 1, 2, \ldots$. In this case, the functions $f(\lambda_1, n_1, 1)$, $f(\lambda_1, n_2, 2)$ appearing in (6.4.4) are given explicitly by

$$\lambda_2 = f(\lambda_1, n_1, 1) = \lambda_1 - n_1^2\pi^2, \quad \lambda_2 = f(\lambda_1, n_2, 2) = -\lambda_1 + n_2^2\pi^2.$$

Equating these and solving we get the defining relations

$$\lambda_1 = \frac{1}{2}(n_1^2 + n_2^2)\pi^2, \quad \lambda_2 = \frac{1}{2}(n_2^2 - n_1^2)\pi^2, \quad n, m \geq 1$$

for the eigenvalue pairs. The corresponding eigenfunctions are of the form $y_1(x_1) = \sin(n_1\pi x_1)$ and $y_2(x_2) = \sin(n_2\pi x_2)$, and these have the required oscillation properties.

The *method of eigencurves* described here will be considered in the next chapter in association with a different two-parameter problem.

6.5 Asymptotic directions of eigencurves

We return now to the topic of §§6.2, 6.3. The result we prove will have as an incidental consequence that the bounds (6.3.1) for the slopes of chords of eigencurves are precise, being attained in an asymptotic sense.

Theorem 6.5.1. Let $p_2(x)$, $a \leq x \leq b$, be positive. Then for each $n = 0, 1, \ldots$ the functions (6.2.5) satisfy

$$f(\lambda_1, n)/\lambda_1 \to \min\{-p_1(x)/p_2(x)\}, \quad \lambda_1 \to +\infty, \tag{6.5.1}$$

$$f(\lambda_1, n)/\lambda_1 \to \max\{-p_1(x)/p_2(x)\}, \quad \lambda_1 \to -\infty. \tag{6.5.2}$$

We suppose the contrary to (6.5.1), that for some sequence

$$\lambda_{1j} \equiv \lambda_1 \to +\infty, \tag{6.5.3}$$

some integral $n \geq 0$, and some $K > 0$, we have

$$f(\lambda_1, n)/\lambda_1 \geq -p_1(x_0)/p_2(x_0) + K, \tag{6.5.4}$$

where $x_0 \in [a, b]$ is the point at which $-p_1(x)/p_2(x)$ attains its minimum. By continuity, there will be an interval $[\xi, \eta] \subset [a, b]$, of positive length, such that

$$f(\lambda_1, n)/\lambda_1 \geq -p_1(x)/p_2(x) + K/2, \quad \xi \leq x \leq \eta. \tag{6.5.5}$$

Hence in the same interval,

$$\lambda_1 p_1(x) + f(\lambda_1, n)p_2(x) \geq K \, \lambda_1 p_2(x)/2$$

and so, if

$$\delta_1 = \min p_2(x), \quad \delta_2 = \min\{-q(x)\} \tag{6.5.6}$$

the minima being over $[\xi, \eta]$, we have

$$\lambda_1 p_1(x) + f(\lambda_1, n)p_2(x) - q(x) \geq K\lambda_1\delta_1/2 + \delta_2, \quad \xi \leq x \leq \eta. \tag{6.5.7}$$

It then follows that as $\lambda_1 \to \infty$ the left-hand side tends uniformly to $+\infty$ for $x \in [\xi, \eta]$. The Sturm comparison theorem then implies that the number of zeros of any associated solutions of (6.1.1) tends to ∞, whereas the oscillation number is to remain fixed at n. This gives a contradiction, and so proves (6.5.1). The proof of (6.5.2) is similar. In the case that $p_2(x)$ is of fixed negative sign, the limits in (6.5.1)–(6.5.2) are to be interchanged.

6.6 The Richardson oscillation theorem for $k = 2$

We use the arguments of the last section to provide a proof of this theorem, somewhat different from that given in Section 5.7 for the case of general k. We suppose the matrix of continuous functions

$$p_{rs}(x_r), \quad r, s = 1, 2, \quad a_r \leq x_r \leq b_r,$$

brought by a linear transformation to one in which the required sign-properties of determinants are automatically fulfilled, assuming that

$$p_{12}(x_1) < 0, \quad a_1 \leq x_1 \leq b_1, \tag{6.6.1}$$

$$p_{22}(x_2) > 0, \quad a_2 \leq x_2 \leq b_2. \tag{6.6.2}$$

We depart from (6.4.2)–(6.4.3) in requiring that both of $p_{11}(x_1)$, $p_{21}(x_2)$ should have zeros in the respective intervals $[a_r, b_r]$, $r = 1, 2$, and that at least one of them should change sign. We have then

Theorem 6.6.1. *Let the $p_{rs}(x_r)$, $r, s = 1, 2$ satisfy the above requirements, and let the forms (5.7.1), with $r = 1, 2$, be negative-definite on the spaces of continuously differentiable functions satisfying the boundary conditions (5.2.2). Then to each pair of non-negative integers n_1, n_2, there are just two eigenvalues with these as oscillation numbers, for which the values of λ_1 have opposite signs.*

In view of (6.6.1)–(6.6.2), the functions (6.4.4) are well-defined, and it is a question of showing that (6.4.5) has precisely one positive and one negative root.

As in Section 5.7, there are three stages in the proof. In the first, we show that

$$f(0, n_2, 2) > 0 > f(0, n_1, 1). \tag{6.6.3}$$

This follows directly from (6.6.1)–(6.6.2) and the assumed negative-definiteness of the forms (5.7.1).

In the second stage, in which we show that these roots exist, we take first the case that both of $p_{11}(x_1)$, $p_{21}(x_2)$ change sign. Since $p_{22}(x_2) > 0$, we have from Theorem 6.5.1 that as $\lambda_1 \to +\infty$,

$$\lim f(\lambda_1, n_2, 2)/\lambda_1 = \min \left(-p_{21}/p_{22} \right) < 0, \tag{6.6.4}$$

while as $\lambda_1 \to -\infty$,

$$\lim f(\lambda_1, n_2, 2)/\lambda_1 = \max \left(-p_{21}/p_{22} \right) > 0; \tag{6.6.5}$$

here the extrema are taken over $a_2 \leq x_2 \leq b_2$.

In a similar way, taking into account that $p_{12}(x_1) < 0$, we have from the duly modified form of Theorem 6.5.1 that, for $\lambda_1 \to +\infty$,

$$\lim f(\lambda_1, n_1, 1)/\lambda_1 = \max \left(-p_{11}/p_{12} \right) > 0, \tag{6.6.6}$$

and, as $\lambda_1 \to -\infty$,

$$\lim f(\lambda_1, n_1, 1))/\lambda_1 = \min \left(-p_{11}/p_{12} \right) < 0. \tag{6.6.7}$$

It follows from (6.6.4), that

$$f(\lambda_1, n_2, 2) - f(\lambda_1, n_1, 1) \to -\infty, \quad \lambda_1 \to +\infty, \tag{6.6.8}$$

and since the left is positive when $\lambda_1 = 0$, we deduce that it vanishes for some positive λ_1. Similarly, we can show that it vanishes for some negative λ_1.

We dispose next of the case that only one of $p_{11}(x_1)$, $p_{21}(x_2)$ changes sign; suppose that $p_{11}(x_1) \geq 0$ (and that $p_{11}(x_1) = 0$ somewhere) while $p_{21}(x_2)$ changes sign. Then (6.6.4)–(6.6.5) still hold but now the right-hand side of (6.6.6) equals zero. Furthermore, (6.6.7) is still in force and so arguing as in the previous case we may deduce that (6.6.8) vanishes for some $\lambda_1 > 0$. A similar discussion applies in the case where $\lambda_1 < 0$.

Finally, it is necessary to show that the positive and negative roots of (6.4.5) are unique. It is sufficient to show that

$$f'(\lambda_1, n_2, 2) - f'(\lambda_1, n_1, 1) \qquad (6.6.9)$$

has the opposite sign to λ_1 when (6.4.5) holds. In view of (6.3.3), (6.6.9) is equal to

$$\int_{a_1}^{b_1} p_{11} y_1^2 \, dx_1 / \int_{a_1}^{b_1} p_{12} y_1^2 \, dx_1 - \int_{a_2}^{b_2} p_{21} y_2^2 \, dx_2 / \int_{a_2}^{b_2} p_{22} y_2^2 \, dx_2. \qquad (6.6.10)$$

By (6.6.1)–(6.6.2), the denominators here are, respectively, negative and positive, and so (6.6.9) has the opposite sign to

$$\det \int_{a_r}^{b_r} p_{rs} y_r^2 \, dx_r. \qquad (6.6.11)$$

We thus need to show that this has the same sign as λ_1 at an eigenvalue.

The proof is similar to that given in a more general case in Section 5.7, and so will be only briefly sketched. We multiply (5.2.1) by the respective non-trivial solutions and integrate over the basic intervals, to get

$$\int_{a_r}^{b_r} (y_r'' - q_r y_r) \, y_r \, dx_r + \sum \lambda_s \int_{a_r}^{b_r} p_{rs} y_r^2 \, dx_r = 0, \quad r = 1, 2. \qquad (6.6.12)$$

Here the first term on the left is negative, being the same as the form appearing in (5.7.1).We then get the result on eliminating λ_2 from (6.6.12).

6.7   Existence of asymptotes

We go back once more to the study of eigencurves for (6.1.1)–(6.1.2), in the case that some linear combination of the $p_s(x)$ is of fixed sign. We continue to take the standardized form in which $p_2(x)$ is positive.

It was shown in Section 6.5 that the eigencurves $\Gamma(n)$, $n = 0, 1, \ldots$ defined by the functions $f(\lambda, n)$ tend to infinity in definite directions as $\lambda \to +\infty$, $\lambda \to -\infty$; the directions did not depend on n, but did depend on whether $\lambda \to \pm\infty$, except in the degenerate case when the eigencurves are straight lines.

The question arises of whether these eigencurves admit not only asymptotic directions, but also asymptotes in the strict geometrical sense; by this we mean that the shortest distance between a point on this curve and the asymptote tends to zero as the point tends to infinity along the curve. The answer is, in general, negative. We take thus up first in a standardized form.

Theorem 6.7.1. Let $p_1(x) \geq 0$, $p_1(x) \not\equiv 0$, $p_2(x) > 0$, $a \leq x \leq b$. Then $f(\lambda_1, n)$ is a strictly increasing function of λ_1. It is bounded as $\lambda_1 \to +\infty$ if and only if $p_1(x)$ vanishes on some sub-interval of (a, b) of positive length.

It follows from Theorem 6.3.1 that $f(\lambda_1, n)$ is non-decreasing in λ_1. It cannot remain constant since $\theta(b; \lambda_1, \lambda_2)$ is a strictly decreasing function of λ_1, if $p_1(x)$ is non-positive and does not vanish identically.

Suppose that $p_1(x) \equiv 0$ on $[a', b']$, where $a \leq a' < b' \leq b$. Then on this interval (6.1.1) takes the form

$$y''(x) + (\lambda_2 p_2(x) - q(x))y(x) = 0, \quad a' \leq x \leq b'; \tag{6.7.1}$$

since $p_2(x) > 0$, it would then follow that if λ_2 were unbounded, the number of zeros of a solution of (6.7.1) would also be unbounded. Since the number of zeros is fixed for $\lambda_2 = f(\lambda_1, n)$, we deduce that λ_2 must be bounded.

Suppose next that λ_2 is bounded, and that $p_2(x)$ does not vanish identically on any sub-interval of (a, b) of positive length. It then follows from Theorem A.6.1 that $\theta(b; \lambda_1, \lambda_2) \to 0$ as $\lambda_1 \to \infty$, whereas we must have $\theta(b; \lambda_1, \lambda_2) = \beta + n\pi > 0$ on Γ_n. This completes the proof.

Notes for Chapter 6

Among the first papers that use the method of eigencurves in the study of a two-parameter Sturm-Liouville problem, we find Hilb (1907) and Richardson (1910). For a current review of such methods, see Binding and Volkmer (1996).

Eigencurves of two-parameter eigenvalue problems associated with formally selfadjoint 2n-th order real linear differential operators determined by regular selfadjoint boundary conditions are considered in Binding and Browne (1989a) (cf., Binding *et al.* [1987]). For bounds on the number of non-real eigenvalues in the abstract two-parameter symmetric case see Binding and Browne (1988) (cf., also Binding and Browne [1981]).

The special case of the study of the stability boundaries of a periodic Hill equation with two parameters was undertaken by Loud (1975). In this vein, but in the case of a doubly periodic differential equation, see Arscott (1987) and each of the references there: Arscott (1964a), Arscott and Sleeman (1968), Erdélyi *et al.* (1953), Ince (1956), Magnus and Winkler (1966), and Sleeman *et al.* (1984). The relationship between the expression of a singly periodic equation in algebraic form and the existence of "multiplicative solutions" is to be found in Arscott and Sleeman (1968). Background material may be found in McLachlan (1947).

Regarding two parameter systems in general: An investigation into the geometric nature of the disconjugacy and non-oscillation domains of a single general Sturm-Liouville equation with two parameters can be found in Mingarelli and Halvorsen (1988). A non-definite analogue of the Haupt-Richardson oscillation theorem (see the Notes for Chapter 5) in the case of a two-parameter system is actually obtained by Faierman (1991, Theorem 1.1). The completeness of the set of eigen-elements of a two-parameter Dirac system was proved in Gadžiev (1979). For a generalization of the standard deficiency index theory of symmetric operators to the multiparameter case, see Browne and Isaev (1988).

6.8 Research problems and open questions

1. Does Faierman (1991, Theorem 1.1) in the two parameter case ($k = 2$) extend to the case $k \geq 3$? It appears that new techniques are required to handle this and all subsequent cases.

2. Determine an oscillation theorem for a two-parameter Dirac system of the form:

$$Jy_r' = \left\{ \sum_{s=1}^{k} \lambda_s A_{rs} + B_r \right\} y_r, \qquad r = 1, 2, \ldots, k$$

in the case $k = 2$, where J is constant, skew-hermitean and non-singular, A_{rs}, B_r are square hermitean matrices; see Atkinson (1964a, Chapters 9.1–9.2 and Chapter 10.9)) for the case $k = 1$.

Chapter 7

Oscillations for Other Multiparameter Systems

7.1 Introduction

In the standard multiparameter theory for arrays

$$\begin{vmatrix} A_{10} & \cdots & A_{1k} \\ \vdots & & \vdots \\ A_{k0} & \cdots & A_{kk} \end{vmatrix} \tag{7.1.1}$$

as applied to Sturm-Liouville problems with several parameters, it is usual to locate differential operators in the first column, while the operators A_{rs}, $1 \le r, s \le k$ in the remaining k columns correspond to multiplication by scalar functions. This accords with the origin of multiparameter Sturm-Liouville theory in the separation of variables, wherein the "separation parameters" appear together with the coefficient-functions rather than with the differential operators. It is also consistent with the operator-formulation in which the first column consists of unbounded operators with compact resolvents, while the remaining columns consist of bounded operators. In an alternative version, one has bounded invertible operators in the first column and compact operators in the last k columns.

The existence of a spectral theory for expressions $S - \lambda T$, where S, T are both differential operators (Pleijel et al.) suggests the study of arrays (7.1.1) in which all the A_{rs} are differential operators, of order zero upward. We examine an example, with $k = 2$, in which each row contains a Sturm-Liouville operator, but not both in the same column.

7.2   An example

We consider the problem

$$\lambda y''(s) + \mu a(s)y(s) + \nu b(s)y(s) = 0, \tag{7.2.1}$$

$$-\lambda c(t)z(t) + \mu z''(t) + \nu d(t)z(t) = 0, \tag{7.2.2}$$

with three homogeneous parameters λ, μ, ν. Here s, t vary over intervals I, J, and we impose, for definiteness, the Dirichlet boundary conditions

$$y = 0, \quad z = 0, \tag{7.2.3}$$

at the end-points of I, J, respectively. The functions a, b, c, d are to be real, and for simplicity continuous.

An eigenvalue will be a triple λ, μ, ν, not all zero, such that (7.2.1)–(7.2.2) both have a solution, not identically zero, satisfying the boundary conditions. Such eigenvalues are considered projectively; a pair of eigenvalues of the form λ, μ, ν, and $c\lambda$, $c\mu$, $c\nu$, are not considered distinct. They form collectively the "point spectrum," or "joint point spectrum" of the problem posed by (7.2.1)–(7.2.2) together with the boundary conditions (7.2.3).

Eigenvalues for which $\lambda = 0$ or $\mu = 0$, for which the differential operators do not appear in either or both of (7.2.1)–(7.2.2) may be termed singular, and may be excluded by suitable assumptions. It is obvious that (7.2.1), for example, will not in general have a nontrivial solution satisfying the boundary conditions if $\lambda = 0$ while μ, ν are not both zero, though this eventuality cannot be excluded in special cases.

Eigenvalues in which λ, μ are both non-zero may be termed "non-singular."

An eigenvalue, or triple λ, μ, ν, will be termed "real" if λ, μ, ν are either real themselves, or are proportional to a triple of real numbers.

We apply the term "eigenfunction" to the product

$$y(s)z(t), \tag{7.2.4}$$

in the event that λ, μ, ν is an eigenvalue.

7.3   Local definiteness

We review the theory developed in Chapter 3 as applied to this situation. This "definiteness" property, or rather hypothesis, plays a central role in the development of a spectral theory of, roughly speaking, self-adjoint type. In particular it serves to ensure the reality of the eigenvalues. To formulate this property in the case of (7.2.1)–(7.2.3), we set up the array of sesquilinear forms associated with the operators in (7.2.1)–(7.2.2), taking into account (7.2.3), thus obtaining the matrix

$$
\begin{vmatrix}
-\int y'^2 \, ds & \int ay^2 \, ds & \int by^2 \, ds \\
-\int cz^2 \, dt & -\int z'^2 \, dt & \int dz^2 \, dt
\end{vmatrix}.
\tag{7.3.1}
$$

We then say that the system (7.2.1)–(7.2.3) is "locally definite"[1] if this matrix has rank two whenever $y(s)z(t)$ is an eigenfunction, that is to say, whenever y, z are non-trivial (real-valued) solutions of the problem (7.2.1)–(7.2.3).

In terms of this array, the eigenvalue λ, μ, ν is determined, projectively, by being orthogonal as a three-vector to the two rows in (7.3.1). The set of such triples $(\lambda,\ \mu,\ \nu)$ is clearly generated by a real triple.

More explicitly, an eigenvalue is given in terms of the non-trivial solutions of (7.2.1)–(7.2.3) by suitably signed minors of the array (7.3.1); λ, μ, ν will be multiples of the three determinants

$$
\begin{vmatrix}
\int ay^2 \, ds & \int by^2 \, ds \\
-\int z'^2 \, dt & \int dz^2 \, dt
\end{vmatrix},
\tag{7.3.2}
$$

$$
\begin{vmatrix}
\int y'^2 \, ds & \int by^2 \, ds \\
\int cz^2 \, dt & \int dz^2 \, dt
\end{vmatrix},
\tag{7.3.3}
$$

$$
\begin{vmatrix}
-\int y'^2 \, ds & \int ay^2 \, ds \\
-\int cz^2 \, dt & -\int z'^2 \, dt
\end{vmatrix}.
\tag{7.3.4}
$$

---

[1]This is actually called "definite" in Section 3.4.

7.4   Sufficient conditions for local definiteness

At the risk of loss of generality, one may ensure local definiteness by imposing simpler but formally more demanding conditions. This may occur in one or both of two ways. Firstly, we may require one specific determinant of the three in (7.3.2)–(7.3.4) be non-zero, for non-trivial solutions of (7.2.1)–(7.2.3); more generally, we can ask that some fixed linear combination of them be non-zero. Secondly, we may require that (7.3.1) have rank two for some more general class of y, z not identically zero; for example, we could impose the rank condition for continuously differentiable y, z satisfying the boundary conditions, but not necessarily the differential equations.

Suppose, for example, we wish to ensure that (7.3.4) is non-zero, or in particular that

$$\int |y'|^2 \, ds \cdot \int |z'|^2 \, dt + \int c|z|^2 \, dt \cdot \int a|y|^2 \, ds > 0. \qquad (7.4.1)$$

This may be ensured crudely by asking that $a(s)$, $c(t)$ have the same fixed signs, so that (7.4.1) will hold for any non-trivial C'-functions, without regard to the boundary conditions. Taking the latter into account, we can derive more sensitive criteria, allowing for a, c to take opposite signs.

Similar remarks apply to the determinants (7.3.2)–(7.3.3), for which we may ask that

$$\int |y'|^2 \, ds \cdot \int d|z|^2 \, dt - \int b|y|^2 \, ds \cdot \int c|z|^2 \, dt \neq 0, \qquad (7.4.2)$$

or that

$$\int b|y|^2 \, ds \cdot \int |z'|^2 \, dt + \int a|y|^2 \, ds \cdot \int d|z|^2 \, dt \neq 0, \qquad (7.4.3)$$

for non-trivial C'-functions satisfying the boundary conditions.

7.5   Orthogonality

Let

$$\lambda_j, \mu_j, \nu_j, \quad j = 1, 2, \qquad (7.5.1)$$

be (projectively) distinct eigenvalues, and y_j, z_j corresponding non-trivial solutions of (7.2.1)–(7.2.2); we may take these solutions to be real-valued. We form the matrix

$$\begin{vmatrix} -\int y_1'y_2'\,ds & \int a y_1 y_2\,ds & \int b y_1 y_2\,ds \\ -\int c z_1 z_2\,dt & -\int z_1'z_2'\,dt & \int d z_1 z_2\,dt \end{vmatrix}. \tag{7.5.2}$$

The rows of this matrix are orthogonal to both the three-vectors (7.5.1), so that this matrix has rank at most one.

In other words, all 2-by-2 determinants formed from the array (7.5.2), in a manner similar to (7.3.2)–(7.3.4), must vanish. These three relationships constitute orthogonalities between the functions

$$y_j(s)z_j(t), \quad j = 1, 2. \tag{7.5.3}$$

These orthogonalities can be extended to suitably differentiable functions of the two variables s, t, not necessarily of the decomposable form (7.5.3).

It follows from the above, together with the hypothesis of local definiteness, that (non-singular) eigenvalues are isolated; we see this by passing to the limit in the orthogonality properties (7.5.2) for distinct eigenvalues (see the proof of Theorem 3.6.1 for details).

7.6 Oscillation properties

We now consider properties similar to the Klein oscillation theorem. It is a question of whether to every pair of non-negative integers m, n there corresponds one, or more, eigenvalues (triples λ, μ, ν) such that (7.2.1)–(7.2.2) have non-trivial solutions satisfying the boundary conditions (7.2.3) and having m, n zeros, respectively, in the interiors of I, J.

We consider the case that

$$a(s) > 0, \quad s \in I, \tag{7.6.1}$$

$$c(t) > 0, \quad t \in J. \tag{7.6.2}$$

This ensures that the determinant (7.3.4) is positive, as in (7.4.1). This implies that for an eigenvalue we have $\nu \neq 0$. We can therefore pass to an inhomogeneous version of eigenvalue in which we take

$$\nu = 1. \tag{7.6.3}$$

We can then use the method of "eigencurves." In the present case, this means that for each equation (7.2.1)–(7.2.2) separately, we form the set in the real

(λ, μ)-plane for which there is a non-trivial solution satisfying the boundary conditions and having the assigned number of zeros in the interior. Intersections of these sets will then give eigenvalues (pairs λ, μ) corresponding to the assigned oscillation numbers.

Substituting (7.6.3) in (7.2.1) we get

$$\lambda y''(s) + \mu a(s)y(s) + b(s)y(s) = 0. \qquad (7.6.4)$$

In view of (7.6.1), and the Sturm oscillation theorem, there will for non-zero λ and non-negative integral m be a unique μ such that (7.6.4) has a non-trivial solution satisfying the boundary conditions and have m zeros in the interior of I. We denote this by

$$\mu = f(\lambda, m), \quad \lambda \neq 0, \quad m \geq 0. \qquad (7.6.5)$$

Similarly, for the equation

$$-\lambda c(t)z(t) + \mu z''(t) + d(t)z(t) = 0, \qquad (7.6.6)$$

we define the function

$$\lambda = g(\mu, n), \quad \mu \neq 0, \quad n \geq 0. \qquad (7.6.7)$$

We are concerned with the intersections of the curves (7.6.5), (7.6.7) and pass to a detailed examination of these curves in turn.

7.7   The curve $\mu = f(\lambda, m)$.

It is easily seen that, for fixed integral $m \geq 0$, f is a continuous function of λ for $\lambda \neq 0$. Moreover, we have

Lemma 7.7.1. *For $\lambda \neq 0$, f is a monotone increasing function of λ.*

Let $\mu = f(\lambda, m)$, $\mu^* = f(\lambda^*, m)$, and let y, y^* be the corresponding solutions of (7.6.4). We deduce that

$$\lambda^*(y^{*''}y - y''y^*) + (\lambda^* - \lambda)\, y''y^* + (\mu^* - \mu)\, ay^*y = 0. \qquad (7.7.1)$$

Integrating over I and passing to the limit as $\lambda^* \to \lambda$ we deduce that

$$d\mu/d\lambda = \int y'^2 \, ds \Big/ \int ay^2 \, ds. \qquad (7.7.2)$$

Next we consider the behavior as $|\lambda| \to \infty$. We denote by α_m the eigenvalue that is such that

$$y'' + \alpha\,a(s)\,y = 0, \qquad\qquad (7.7.3)$$

has a solution satisfying the boundary conditions, with precisely m zeros in the interior of I. We claim then

Lemma 7.7.2. *We have*

$$f(\lambda, m) \sim \alpha_m \lambda \quad \text{as } |\lambda| \to \infty. \qquad\qquad (7.7.4)$$

This follows on writing (7.6.4) in the form

$$y'' + \{(\mu/\lambda)\,a(s) + b(s)/\lambda\}\,y = 0. \qquad\qquad (7.7.5)$$

Finally, we must consider small values of λ. We observe that the following lemmas maybe considered as special cases of Theorem 6.5.1. It is instructive to give the simplified arguments below.

Lemma 7.7.3. *As $\lambda \to 0^+$, we have*

$$f(\lambda, m) \to \inf\{-b(s)/a(s)\}, \quad s \in I. \qquad\qquad (7.7.6)$$

Suppose first that for some $\varepsilon > 0$ and arbitrarily small $\lambda > 0$ we have

$$f(\lambda, m) \geq \inf\{-b(s)/a(s)\} + \varepsilon. \qquad\qquad (7.7.7)$$

It will then follow that for some $\delta > 0$ we have

$$f(\lambda, m) + b(s)/a(s) > \varepsilon/2,$$

in an interval of length δ, so that

$$\{\mu a(s) + b(s)\}/\lambda \geq \varepsilon\,a(s)\,/\,(2\lambda),$$

in such an interval. It then follows from the Sturm comparison theorem that the number of zeros of y in this interval becomes unbounded as $\lambda \to +0$, contrary to the definition of $f(\lambda, m)$.

Suppose again that

$$f(\lambda, m) \leq \inf\{-b(s)/a(s)\} - \varepsilon,$$

for some $\varepsilon > 0$ and arbitrarily small $\lambda > 0$. It then follows that

$$\{\mu a(s) + b(s)\}/\lambda \to -\infty$$

uniformly in I, and this again contradicts the definition of $f(\lambda, m)$.

In a similar way, we have

Lemma 7.7.4. *As $\lambda \to 0^-$,*

$$f(\lambda, m) \to \sup\{-b(s)/a(s)\}, \quad s \in I. \tag{7.7.8}$$

The proof is similar. We suppose that for some $\varepsilon > 0$ and arbitrarily small $\lambda < 0$ we have

$$f(\lambda, m) < \sup\{-b(s)/a(s)\} - \varepsilon,$$

so that

$$f(\lambda, m) < -b(s)/a(s) - \varepsilon/2$$

in some interval of length $\delta > 0$. We then deduce that

$$\mu a(s) + b(s) < -\varepsilon a(s)/2,$$

or

$$\{\mu a(s) + b(s)\}/\lambda > \varepsilon a(s)/(2|\lambda|),$$

in such an interval. This would imply that y has an arbitrarily large number of zeros.

If again

$$f(\lambda, m) > \sup\{-b(s)/a(s)\} + \varepsilon$$

for some $\varepsilon > 0$ and arbitrarily small $\lambda < 0$, we would get

$$\{\mu a(s) + b(s)\} \to -\infty,$$

uniformly in I, which again contradicts the requirements on the zeros of a solution.

Lemma 7.7.5. *Assume that $b(t) \not\equiv 0$ and $b(t)$ takes on both signs in I. Let the eigenvalues of*

$$y'' + \beta\, b(s)\, y = 0, \tag{7.7.9}$$

with the same boundary conditions be

$$\cdots < \beta_{-1} < \beta_{-0} < 0 < \beta_0 < \beta_1 < \ldots \tag{7.7.10}$$

Then

$$f(1/\beta_m, |m|) = 0. \tag{7.7.11}$$

Indeed, if this were not the case then a Sturmian argument shows that the β_m could not be eigenvalues of (7.7.9) and the first of (7.2.3).

7.8 The curve $\lambda = g(\mu, n)$

We now choose λ in (7.6.6), that is to say, in

$$-\lambda c(t)z + \mu z'' + d(t)z = 0,$$

for given $\mu \neq 0$ and integral $n \geq 0$, so that z should satisfy the boundary conditions and have n zeros in the interior of J. Again g is a continuous function of μ. The following lemmas are parallel to those of §7.7, and so their proofs are left to the reader.

Lemma 7.8.1. *For $\mu \neq 0$, g is a decreasing function of μ.*

We have in fact

$$d\lambda/d\mu = -\int z'^2 \, ds / \int cz^2 \, ds. \tag{7.8.1}$$

We define next γ_n as the eigenvalue such that

$$z'' + \gamma_n \, c(t) \, z = 0 \tag{7.8.2}$$

has a solution satisfying the boundary conditions and having just n zeros in the interior of J, and have

Lemma 7.8.2. *As $|\mu| \to \infty$,*

$$g(\mu, n) \sim -\gamma_n \mu. \tag{7.8.3}$$

Finally, we must consider the behavior as $\mu \to 0$.

Lemma 7.8.3. *As $\mu \to 0^+$,*

$$g(\mu, n) \to \sup\{d(t)/c(t)\} \tag{7.8.4}$$

Lemma 7.8.4. *As $\mu \to 0^-$,*

$$g(\mu, n) \to \inf\{d(t)/c(t)\}. \tag{7.8.5}$$

Lemma 7.8.5. *Assume that $d(t) \not\equiv 0$ and let $d(t)$ take on both signs in J. Let the eigenvalues of*

$$z'' + \delta d(t)z = 0, \tag{7.8.6}$$

with the same boundary conditions be

$$\cdots < \delta_{-1} < \delta_{-0} < 0 < \delta_0 < \delta_1 < \cdots \tag{7.8.7}$$

Then

$$g(1/\delta_n, |n|) = 0. \tag{7.8.8}$$

We now consider the boundary problem (7.2.1)–(7.2.3), taking it that (7.6.1)–(7.6.2). From these lemmas, we deduce an oscillation theorem according to which the eigenvalues may be associated with the numbers of zeros of the eigenfunctions $y(s)z(t)$.

Theorem 7.8.1. *For every sufficiently large n, there is an m such that there is at least one eigenvalue (λ, μ) of (7.6.4), (7.6.6), (7.2.3) whose eigenfunction $y(s)z(t)$ is such that $y(s)$ has just m zeros in I and $z(t)$ has just n zeros in J.*

It suffices to show that the eigencurves $\mu = f(\lambda, m)$, $\lambda = g(\mu, n)$ intersect for all sufficiently large values of n, m, or equivalently, that $\mu - f(g(\mu, n), m) = 0$ always has a solution for such n, m.

To this end, let $\mu > 0$. By Lemma 7.8.2, we can choose n sufficiently large so that $g(\mu, n) \leq -\gamma_n \mu/2$. In addition, by Lemma 7.7.2 we can make $f(-\gamma_n \mu/2, m) \leq -\gamma_n \mu \alpha_m/4$ for all sufficiently large m. Combining these estimates with Lemma 7.7.1 we get that for $\mu > 0$ and n, m sufficiently large

$$\mu - f(g(\mu, n), m) \geq \mu(1 + \gamma_n \alpha_m/4) > 0.$$

For $\mu < 0$, a similar argument applies. For example, $g(\mu, n) \geq \gamma_n |\mu|/2$ and so $f(g(\mu, n), m) \geq f(\gamma_n |\mu|/2, m)$ since f increases in its first variable. In addition, for all sufficiently large n, m we also have $f(\gamma_n |\mu|/2, m) \geq \gamma_n |\mu| \alpha_m/4$. Hence,

$$\mu - f(g(\mu, n), m) \leq \mu(1 - \gamma_n \alpha_m |\mu|/4) < 0.$$

It follows that for all sufficiently large n, m the equation $\mu = f(g(\mu, n), m)$ has a solution $\mu > 0$ and the result follows.

Example We show that Theorem 7.8.1 cannot be true for all non-negative integer values of n, m, by exhibiting a counterexample in the case $n = 0, m = 0^2$.

Let $a(s) = 20$, $c(t) = 20$ for all $s \in I = [-1, 1], t \in J = [-1, 1]$. Define $b(s) = s$, $d(t) = t$ for all such $s \in I, t \in J$. Then, by Lemma 7.7.3 we must have $f(\lambda, m) \to -0.05$ as $lambda \to 0^+$ while by Lemma 7.7.4, there holds $f(\lambda, m) \to 0.05$ as $lambda \to 0^-$. In addition, Lemma 7.8.3 tells us that $g(\mu, n) \to 0.05$ as $\mu \to 0^+$ while $g(\mu, n) \to -0.05$ as $\mu \to 0^-$ (by Lemma 7.8.4).

Next, a simple (Maple) calculation of the first positive (resp. first negative) eigenvalue $\sigma_0^{\pm} := 1/\lambda_0^{\pm}$ of the Airy type eigenvalue problem $y''(s) + \sigma s y(s) = 0$, $y(-1) = 0, y(1) = 0$ show that numerically, $\lambda_0^{\pm} \approx \pm 0.07798$.

Similarly, we show that $\mu_0^{\pm} \approx \pm 0.07798$ where these represent the (reciprocals of the) first positive and negative eigenvalues of the problem $z''(t) + \sigma t z(t) = 0$,

---

[2]This example is inspired by an analogous construction of Professor Volkmer for discontinuous coefficients (personal communication).

$z(-1) = 0, z(1) = 0$, where now $\sigma = 1/\mu$. These results together imply that
the eigencurves $\mu = f(\lambda, 0), \lambda = g(\mu, 0)$ both intersect the respective axes at
the values ± 0.07798 (as there is symmetry in the problem).

Applying Lemma 7.7.1 and Lemma 7.8.1 to the foregoing we get that the eigencurves given by $\lambda = g(\mu, 0)$, $\mu = f(\lambda, 0)$ can therefore never intersect each other
in this case, and thus there cannot be any solutions satisfying the conclusions
of the Theorem.

Notes for Chapter 7

This chapter was written by Atkinson certainly before 1992 judging from the
dated digital material at hand. Indeed, basing ourselves on the manuscripts in
our possession the results within this chapter date back to the early 1980s although references to it can be found in all four Table of Contents (some of which
date to the mid-1970s). Very similar results were obtained later by Faierman
and Mennicken (2005); there one finds a slight sharpening of Theorem 7.8.1
above, cf., Faierman and Mennicken (2005, Theorem 3.2). The L^2 completeness of the eigenfunctions in special cases of this same problem is considered in
Faierman *et al.* (2008) using the method of partial differential equations.

In extant table of contents this chapter had at least six more sections present
for which no material was found. The headings for the remaining sections were
7.9 Dirac systems; 7.10 Ma systems;[3] 7.11 Non-standard arrays of operators;
7.12 Interface conditions; 7.13 Parameter in the boundary conditions; and 7.14
Orthogonal polynomials.

Regarding such "other" types of multiparameter systems there appears to be
only one paper by Binding and Volkmer (2001a) that deals with the existence
of eigenvalues for indefinite one-parameter Dirac systems.

On the subject of eigenvalue problems with a parameter in the boundary conditions, there has been a flurry of activity lately. We cite the works of Allahverdiev
and Isaev (1981), Browne and Sleeman (1979), Bhattacharyya et al. (2001),
(2002), and Binding *et al.* (1994) as representative.

---

[3]Section 7.10 likely would have referred to the systems considered by Stephen Ma in his
Ph.D. thesis of 1972 under the supervision of Atkinson although once again nothing else is
known. Still less can be gathered about the remaining sections.

7.9   Research problems and open questions

1. What can one say about the uniform convergence of the expansion considered in Faierman *et al.* (2008) for the system considered here?

2. Improve Theorem 7.8.1. For example, find bounds on n, m depending on the coefficients alone.

3. Answer the questions raised in Faierman and Mennicken (2005, Problem 4.8, p. 1559) for the system considered in this chapter.

4. Formulate an oscillation theorem for *Ma systems*, Ma (1972), in the case of $k > 2$ parameters.

Chapter 8

Distribution of Eigenvalues

8.1 Introduction

For the classical Sturm-Liouville problem, given by

$$y'' + \{\lambda\, p(x) - q(x)\}y = 0, \quad a \le x \le b, \tag{8.1.1}$$

with continuous p, q, positive p and the usual Sturmian boundary conditions,

$$y(a)\cos\alpha = y'(a)\sin\alpha, \quad y(b)\cos\beta = y'(b)\sin\beta, \tag{8.1.2}$$

the asymptotic behavior of the eigenvalues was among the earliest solved problems.

In this chapter, our concern is to extend to the multiparameter case the following statements concerning the Sturm-Liouville problem (8.1.1)–(8.1.2):

(i) if $p(x)$ is positive in (a, b), the n-th eigenvalue in ascending order is of magnitude n^2, and the eigenvalues accumulate only at $+\infty$,

(ii) if $p(x)$ changes sign, but a certain quadratic form

$$\int_a^b y\{y'' - q(x)y - \mu p(x)y\}\, dx \tag{8.1.3}$$

is negative-definite, there is both an ascending and descending sequence of eigenvalues, the n-th in each case being of order n^2, eigenvalues accumulating both at $+\infty$ and at $-\infty$.

The statement about the order of magnitude of eigenvalues may, in view of oscillation theorems, be re-phrased by saying that the magnitude of an eigenvalue is of the order of the square of the oscillation number. It is fairly clear what the extension of this statement to the multiparameter case should be. This will be dealt with in Section 8.2; the result will deal only with the order of magnitude, and does not amount to an asymptotic formula.

The remaining statements in (i), (ii) had to do with the limit-points of eigenvalues, in particular the fact that the eigenvalues had no finite limit-point. We have already extended the latter property to the multiparameter case, under suitable hypotheses, in Section 3.6. It is now a question of identifying, so to speak, infinite limit-points of the spectrum. In the one-parameter case, these are $+\infty$, if $p(x)$ is positive, $-\infty$ if it is negative, and both if $p(x)$ takes both signs. Such limit-points may be considered, at least provisionally, as forming a sort of essential spectrum.

In the one-parameter case, it is natural to distinguish between the points $+\infty$, $-\infty$ as limit-points of eigenvalues. This means that we consider the parameter as taking its values not merely on the real line, but on the real line extended by adjoining the two points $+\infty$, $-\infty$; when we endow these points at infinity with neighborhoods in the obvious way, the real line so extended becomes a topological space, distinct from the real line, and also from the real line extended by a single point at infinity, in which $+\infty$, $-\infty$ are identified. Similarly, in the multiparameter case, in which an eigenvalue is specified by a set of real numbers $\lambda_1, \ldots, \lambda_k$, and so by a point in R^k, we do not adopt a one-point compactification, in which all points at infinity are mutually identified. Instead, we consider what we term "asymptotic directions," roughly speaking directions in which some sequence of eigenvalues tends to a point at infinity. The set of such asymptotic directions forms what we may again consider as an essential spectrum.

The use of points at infinity may be avoided, in the usual way, by using a homogeneous formulation. Thus, in the one-parameter case, we consider in place of the pair $(1, \lambda)$, an equivalence class of pairs of the form $(\sigma, \sigma\lambda)$, $\sigma > 0$. Again, such a pair may be represented by a normalized pair,

$$1/\sqrt{(1+\lambda)}, \quad \lambda/\sqrt{(1+\lambda)},$$

and so by a point on a circle. In the multiparameter case, an eigenvalue $(\lambda_1, \ldots, \lambda_k)$ may be represented by an equivalence class of $(k+1)$-tuples

$$(\sigma, \sigma\lambda_1, \ldots, \sigma\lambda_k),$$

$\sigma > 0$, or alternatively by a point on a sphere S^k. The essential spectrum is then the set of cluster-points on this sphere, in the usual topology.

Points of the spectrum may be represented in a less explicit but more basic way in terms of sets of quadratic forms instead of eigenvalues. In the notation

(3.4.1), we have a k-by-$(k+1)$ matrix of forms $\Phi_{rs}(y_r)$. Taking the y_r to be a non-trivial solution of the boundary-value problem, the k-by-k minors obtained by deleting any of the $k+1$ columns are, when suitably signed, proportional to $1, \lambda_1, \ldots, \lambda_k$.

8.2 A lower order-bound for eigenvalues

Passing now to the multiparameter case, we aim to show that in the situation of either the Klein or the Richardson oscillation theorems, Theorems 5.5.1 or 5.7.1, the order of magnitude of the eigenvalue $(\lambda_1, \ldots, \lambda_k)$ is given by the sum of squares of the oscillation numbers $(n_1, \ldots, n_k)$. We first take up the simpler part of this statement, involving a lower order-bound for $\sum |\lambda_s|$. This follows easily from the Sturm comparison theorem, and is actually independent of whether the Klein or the Richardson case is involved; definiteness conditions will be brought into play subsequently, to yield upper bounds for the magnitude of eigenvalues.

Theorem 8.2.1. *Let $p_{rs}(x_r)$, $q_r(x_r)$, $s = 1, \ldots, k$, $a_r \leq x_r \leq b_r$, be real and continuous. Suppose that for an infinity of sets of values of the non-negative integers $n_1, \ldots, n_k$ there exist sets of values of $\lambda_1, \ldots, \lambda_k$ such that the equations (5.2.1) have non-trivial solutions with n_r zeros in (a_r, b_r), respectively. Then there exist $A_1 > 0$, A_2 such that*

$$\sum |\lambda_s| \geq A_1 \sum n_r^2 + A_2. \tag{8.2.1}$$

Proof. We write
$$A = \max\{|p_{rs}(x_r)|, |q_r(x_r)|\}, \tag{8.2.2}$$

taken over all r, s and x_r. We compare (5.2.1), respectively, with

$$z_r''(x_r) + A\left\{\sum |\lambda_s| + 1\right\} z_r(x_r) = 0. \tag{8.2.3}$$

The Sturm comparison theorem ensures that a solution of (8.2.3) must have at least $n_r - 1$ zeros in (a_r, b_r). Hence, using this theorem again, we have

$$(b_r - a_r)\sqrt{\left\{A\left(\sum |\lambda_s| + 1\right)\right\}} \geq (n_r - 1)\pi, \quad r = 1, \ldots, k. \tag{8.2.4}$$

These imply bounds of the form

$$\sum |\lambda_s| + 1 \geq A_3(n_r - 1)^2, \quad r = 1, \ldots, k, \tag{8.2.5}$$

where $A_3 > 0$. Since $(n_r - 1)^2 \geq n_r^2/2 - 1$, we have on summing that

$$k \sum |\lambda_s| + k \geq (A_3/2) \sum n_r^2 - k.$$

This is equivalent to a result of the required form (8.2.1).

8.3   An upper order-bound under condition (A)

We now supplement the last result by a bound in the opposite sense, for eigenvalues of (5.2.1-2), under conditions ensuring the Klein oscillation theorem.

Theorem 8.3.1. *Assume the stronger Condition (A) in Theorem 5.5.1. Then the unique eigenvalue $\lambda_1, \ldots, \lambda_k$ associated with oscillation numbers $n_1, \ldots, n_k$ satisfies bounds of the form*

$$\sum |\lambda_s| \geq A_1 \sum n_r^2 + A_2, \qquad (8.3.1)$$

$$\sum |\lambda_s| \leq B_1 \sum n_r^2 + B_2, \qquad (8.3.2)$$

where $A_1 > 0$, $B_1 > 0$, and A_2, B_2 are constants.

The existence of eigenvalues is assured in this case by the Klein oscillation theorem. Since (8.3.1) was established in the last section, we have only to prove (8.3.2).

We suppose an eigenvalue $\lambda_1, \ldots, \lambda_k$ represented in polar form as

$$\lambda_s = \chi \mu_s, \quad s = 1, \ldots, k, \qquad (8.3.3)$$

where

$$\chi = \sum |\lambda_s|, \qquad (8.3.4)$$

$$\sum |\mu_s| = 1. \qquad (8.3.5)$$

The μ_s are uniquely fixed by the λ_s, except when $\chi = 0$. It is sufficient to prove the result for $\chi > \chi_0$, say, where $\chi_0 > 0$ is suitably large, if we require that $B_2 > \chi_0$.

We now apply Theorem 4.3.2, by which there is a fixed $\delta > 0$ such that for any set μ_s, satisfying (8.3.5), there is a t, $1 \leq t \leq k$, such that either

$$\sum \mu_s p_{ts}(x_t) \geq \delta, \quad a_t \leq x_t \leq b_t, \qquad (8.3.6)$$

or else

$$\sum \mu_s p_{ts}(x_t) \leq -\delta, \quad a_t \leq x_t \leq b_t, \qquad (8.3.7)$$

Thus, with A as in (8.2.2), we have either

$$\sum \lambda_s p_{ts}(x_t) - q_t(x_t) \geq \chi\delta - A, \qquad (8.3.8)$$

or

$$\sum \lambda_s p_{ts}(x_t) - q_t(x_t) \leq -\chi\delta + A, \qquad (8.3.9)$$

for some t, throughout $[a_t, b_t]$.

We show next that (8.3.9) cannot be the case, if χ_0 is suitably large. To this end, we introduce the real numbers Λ_r, $r = 1, \ldots, k$, as the least eigenvalues of the one-parameter problem

$$y_r''(x_r) + \Lambda\, y_r(x_r) = 0, \qquad (8.3.10)$$

subject to the boundary conditions (5.2.2). It then follows that for an eigenvalue of (5.2.1-2) we cannot have

$$\sum \lambda_s p_{ts}(x_t) - q_t(x_t) < \Lambda_t, \quad a_t \leq x_t \leq b_t, \qquad (8.3.11)$$

for any t. Thus (8.3.9) is excluded if

$$-\chi_0\delta + A < \Lambda_t.$$

We conclude that if

$$\chi_0\delta - A > -\Lambda_t, \quad t = 1, \ldots, k, \qquad (8.3.12)$$

then (8.3.8) must hold in $[a_t, b_t]$ for at least one t.

We then use the Sturm comparison theorem to compare (5.2.1), in the case $r = t$, with

$$z_t'' + (\chi\delta - A)z_t = 0.$$

We impose the condition that

$$(b_r - a_r)\sqrt{(\chi_0\delta - A)} > 4\pi, \quad r = 1, \ldots, k. \qquad (8.3.13)$$

Then the number of zeros of z_t in (a_t, b_t) is not less than

$$(b_t - a_t)\sqrt{(\chi\delta - A)}/\pi - 1,$$

and so

$$n_t \geq (b_t - a_t)\sqrt{(\chi\delta - A)}/\pi - 2 \geq (b_t - a_t)\sqrt{(\chi\delta - A)}/(2\pi),$$

by (8.3.13). It follows that, for some t,

$$\chi\delta \leq A + \{2\pi n_t/(b_t - a_t)\}^2.$$

This clearly implies a result of the form (8.3.2). This concludes the proof of Theorem 8.3.1.

Example For example, we let $k = 2$, $[a_r, b_r] = [0, 1]$ and $p_{11} = 1$, $p_{22} = 1$, $p_{21} = 1$, $p_{12} = -1$. Consider the Dirichlet problem associated with the resulting equations. Then

$$y_1''(x_1) + (\lambda_1 - \lambda_2)y_1(x_1) = 0, \quad y_2''(x_2) + (\lambda_1 + \lambda_2)y_2(x_2) = 0$$

has the eigenvalues given by pairs (λ_1, λ_2) where $\lambda_1 - \lambda_2 = n_1^2\pi^2$ and $\lambda_1 + \lambda_2 = n_2^2\pi^2$ for $n_1, n_2 = 1, 2, \ldots$. From this, we find that

$$\lambda_1 = \frac{1}{2}(n_1^2 + n_2^2)\pi^2, \quad \lambda_2 = \frac{1}{2}(n_2^2 - n_1^2)\pi^2, \quad n, m \geq 1$$

as required by the theorem.

8.4   An upper bound under condition (B)

We show that the same result for the order of magnitude holds for the second definiteness condition, in the form that the determinants (4.6.1) have the same fixed sign, while the full determinant $\det p_{rs}(x_r)$ changes sign, the latter means that the first definiteness condition does not hold. Actually, the second definiteness condition requires also the sign-definiteness of certain quadratic forms, and the Richardson oscillation theorem then ensures the existence of just two eigenvalues of (5.2.1-2) for each set of oscillation numbers. However, we shall not postulate this sign-definiteness condition for the quadratic forms associated with the differential operators. Instead, we rely on Theorem 5.9.1, which ensures the existence of at least two eigenvalues for sufficiently large sets of oscillation numbers.

Theorem 8.4.1. Let the $p_{rs}(x_r)$, $q_r(x_r)$ be continuous in $[a_r, b_r]$, r, $s = 1, \ldots, k$, let $\det p_{rs}(x_r)$ change sign, and let the determinants (4.6.1) all be positive for some fixed set $\rho_1, \ldots, \rho_k$. Then the result (8.3.2) holds, with $B_1 > 0$, $B_2 > 0$, for the eigenvalues $\lambda_1, \ldots, \lambda_k$ associated with every set of oscillation numbers $n_1, \ldots, n_k$ such that

$$\sum n_r^2 \geq C, \tag{8.4.1}$$

for some $C > 0$.

If the second definiteness condition holds in its complete form, the assertion will be that (8.3.2) holds for the two eigenvalues associated with any set of oscillation numbers.

In proving the theorem, we use an argument similar to that of Section 5.6. Since $\det p_{rs}(x_r)$ changes sign, alternative (i) of Theorem 4.6.1 holds. Using Theorem 4.6.2, with reversed sign in (4.6.12), which we may do, we have that there exist positive δ, η such that for any set $\lambda_1, \ldots, \lambda_k$ we have

$$\sum \lambda_s p_{rs}(x_r) \geq \delta \sum |\lambda_s| \qquad (8.4.2)$$

for some r and some x_r-interval of length η. Hence, with A as in (8.2.2), we have in this interval

$$\sum \lambda_s p_{rs}(x_r) - q_r(x_r) \geq \delta \sum |\lambda_s| - A. \qquad (8.4.3)$$

Increasing C if necessary, we may assert that if $\sum n_r^2 \geq C$, we have

$$\delta \sum |\lambda_s| > 2A,$$

by (8.2.1). It then follows from (8.4.2) that

$$\sum \lambda_s p_{rs}(x_r) - q_r(x_r) \geq (\delta/2) \sum |\lambda_s|,$$

in an interval of length η. The Sturm comparison theorem then implies that

$$n_r \geq \pi^{-1} \eta \sqrt{\{\delta \sum |\lambda_s|/2\}} - 2,$$

or

$$\sum |\lambda_s| \leq 2\pi^2 \eta^{-2} \delta^{-1} (n_r + 2)^2 \leq 4\pi^2 \eta^{-2} \delta^{-1} (n_r^2 + 4).$$

This clearly implies (8.3.2) and completes our proof.

8.5 Exponent of convergence

This provides a rather crude measure of the overall density of the spectrum, which is useful in connection with eigenfunction expansions. For any k-parameter eigenvalue problem such as those of the last two sections, we introduce the norm

$$|\lambda| = \sum |\lambda_s| \qquad (8.5.1)$$

of an (inhomogeneous) eigenvalue $(\lambda_1, \ldots, \lambda_k)$. The exponent of convergence of the eigenvalues is then defined as the greatest lower bound σ, say, of numbers γ such that

$$\sum \{1 + |\lambda|\}^{-\gamma} < \infty. \tag{8.5.2}$$

The series is to converge when $\gamma > \sigma$, but not necessarily when $\gamma = \sigma$.

In the ordinary one-parameter Sturm-Liouville case, when the n-th eigenvalue is of order n^2, it is evident that $\sigma = 1/2$. The case of general k is covered by the last two sections for non-singular cases.

Theorem 8.5.1. *Under the conditions of Theorem 8.3.1 or 8.4.1, the exponent of convergence of the eigenvalues is $k/2$.*

It is a question of the convergence of the multiple series

$$\sum \cdots \sum (n_1^2 + \cdots + n_k^2)^{-\gamma},$$

taken over all sets of integral $n_1, \ldots, n_k$, not all zero.

In particular, we have

$$\sum \{1 + |\lambda|\}^{-1} < \infty, \quad \text{if } k = 1,$$

and

$$\sum \{1 + |\lambda|\}^{-2} < \infty, \quad \text{if } k = 1, 2, 3.$$

8.6 Approximate relations for eigenvalues

In the case of the standard one-parameter problem

$$y'' + \{\lambda p(x) - q(x)\} y = 0, \quad a \le x \le b,$$

one has under fairly general conditions a relation of the form

$$n\pi \sim \sqrt{\lambda_n} \int_a^b \sqrt{p(x)} \, dx, \tag{8.6.1}$$

in the case that $p(x) > 0$, and a similar one in the case that $p(x)$ changes sign. This relationship is readily inverted to express λ_n asymptotically in terms of n for large n. Not surprisingly, the multiparameter case is more difficult. The analogue of (8.6.1) relates each of the oscillation numbers n_r to a function of the $(\lambda_1, \ldots, \lambda_k)$. This multivariate mapping is not explicitly invertible when $k > 1$. Another point is that not all the oscillation numbers corresponding to a large eigenvalue need be large.

Nevertheless, the analogue of (8.6.1) provides some information. The application of Theorem A.3.2 (or Theorem A.2.1) to (5.2.1) suggests the introduction of the functions

$$F_r(\lambda_1, \ldots, \lambda_k) = \int_{a_r}^{b_r} Re\sqrt{\{\sum \lambda_s p_{rs}(x_r)\}}\, dx_r, \quad r = 1, \ldots, k; \quad (8.6.2)$$

as previously, the square root is taken to be non-negative when it is real. These are continuous functions of the real variables $\lambda_1, \ldots, \lambda_k$, with non-negative values; they are homogeneous functions, of degree $1/2$. They represent approximately, under suitable conditions, the increases in the associated Prüfer angles (5.3.3) over (a_r, b_r). For specific results we introduce, for any positive integral n, the quantity

$$\eta_n = \max |p_{rs}(x_{r2}) - p_{rs}(x_{r1})|, \quad (8.6.3)$$

the maximum being subject to

$$|x_{r2} - x_{r1}| \le (b_r - a_r)/n, \quad (8.6.4)$$

and being extended over all r, s. We write also

$$A = \max |q_r(x_r)|, \quad a_r \le x_r \le b_r, \quad 1 \le r \le k. \quad (8.6.5)$$

From Theorem A.3.2 we then have immediately

Theorem 8.6.1. *Let the $p_{rs}(x_r)$, $q_r(x_r)$ be continuous. Then the number of zeros in $(a_r, b_r]$ of a non-trivial solution of the r-th equation in (5.2.1) differs from $\pi^{-1} F_r(\lambda_1, \ldots, \lambda_k)$ by not more than*

$$n + \pi^{-1}(b_r - a_r)\left\{\sqrt{\left(\eta_n \sum |\lambda_s|\right)} + 3\sqrt{A}\right\}. \quad (8.6.6)$$

We deduce asymptotic estimates concerning eigenvalues, associated with large oscillation numbers.

Theorem 8.6.2. *Let the $p_{rs}(x_r)$, $q_r(x_r)$ be continuous and let $\det p_{rs}(x_r)$ have fixed sign. Then the unique eigenvalue of (5.2.1–5.2.2) associated with oscillation numbers $n_1, \ldots, n_k$ satisfies*

$$n_r = \pi^{-1} F_r(\lambda_1, \ldots, \lambda_k) + o\left\{\sqrt{\sum |\lambda_s|}\right\}, \quad r = 1, \ldots, k, \quad (8.6.7)$$

as

$$\sum n_r^2 \to \infty. \quad (8.6.8)$$

The same is true of all eigenvalues associated with these oscillation numbers if $\det p_{rs}(x_r)$ changes sign, and the array $p_{rs}(x_r)$, r, $s = 1,\ldots,k$, satisfies condition (B).

We have to show that for any $\varepsilon > 0$ there is a C such that if

$$\sum n_r^2 > C \tag{8.6.9}$$

then

$$|\pi n_r - F_r(\lambda_1,\ldots,\lambda_k)| \leq \varepsilon\sqrt{\sum|\lambda_s|}. \tag{8.6.10}$$

By Theorem 8.2.1, we can choose C so that (8.6.9) implies that

$$(b_r - a_r)3\sqrt{A} < (\varepsilon/3)\sum|\lambda_s|, \quad r = 1,\ldots,k. \tag{8.6.11}$$

We choose the integer n so that

$$(b_r - a_r)\sqrt{\eta_n} < \varepsilon/3, \quad r = 1,\ldots,k. \tag{8.6.12}$$

The desired result (8.6.10) is then ensured if C in (8.6.9) is so large that, in addition to (8.6.11), we have

$$\pi(n+1) < (\varepsilon/3)\left\{\sum|\lambda_s|\right\}^{1/2}. \tag{8.6.13}$$

The result of the last theorem can also be expressed with error-terms dependent on the oscillation numbers, rather than on the eigenvalue. We have

Theorem 8.6.3. *Under the assumptions of Theorem 8.6.2, the eigenvalue or eigenvalues associated with oscillation numbers $n_1,\ldots,n_k$ satisfy*

$$F_r(\lambda_1,\ldots,\lambda_k) = \pi n_r + o\{\sum n_t\}, \quad r = 1,\ldots,k. \tag{8.6.14}$$

This follows from the last theorem, together with Theorem 8.3.1, if the array $p_{rs}(x_r)$ satisfies condition (A), and Theorem 8.4.1 if it satisfies condition (B) and $\det p_{rs}(x_r)$ changes sign.

8.7   Solubility of certain equations

One may view (8.6.14) as a set of equations to determine approximately the eigenvalue $\lambda_1,\ldots,\lambda_k$ when the oscillation numbers $n_1,\ldots,n_k$ are given. This raises the question of whether the equations

$$F_r(\mu_1, \ldots, \mu_k) = \xi_r, \quad r = 1, \ldots, k, \qquad (8.7.1)$$

are soluble, uniquely or at all, when $\xi_1, \ldots, \xi_k$ are given. As usual, we have different results for the cases of conditions (A) and (B).

On the first, we have

Theorem 8.7.1. *Let the $p_{rs}(x_r)$ be continuous, and let $\det p_{rs}(x_r)$ have fixed sign. Then to each set*

$$\xi_r > 0, r = 1, \ldots, k, \qquad (8.7.2)$$

there is precisely one set $\mu_1, \ldots, \mu_k$ such that (8.7.1) holds.

Taking first the question of uniqueness, we suppose that there is a also a distinct set $\mu'_1, \ldots, \mu'_k$ such that

$$F_r(\mu'_1, \ldots, \mu'_k) = \xi_r, \quad r = 1, \ldots, k. \qquad (8.7.3)$$

By Theorem 4.3.1, there is an r such that

$$\sum (\mu'_s - \mu_s) p_{rs}(x_r) \qquad (8.7.4)$$

has fixed sign in $[a_r, b_r]$. Suppose first that (8.7.4) is positive for some r. We then have

$$\mathrm{Re} \sqrt{\sum \mu'_s p_{rs}(x_r)} \geq \mathrm{Re} \sqrt{\sum \mu_s p_{rs}(x_r)}, \qquad (8.7.5)$$

with strict inequality whenever the right is positive. The latter case must occur for some x_r, by the assumption (8.7.2). Hence

$$F_r(\mu'_1, \ldots, \mu'_k) > F_r(\mu_1, \ldots, \mu_k) \qquad (8.7.6)$$

in contradiction to (8.7.3). A similar argument deals with the case that (8.7.4) is negative in $[a_r, b_r]$.

We now have to show that the equations (8.7.1) have at least one solution, if the ξ_r are all positive. This depends on the Klein oscillation theorem. For any $\chi > 0$, we take n_r to be the smallest integer with

$$n_r \geq \chi \xi_r / \pi, \quad r = 1, \ldots, k. \qquad (8.7.7)$$

For boundary conditions (5.2.2), chosen at will, and zero $q_r(x_r)$, we take $\lambda_1, \ldots, \lambda_k$ to be the eigenvalue associated with the oscillation numbers $n_1, \ldots, n_k$; this exists by Theorem 5.5.1. By Theorem 8.6.3, we then have, as $\chi \to \infty$,

$$F_r(\lambda_1, \ldots, \lambda_k) = \chi \xi_r + o(\chi^{1/2}),$$

and so, since the F_r are of degree $1/2$,

$$F_r(\lambda_1/\chi^2, \ldots, \lambda_k/\chi^2) \to \xi_r, \quad r = 1, \ldots, k. \tag{8.7.8}$$

By Theorem 8.3.1, the quantities λ_s/χ^2 are bounded as $\chi \to \infty$, and so we can make $\chi \to \infty$ through a sequence which makes them all converge; we then have the result on denoting the limits by μ_s, $s = 1, \ldots, k$, since the F_r are continuous.

The above proof of the uniqueness of the solution of (8.7.1) fails if some or all of the ξ_r are zero. In the latter case, we have as solution any set $(\mu_1, \ldots, \mu_k)$ such that the k functions $\sum \mu_s p_{rs}(x_r)$ are all non-positive. Such sets exist, by Theorem 4.4.1.

The companion result for the case of condition (B) is

Theorem 8.7.2. *Let the $p_{rs}(x_r)$ be continuous, let $\det p_{rs}$ change sign, and let the determinants (4.6.1) all be positive, for some set $\rho_1, \ldots, \rho_k$. Then for given positive $\xi_1, \ldots, \xi_k$, the equations (8.7.1) have precisely two solutions, one satisfying each of the side conditions*

$$\sum \rho_s \mu_s > 0, \tag{8.7.9}$$

$$\sum \rho_s \mu_s < 0. \tag{8.7.10}$$

Using Theorem 4.11.2, we suppose a linear transformation performed so that the (4.6.1) are positive with $\rho_1 = 1$, $\rho_2 = \cdots = \rho_k = 0$. We prove first that (8.7.1) cannot have more then one solution with $\mu_1 > 0$, nor more than one with $\mu_1 < 0$, and that it has none with $\mu_1 = 0$.

We dispose first of the last assertion. If $\mu_1 = 0$, and (8.7.1)–(8.7.2) hold, then $\mu_2, \ldots, \mu_k$ cannot all be zero and so, by Theorem 4.7.1, there is an r such that

$$\sum \mu_s p_{rs}(x_r) < 0, \tag{8.7.11}$$

in $[a_r, b_r]$. We then have $F_r(0, \mu_2, \ldots, \mu_k) = 0$, in contradiction to (8.7.2).

Suppose next that (8.7.1) has two distinct solutions in which the first component is positive, given by (8.7.1) and (8.7.3). Without loss of generality, we suppose that

$$0 < \mu_1 \le \mu_1'. \tag{8.7.12}$$

We write

$$\mu_s'' = \mu_s'(\mu_1/\mu_1'), \quad s = 1, \ldots, k. \tag{8.7.13}$$

Since the F_r are of degree $1/2$, we have, for $r = 1, \ldots, k$,

$$F_r(\mu_1', \ldots, \mu_k') = \sqrt{(\mu_1'/\mu_1)} F_r(\mu_1'', \mu_2'', \ldots, \mu_k''). \tag{8.7.14}$$

If

$$\mu_s'' = \mu_s, \quad s = 2, \ldots, k, \tag{8.7.15}$$

then we cannot have $\mu_1 = \mu_1'$, for then the sets μ_s, μ_s' would not be distinct. Thus if (8.7.15) holds, we must have $\mu_1 < \mu_1'$, and then (8.7.14) implies (8.7.6), giving a contradiction to (8.7.1)–(8.7.3). Hence we may pass to the case that (8.7.15) do not all hold. By Theorem 4.7.1, there is an r such that

$$\sum (\mu_s'' - \mu_s) p_{rs}(x_r)$$

takes positive values only. This implies that

$$F_r(\mu_1, \mu_2'', \ldots, \mu_k'') > F_r(\mu_1, \mu_2, \ldots, \mu_k),$$

since the latter is positive. In view of (8.7.14), this gives (8.7.6) again. Thus (8.7.1), subject to (8.7.2), will not have more than one solution with $\mu_1 > 0$; the case $\mu_1 < 0$ is considered similarly.

Finally, we have to show that these solutions actually exist. The proof is similar to that of the corresponding property in the case of Theorem 8.7.1. We rely on this occasion on the Richardson oscillation theorem, Theorem 5.7.1, and Theorem 8.4.1; the conditions (5.7.1) may be ensured by choosing either $q_r(x_r) > 0$ or $q_r(x_r) \equiv 0$; $\alpha_r = 0$, $\beta_r = \pi$, $r = 1, \ldots, k$. Alternatively, the argument of Section 5.9 may be used.

Notes for Chapter 8

For results regarding the approximation of the eigenvalues of one two-parameter system of second-order differential equations by means of another such under condition (A), see Faierman (1977). On the other hand, Shibata (1996a), (1996b), (1997d), (1998b), and (1999) considered the more general problem of spectral asymptotics of multiparameter nonlinear second-order differential equations on a finite interval, where the parameters appear linearly. Of course, the one-parameter case is the most studied one with the latest contributions being in the indefinite cases. In this respect, see the papers by Atkinson and Mingarelli (1987), Tumanov (2000), (2001), D'yachenko (2000), and Binding *et al.* (2002b), (2003) as typical.

8.8   Research problems and open questions

1. Can it be shown that there is a universal constant C such that $A_1 = B_1 = C$ in Theorem 8.3.1?

2. What is the rate of decay of the o-term in (8.6.14)? Normally one imposes additional smoothness upon the coefficients of the multiparameter problem.

3. What is the rate of decay of the o-term in Theorem 8.6.2?

Chapter 9

The Essential Spectrum

9.1   Introduction

In Chapter 3, we interpreted the term "spectrum" in connection with a multiparameter boundary-value problem as meaning simply the collection of eigenvalues, considered inhomogeneously as k-tuples. We showed there, and in Chapter 5, that under suitable conditions eigenvalues were of finite multiplicity and that the set of eigenvalues had no finite limit-points.The purpose of this chapter is to take account of, so to speak, infinite limit-points of the set of eigenvalues. The treatment will be confined to the non-singular Sturm-Liouville setting of Chapter 3.

In the one-parameter case, the situation may be summed up very briefly. In the standard "orthogonal" case, in which the coefficient of λ in the equation has fixed positive sign, the eigenvalues can accumulate only at $+\infty$. In the "polar" case, in which the coefficient of λ changes sign, the eigenvalues accumulate both at $+\infty$ and at $-\infty$. These two types of limiting behavior for the eigenvalues are naturally to be regarded as distinct; we shall not confuse them by adopting a "one-point compactification" of the real line, nor by considering the real line as a subset of the closed complex plane.

For the multiparameter case, in which an eigenvalue is a real k-tuple, a sequence of eigenvalues can tend to infinity in various directions. Our main objective is to determine possible such directions. Naturally, we discriminate between different directions, or "points at infinity," in a particular sense.

In the treatment to be used in the sequel, points at infinity will be characterized by unit vectors in the directions concerned. They will thus be determined, like the eigenvalues, by k-tuples of scalars. This formulation, though economical, is open to the objection that the same notation is being used, with two interpretations, for two classes of what is really a single set, a properly interpreted notion of "spectrum." One can avoid this objection, although we shall not do this, by going over to a formulation in which points of the spectrum are real $(k + 1)$-tuples, not all zero. We identify two such $(k + 1)$-tuples if they differ only by a positive scale factor; they may be identified with points of a sphere.

In such a formulation, an eigenvalue $\lambda_1, \ldots, \lambda_k$ is represented by the $(k + 1)$-tuple

$$(1, \lambda_1, \ldots, \lambda_k), \tag{9.1.1}$$

or by its normalized representative on a unit sphere S; a "point at infinity," in the direction $(\mu_1, \ldots, \mu_k)$, is represented by

$$(0, \mu_1, \ldots, \mu_k). \tag{9.1.2}$$

We are concerned here with points (9.1.2), which are the limits of a sequence of points (9.1.1) when the latter are normalized.

9.2   The essential spectrum

For a multiparameter Sturm-Liouville problem, such as (5.2.1)–(5.2.2), we introduce the notion of an "asymptotic direction." By this we mean a real k-tuple

$$\mu_1, \ldots, \mu_k \tag{9.2.1}$$

not all zero, such that for some infinite sequence of eigenvalues

$$\lambda_1^{(m)}, \ldots, \lambda_k^{(m)}, \quad m = 1, 2, \ldots \tag{9.2.2}$$

and some unbounded sequence of positive numbers

$$\chi^{(m)} \to \infty, \quad m = 1, 2, \ldots, \tag{9.2.3}$$

we have

$$\lambda_s^{(m)} / \chi^{(m)} \to \mu_s, \quad s = 1, \ldots, k. \tag{9.2.4}$$

Clearly, the property of being an asymptotic direction is preserved if all the μ_s are multiplied by the same positive number. We shall say that (9.2.1) is "normalized" if

$$\sum |\mu_s| = 1. \tag{9.2.5}$$

The set of all asymptotic directions will be termed the "essential spectrum," in the case of the non-singular eigenvalue problem (5.2.1)–(5.2.2). In non-singular cases, for example, if one of the basic intervals (a_r, b_r) were infinite, in one or both directions, the notion of the essential spectrum must be modified; indeed, this is true also of the notion of the spectrum as a collection of eigenvalues.

The set of normalized asymptotic directions will be termed the "normalized essential spectrum," and will be denoted by S. One may view it as follows. We form "normalized eigenvalues," $(k+1)$-tuples of the form

$$\left\{1+\sum|\lambda_s|\right\}^{-1}, \lambda_1\left\{1+\sum|\lambda_s|\right\}^{-1}, \ldots, \lambda_k\left\{1+\sum|\lambda_s|\right\}^{-1}, \qquad (9.2.6)$$

where $\lambda_1, \ldots, \lambda_k$ is an eigenvalue. Assuming, as is the case for (5.2.1)–(5.2.2), that the eigenvalues have no finite limit-point, the cluster-points of the set (9.2.6) will all lie on the "great circle," formed by the subset of the "sphere" $\sum|\mu_s| = 1$ for which $\mu_0 = 0$. On suppressing the first zero coordinate for points of this set, we obtain the set S.

We have trivially

Theorem 9.2.1. *The normalized essential spectrum is closed.*

For the set of cluster-points of any set is closed. In the one-parameter case, the normalized essential spectrum consists of the point $+1$ in the usual orthogonal case, and of the pair $+1$, -1 in the polar case. Its determination in the multiparameter case will occupy the next few sections.

9.3 Some subsidiary point-sets

Let $p_{rs}(x_r)$, $a_r \leq x_r \leq b_r$, $r, s = 1, \ldots, k$, be any array of continuous real-valued functions. We denote by S_1 the collection of real k-tuples $(\mu_1, \ldots, \mu_k)$ such that

$$\sum_{s=1}^{k} \mu_s p_{rs}(x_{r0}) \geq 0, \quad r = 1, \ldots, k, \qquad (9.3.1)$$

for some set

$$x_{r0} \in [a_r, b_r], \quad r = 1, \ldots, k; \qquad (9.3.2)$$

the set $x_{10}, \ldots, x_{k0}$ may vary with choice of the μ_s. We denote by S_2 the collection of real k-tuples $(\mu_1, \ldots, \mu_k)$ such that

$$\sum \mu_s p_{rs}(x_{r0}) > 0, \quad r = 1, \ldots, k, \qquad (9.3.3)$$

for some set (9.3.2). The subsets of S_1, S_2 formed by k-tuples, which are normalized in the sense (9.2.5), will be denoted by S_{10}, S_{20}, respectively.

It is clear that

$$S_2 \subseteq S_1, \quad S_{20} \subseteq S_{10}. \tag{9.3.4}$$

We shall show later that $S_{20} \subseteq S \subseteq S_{10}$, under certain conditions, and will ultimately identify S with S_{10}. Before doing this, we need some elementary remarks, based on Chapter 4, and related considerations.

We have first

Theorem 9.3.1. *The set S_1 is closed, while S_2 is open. The set S_2, and so also S_{20}, S_{10} is non-empty if $\det p_{rs}(x_r)$ does not vanish identically.*

For the first assertion, we suppose that we have a convergent sequence of elements of S_1,

$$(\mu_{1m}, \ldots, \mu_{km}) \to (\mu_1, \ldots, \mu_k) \tag{9.3.5}$$

with

$$\sum \mu_{sm} p_{rs}(x_{rm}) \geq 0, \quad r = 1, \ldots, k, \quad m = 1, 2, \ldots. \tag{9.3.6}$$

By restriction to a subsequence, we can take it that the x_{rm} converge as $m \to \infty$ to points $x_{r0} \in [a_r, b_r]$, $r = 1, \ldots, k$. In view of the continuity of the $p_{rs}(x_r)$, we then get in the limit (9.3.1), as asserted.

In proving that S_2 is open, we do not need to assume the $p_{rs}(x_r)$ continuous. If (9.3.3) holds, it clearly holds if the μ_s are replaced by values in a suitable neighborhood of the μ_s.

Passing to the last remark of the theorem, we suppose that for some set (9.3.2) we have

$$\det p_{rs}(x_{r0}) \neq 0. \tag{9.3.7}$$

We can then choose the μ_s so that the left-hand sides in (9.3.3) take any assigned values, in particular a set of values that are all positive. We prove next a less trivial property.

Theorem 9.3.2. *The set S_1 includes the closure of S_2; it coincides with the closure of S_2 if either $\det p_{rs}(x_r)$ has fixed sign, or again if $\det p_{rs}(x_r)$ changes sign and there is a set $\rho_1, \ldots, \rho_k$ such that the determinants (4.6.1) are all positive.*

For the first statement, we assume that (9.3.5) relates to a convergent sequence of elements of S_2, so that (9.3.6) holds with strict inequality. As before, we may take it that $x_{rm} \to x_{r0}$ as $m \to \infty$. We then get (9.3.1) in the limit, assuming the $p_{rs}(x_r)$ continuous.

Suppose next that $\det p_{rs}(x_r)$ has fixed sign, and that

$$(\mu_1, \ldots, \mu_k) \in S_1,$$

so that (9.3.1) holds. By (9.3.7), we can determine $\nu_1, \ldots, \nu_k$ so that

$$\sum \nu_s p_{rs}(x_{r0}) = 1, \quad r = 1, \ldots, k. \tag{9.3.8}$$

For any $\varepsilon > 0$, we then have

$$\sum (\mu_s + \varepsilon \nu_s) p_{rs}(x_{r0}) > 0, \quad r = 1, \ldots, k, \tag{9.3.9}$$

so that

$$(\mu_1 + \varepsilon \nu_1, \ldots, \mu_k + \varepsilon \nu_k) \in S_2, \quad \varepsilon > 0. \tag{9.3.10}$$

Thus $(\mu_1, \ldots, \mu_k)$ is the limit of a sequence of points in S_2, as was to be proved.

We suppose next that (9.3.1) holds, that $\det p_{rs}(x_r)$ changes sign, and that the determinants (4.6.1) take only positive values. Appealing to Theorem 4.6.1, we have that since $\det p_{rs}(x_r)$ changes sign, only case (i) of this theorem is admissible. Thus, if the μ_s are not all zero, we must have (4.6.3) for some r, x_r and, by considering the set $-\mu_s$, we must have (4.6.3) with reversed sense, for some r, x_r. We may therefore take it that in (9.3.1) we have strict inequality in at least one case.

It will be sufficient to discuss the case that we have strict inequality in (9.3.1) with $r = 1$. We can then determine a set $\nu_1, \ldots, \nu_k$ such that

$$\sum_1^k \nu_s \rho_s = 1, \tag{9.3.11}$$

say, while (9.3.8) hold for $r = 1, \ldots, k$. It then follows that (9.3.9) holds for suitably small $\varepsilon > 0$, and the result follows as before.

For completeness, we note that the result holds also for the point $(0, \ldots, 0) \in S_1$. This follows from the fact that S_2 is non-empty, by Theorem 9.3.1.

We note that if S_1 is the closure of S_2, then S_{10} is the closure of S_{20}. This may be shown by normalizing the approximating family (9.3.10).

9.4   The essential spectrum under condition (A)

In this and the next section, we identify the essential spectrum of a non-singular Sturm-Liouville problem, under the two principal definiteness conditions. We assume in both cases that the $p_{rs}(x_r)$, $q_r(x_r)$ are continuous.

Theorem 9.4.1. *Let* $\det p_{rs}(x_r)$ *have fixed sign. Then the normalized essential spectrum of* (5.2.1)–(5.2.2) *coincides with the set* S_{10} *of normalized k-tuples satisfying* (9.3.1) *for some set* (9.3.2).

We shall prove that

$$S_{20} \subseteq S \subseteq S_{10}. \tag{9.4.1}$$

Since S is closed, and S_{10} is the closure of S_{20}, this will prove the result.

The second of (9.4.1) is easily proved, and does not depend on the hypothesis that $\det p_{rs}(x_r)$ has fixed sign. We suppose that the point $(\mu_1, \ldots, \mu_k) \in S$ but is not in S_{10}, so that for some r (9.3.1) does not hold for any x_{r0}. We have then

$$\sum_1^k \mu_s p_{rs}(x_r) \leq -\delta, \quad a_r \leq x_r \leq b_r, \tag{9.4.2}$$

for some $\delta > 0$. If nevertheless there is a sequence of eigenvalues (9.2.2) satisfying (9.2.3)–(9.2.5), it will follow from (9.4.2) that

$$\sum_1^k \lambda_s{}^{(m)} p_{rs}(x_r) - q_r(x_r) \to -\infty, \quad m \to \infty \tag{9.4.3}$$

uniformly in $[a_r, b_r]$. In view of Theorem A.6.1, this is in conflict with the boundary conditions.

It remains to prove the first of (9.4.1), or, what comes to the same thing, that every k-tuple satisfying (9.3.3) for some (9.3.2) is an asymptotic direction.

It will be sufficient to prove the following. Let the normalized set $\mu_1, \ldots, \mu_k$ satisfy (9.3.3) for some set (9.3.2). Then for any $\eta > 0$, there is a $K_0(\eta; \mu_1, \ldots, \mu_k)$ such that, if

$$\rho > K_0(\eta; \mu_1, \ldots, \mu_k), \tag{9.4.4}$$

then the ball

$$\sum_1^k |\lambda_s - \rho\mu_s| \leq \rho\eta \tag{9.4.5}$$

contains an eigenvalue of (5.2.1)–(5.2.2).

The proof uses ideas similar to those of Sections 5.3. We define the Prüfer angles θ_r as in (5.3.1)–(5.3.5), and have to prove that the equations

$$\theta_r(b_r; \lambda_1, \ldots, \lambda_k) = \beta_r + n_r\pi, \quad r = 1, \ldots, k, \tag{9.4.6}$$

have a solution in (9.4.5), for suitable integers n_r, which may of course depend on the choice of ρ, subject to (9.4.4). We choose the n_r to be the smallest integers satisfying

$$\theta_r(b_r; \rho\mu_1, \ldots, \rho\mu_k) \leq \beta_r + n_r\pi, \quad r = 1, \ldots, k. \tag{9.4.7}$$

We seek to minimize the function

$$\psi(\lambda_1, \ldots, \lambda_k) = \sum_1^k \{\beta_r + n_r\pi - \theta_r(b_r; \lambda_1, \ldots, \lambda_k)\}^2, \tag{9.4.8}$$

as $(\lambda_1, \ldots, \lambda_k)$ ranges over (9.4.5).

Suppose first that this minimum is attained at a point of the interior of (9.4.5). Then by equating to zero the partial derivatives of ψ we have

$$2 \sum_{s=1}^k (\partial\theta_r/\partial\lambda_s)(\beta_r + n_r\pi - \theta_r) = 0, \quad r = 1, \ldots, k.$$

Since $\det p_{rs}(x_r)$ is to have fixed sign, the Jacobian (5.4.8) is not zero, and so we have (9.4.6), as required.

We must therefore eliminate the possibility that the minimum of ψ is achieved on the boundary of (9.4.5). For this it will be sufficient to arrange that the value on the boundary exceeds its value at the point $(\rho\mu_1, \ldots, \rho\mu_k)$, or that

$$\psi(\rho(\mu_1 + \eta\nu_1), \ldots, \rho(\mu_k + \eta\nu_k)) > \psi(\rho\mu_1, \ldots, \rho\mu_k),$$

for all normalized sets $\nu_1, \ldots, \nu_k$. By our prescription for the n_r in (9.4.7), it will be sufficient to ensure that

$$\psi(\rho(\mu_1 + \eta\nu_1), \ldots, \rho(\mu_k + \eta\nu_k)) \geq k\pi^2. \tag{9.4.9}$$

This in turn can be replaced by an inequality concerning the zeros of solutions of (5.2.1). It is clear from (9.4.8) that it will be sufficient to show that

$$|\beta_r + n_r\pi - \theta_r(b_r; \rho(\mu_1 + \eta\nu_1), \ldots, \rho(\mu_k + \eta\nu_k))| \geq k\pi,$$

for at least one r. Let $N_r(\lambda_1, \ldots, \lambda_k)$ denote the number of zeros in $(a_r, b_r]$ of a solution of (5.2.1) with the initial data (5.3.2); by Theorem 5.3.1 and (5.2.3) we have then

$$|\theta_r(b_r; \lambda_1, \ldots, \lambda_k) - \pi N_r(\lambda_1, \ldots, \lambda_k)| < \pi.$$

We have also, by our determination of the n_r,

$$|\pi N_r(\rho\mu_1, \ldots, \rho\mu_k) - (\beta_r + n_r\pi)| < 2\pi.$$

Hence (9.4.9) will be ensured if we have

$$|\pi\, N_r(\rho(\mu_1+\eta\nu_1),\ldots,\rho(\mu_k+\eta\nu_k)) - \pi\, N_r(\rho\mu_1,\ldots,\rho\mu_k)| \geq (k+3)\pi, \quad (9.4.10)$$

for at least one r, for any normalized set of ν_s, and for sufficiently large ρ.

We now use the results of Section 8.6, which show that

$$N_r(\rho(\mu_1+\eta\nu_1),\ldots,\rho(\mu_k+\eta\nu_k)) - N_r(\rho\mu_1,\ldots,\rho\mu_k) =$$
$$= \pi^{-1}\sqrt{\rho}\{F_r(\mu_1+\eta\nu_1,\ldots,\mu_k+\eta\nu_k) - F_r(\mu_1,\ldots,\mu_k)\} + o(\sqrt{\rho}). \quad (9.4.11)$$

Here we note that $F_r(\mu_1,\ldots,\mu_k) > 0$, $r = 1,\ldots,k$, since by hypothesis $(\mu_1,\ldots,\mu_k) \in S_2$. It thus follows from Theorem 8.6.1 that

$$F_r(\mu_1+\eta\nu_1,\ldots,\mu_k+\eta\nu_k) \neq F_r(\mu_1,\ldots,\mu_k) \quad (9.4.12)$$

for at least one r. Thus, by continuity, there is a $\delta > 0$ such that

$$|F_r(\mu_1+\eta\nu_1,\ldots,\mu_k+\eta\nu_k) - F_r(\mu_1,\ldots,\mu_k)| \geq \delta, \quad (9.4.13)$$

for at least one r, for every normalized set of ν_s. It then follows from (9.4.11) that, for sufficiently large ρ, we have

$$|N_r(\rho(\mu_1+\eta\nu_1),\ldots,\rho(\mu_k+\eta\nu_k)) - N_r(\rho\mu_1,\ldots,\rho\mu_k)| \geq \pi^{-1}\delta\sqrt{\rho}, \quad (9.4.14)$$

for at least one r. This establishes (9.4.10), and completes the proof of Theorem 9.4.1.

As a consequence, we deduce the "essentiality" of the essential spectrum.

Theorem 9.4.2. *If* $\det p_{rs}(x_r)$ *has fixed sign, the essential spectrum is independent of the quantities* α_r, β_r *in the boundary conditions (5.2.2).*

9.5   The essential spectrum under condition (B)

Similar results hold in this case also.

Theorem 9.5.1. *Let* $\det p_{rs}(x_r)$ *change sign, and let there be a set* $\rho_1,\ldots,\rho_k$ *such that the determinants (4.6.1) have fixed positive sign. Then the normalized essential spectrum coincides with the set* S_{10}, *and is independent of the boundary-angles* α_r, β_r.

As before, the result will follow from Theorems 9.2.1 and 9.3.2 if we prove the inclusions (9.4.1). The second of these was proved in the last section; the proof given there made no hypotheses concerning the sign of determinants. It is thus sufficient to establish the inclusion $S_{20} \subset S$.

We suppose as usual that a linear transformation has been performed on the parameters so that (4.6.1) are all positive with $\rho_1 = 1$, $\rho_2 = \cdots = \rho_k = 0$.

Assume that $(\mu_1, \ldots, \mu_k) \in S_{20}$; by Theorem 4.7.1, we must have that $\mu_1 \neq 0$. We take as representative the case that $\mu_1 > 0$, and wish to show that for large $\rho > 0$, suitable neighborhoods of $(\rho\mu_1, \ldots, \rho\mu_k)$ will contain eigenvalues of (5.2.1)–(5.2.2). In place of the ball (9.4.5), we take now the "cylinder"

$$|\lambda_1 - \rho\mu_1| \leq \rho\eta_1, \quad \sum_2^k |\lambda_s - \rho\mu_s| \leq \rho\eta_2. \tag{9.5.1}$$

We claim that there exist arbitrarily small positive η_1, η_2, such that (9.5.1) contains an eigenvalue, for large ρ, say, if

$$\rho > K_1 = K_1(\eta_1, \eta_2; \mu_1, \ldots, \mu_k). \tag{9.5.2}$$

On η_1, η_2, we impose first the requirement that

$$(\mu_1 + \nu_1, \ldots, \mu_k + \nu_k) \in S_2, \tag{9.5.3}$$

for all sets $\nu_1, \ldots, \nu_k$ such that

$$|\nu_1| \leq \eta_1, \quad \sum_2^k |\nu_s| \leq \eta_2. \tag{9.5.4}$$

This is possible since S_2 is open. In view of Theorem 4.7.1, we must then have in particular that

$$\eta_1 < \mu_1. \tag{9.5.5}$$

Let $\delta > 0$ be such that the inequality

$$\sum_2^k \nu_s p_{rs}(x_r) \geq \delta \sum_2^k |\nu_s|, \quad a_r \leq x_r \leq b_r, \tag{9.5.6}$$

holds for some r, and the inequality

$$\sum_2^k \nu_s p_{rs}(x_r) \leq -\delta \sum_2^k |\nu_s|, \quad a_r \leq x_r \leq b_r, \tag{9.5.7}$$

for some other r-value. Theorem 4.7.2 ensures the existence of such a δ; the values of r concerned may depend on the choice of $\nu_1, \ldots, \nu_k$. We denote by A some bound for $|p_{rs}(x_r)|$, valid for all r, s, x_r; we impose on η_1, η_2 the further restriction that

$$0 < \eta_1 A < \delta \eta_2/2. \tag{9.5.8}$$

We next assemble some consequences of these restrictions. Since the functions F_r, defined in (8.6.2), are positive and continuous on S_2, there will be a $\zeta_1 > 0$ such that, if (9.5.4) holds,

$$F_r(\mu_1 + \nu_1, \ldots, \mu_k + \nu_k) \geq \zeta_1, \quad r = 1, \ldots, k. \tag{9.5.9}$$

We impose (9.5.8) for the consequence that if

$$|\nu_1| \leq \eta_1, \quad \sum_2^k |\nu_s| = \eta_2, \tag{9.5.10}$$

then

$$\left| \sum_1^k \nu_s p_{rs}(x_r) \right| \geq \delta \eta_2/2, \quad a_r \leq x_r \leq b_r, \tag{9.5.11}$$

for either of the r-values satisfying (9.5.6)–(9.5.7); since $\sum_1^k \nu_s p_{rs}(x_r)$ will then have fixed sign, we have

$$F_r(\mu_1 + \nu_1, \ldots, \mu_k + \nu_k) \neq F_r(\mu_1, \ldots, \mu_k)$$

for such r. By continuity, there is a $\zeta_2 > 0$ such that if (9.5.10) holds, then

$$|F_r(\mu_1 + \nu_1, \ldots, \mu_k + \nu_k) - F_r(\mu_1, \ldots, \mu_k)| \geq \zeta_2, \tag{9.5.12}$$

for some r with $2 \leq r \leq k$.

We now follow an argument resembling that of Section 5.7. We consider the partial eigenvalue problem

$$\theta_r(b_r; \lambda_1, \ldots, \lambda_k) = \beta_r + n_r \pi, \quad r = 2, \ldots, k, \tag{9.5.13}$$

where the integers $n_2, \ldots, n_k$ are the least admissible in (9.4.7), with $r = 2, \ldots, k$; we do not fix n_1. By Section 5.5, this fixes $\lambda_2, \ldots, \lambda_k$ uniquely as continuous functions of the real variable λ_1, in view of the hypothesis that

$$\det_{2 \leq r, s, \leq k} p_{rs}(x_r) > 0;$$

we denote these functions by $\lambda_2(\lambda_1), \ldots, \lambda_k(\lambda_1)$.

We consider now the curve

$$(\lambda_1, \lambda_2(\lambda_1), \ldots, \lambda_k(\lambda_1)), \tag{9.5.14}$$

as λ_1 describes the interval

$$\rho(\mu_1 - \eta_1) \leq \lambda_1 \leq \rho(\mu_1 + \eta_1). \tag{9.5.15}$$

We claim that for large $\rho > 0$, this curve lies strictly inside the cylinder (9.5.1), except for the end-points, which lie on the "faces"

$$\lambda_1 = \rho(\mu_1 - \eta_1), \quad \sum_2^k |\lambda_s - \rho\mu_s| \leq \rho\eta_2, \qquad (9.5.16)$$

$$\lambda_1 = \rho(\mu_1 + \eta_1), \quad \sum_2^k |\lambda_s - \rho\mu_s| \leq \rho\eta_2. \qquad (9.5.17)$$

We prove first that this is so for the "mid-point" of (9.5.14). The argument is very similar to that of the last section. By a slight modification of (9.4.14), we have

$$|N_r(\rho(\mu_1 + \nu_1), \ldots, \rho(\mu_k + \nu_k)) - N_r(\rho\mu_1, \ldots, \rho\mu_k)| \geq \zeta_2\sqrt{\rho}/(2\pi), \quad (9.5.18)$$

if (9.5.10) holds, and ρ is suitably large, for some r, $2 \leq r \leq k$. Applying this with $\nu_1 = 0$, we shall have

$$|\beta_r + n_r\pi - \theta_r(b_r; \rho\mu_1, \rho(\mu_2 + \nu_2), \ldots, \rho(\mu_k + \nu_k))| > k\pi, \qquad (9.5.19)$$

by reasoning similar to that applied to (9.4.10). The argument of Section 9.4 then shows that, for large ρ, (9.5.13) have a unique solution with

$$\lambda_1 = \rho\mu_1, \quad \sum_2^k |\lambda_s - \rho\mu_s| < \rho\eta_2.$$

Thus, for large ρ, the curve (9.5.14) lies inside the cylinder (9.5.1) when $\lambda_1 = \rho\mu_1$.

We next argue that, as λ_1 varies subject to (9.5.15), the curve cannot meet the "sides" of the cylinder, that is to say, the set

$$|\lambda_1 - \rho\mu_1| \leq \rho\eta_1, \quad \sum_2^k |\lambda_s - \rho\mu_s| = \rho\eta_2.$$

This again is a consequence of (9.5.18), more precisely of (9.5.19) with $\rho\mu_1$ replaced by $\rho(\mu_1 + \nu_1)$.

Let now $(\lambda_1', \ldots, \lambda_k')$, $(\lambda_1'', \ldots, \lambda_k'')$ denote the end-points of (9.5.14), lying, respectively, in the faces (9.5.16)–(9.5.17). We have by construction

$$\theta_r(b_r; \lambda_1', \ldots, \lambda_k') = \theta_r(b_r; \lambda_1'', \ldots, \lambda_k''), \quad r = 2, \ldots, k. \qquad (9.5.20)$$

If we show that

$$\theta_1(b_1; \lambda_1', \ldots, \lambda_k') - \theta_1(b_1; \lambda_1'', \ldots, \lambda_k'') \geq \pi, \qquad (9.5.21)$$

it will follow that (9.5.1) contains a solution of (9.4.6), for some integer n_1. This will complete our proof.

We show first that for some $\zeta_3 > 0$ and all $\rho > 0$ we have

$$F_r(\lambda_1', \ldots, \lambda_k') - F_r(\lambda_1'', \ldots, \lambda_k'') \geq \zeta_3\sqrt{\rho}, \qquad (9.5.22)$$

for some r, $1 \leq r \leq k$. For by homogeneity,

$$F_r(\lambda_1', \ldots, \lambda_k') = \sqrt{(\lambda_1'/\lambda_1'')} F_r(\lambda_1'', \lambda_2^\dagger, \ldots, \lambda_k^\dagger),$$

where

$$\lambda_1''/\lambda_1' = (\mu_1 - \eta_1)/(\mu_1 + \eta_1), \ldots, \lambda_s^\dagger = (\lambda_1''/\lambda_1')\lambda_s', \quad s = 2, \ldots, k.$$

Hence the left of (9.5.22) may be written

$$\{1 - \sqrt{(\lambda_1''/\lambda_1')}\} F_r(\lambda_1', \ldots, \lambda_k') + \{F_r(\lambda_1'', \lambda_2^\dagger, \ldots, \lambda_k^\dagger) - F_r(\lambda_1'', \lambda_2'', \ldots, \lambda_k'')\}.$$

By Theorem 4.7.1, there is an r such that

$$\sum_2^k (\lambda_s^\dagger - \lambda_s'') p_{rs}(x_r) \geq 0, \quad a_r \leq x_r \leq b_r,$$

and so

$$F_r(\lambda_1'', \lambda_2^\dagger, \ldots, \lambda_k^\dagger) \geq F_r(\lambda_1'', \lambda_2'', \ldots, \lambda_k'').$$

We thus get, for some r,

$$F_r(\lambda_1', \ldots, \lambda_k') - F_r(\lambda_1'', \ldots, \lambda_k'') \geq \zeta_1\sqrt{\rho}\{1 - (\mu_1 - \eta_1)/(\mu_1 + \eta_1)\},$$

where ζ_1 is as in (9.5.9). This is equivalent to (9.5.22).

By Section 8.5, we then conclude that, for large ρ, we have

$$\theta_r(b_r; \lambda_1', \ldots, \lambda_k') - \theta_r(b_r; \lambda_1'', \ldots, \lambda_k'') \geq \zeta_3\sqrt{\rho}/(2\pi),$$

for some r. By (9.5.20), we cannot have $r = 2, \ldots, k$, and so we must have $r = 1$. This justifies (9.5.21) for large ρ, and completes the proof of Theorem 9.5.1.

9.6   Dependence on the underlying intervals

From our identification of the essential spectrum in the last two sections, it was immediate that the essential spectrum did not depend on the choice of the one-point or separated boundary conditions (5.2.2). There is on more simple deduction to be made. Heuristically expressed, we have

Theorem 9.6.1. *If the basic intervals $[a_r, b_r]$, are expanded, with preservation of the hypotheses of either Theorem 9.4.1 or Theorem 9.5.1, the new essential spectrum will include the original one.*

For the set of non-zero k-tuples $\mu_1, \ldots, \mu_k$ for which a set (9.3.2) exists satisfying (9.3.1) does not diminish if the intervals $[a_r, b_r]$ are enlarged.

9.7 Nature of the essential spectrum

We make here some general remarks of a geometrical character, and we assume throughout that the $p_{rs}(x_r)$, $q_r(x_r)$ are continuous.

Theorem 9.7.1. *If $\det p_{rs}(x_r)$ has fixed sign, the essential spectrum is connected.*

By Theorem 9.4.1, the (unnormalized) essential spectrum coincides with the set of non-zero elements of S_1, and so we have to show that the latter set S_1', say, is connected. Let $\mu_1, \ldots, \mu_k$ be a set of numbers, not all zero, satisfying (9.3.1) for some set (9.3.2). By Theorem 4.4.1, there is a set $\mu_1', \ldots, \mu_k'$ such that

$$\sum_1^k \mu_s' p_{rs}(x_r) > 0, \quad a_r \le x_r \le b_r, \ r = 1, \ldots, k. \tag{9.7.1}$$

It then follows that, for $0 < \tau \le 1$,

$$\sum_1^k \{(1-\tau)\mu_s + \tau\mu_s'\} p_{rs}(x_{r0}) > 0, \ r = 1, \ldots, k, \tag{9.7.2}$$

so that the set

$$((1-\tau)\mu_1 + \tau\mu_1', \ldots, (1-\tau)\mu_k + \tau\mu_k'), \quad 0 \le \tau \le 1, \tag{9.7.3}$$

is non-zero and is in S_1; in fact it is in S_2 for $0 < \tau \le 1$. Thus any element of S_1' is connected in S_1' to some fixed element $(\mu_1', \ldots, \mu_k')$; hence the essential spectrum is connected, and indeed starlike.

Notes for Chapter 9

Of the few papers that study asymptotic directions in this and more general scenarios, we cite Binding and Browne (1980) who consider the question in a general setting, Binding *et al.* (1992) for the general two-parameter Hilbert space case, Shibata (1996a) for nonlinear multiparameter problems, and Volkmer (1996) undertakes a study of the asymptotic directions in the general multiparameter case in Hilbert space.

9.8   Research problems and open questions

1. Is there a version of Theorem 9.4.2 in singular cases?

2. Determine the validity of Theorem 9.7.1 if the determinant there changes sign or vanishes identically on some subinterval of $[a_r, b_r]$ for some r.

Chapter 10

The Completeness of Eigenfunctions

10.1   Introduction

We pass now from the description of the properties of individual eigenvalues and eigenfunctions to the principal property enjoyed by the eigenfunctions as a set. This is represented by the theory of "eigenfunction expansions"; an "arbitrary" function, of the several variables involved, can be expanded as a series of eigenfunctions, products of solutions of the individual ordinary differential equations that make up the eigenvalue problem. Since the eigenfunctions are mutually orthogonal, with respect to a determinantal weight-function, the coefficients in such an expansion are easily found, given the function to be expanded. The difficulty is, of course, to show that the function being "expanded" is indeed represented by the series of eigenfunctions, even when the latter series is known to be convergent.

In this chapter, we are concerned with proofs of the completeness of the eigenfunctions of a multiparameter Sturm-Liouville problem involving finite intervals. It will be a question of deducing this completeness from the corresponding property for some related eigenvalue problem. Our procedure in the present chapter will be to appeal to completeness properties for some associated multiparameter problems.

In the first approach, we use the method, common in the one-parameter case, of reduction to Fredholm integral equations with symmetric kernel; here the kernel is given by the Green's function of the differential equation and boundary conditions. The completeness, under suitable hypotheses, of the eigenfunctions of a multiparameter set of Fredholm integral equations is covered by a similar theory for sets of compact operators (Chapter 11 of Atkinson [1972]).

10.2   Green's function

As a preliminary to the method of reduction to a set of integral equations, we assemble here standard results concerning the Green's function of a Sturm-Liouville problem.

Theorem 10.2.1. *Let $q(x)$, $a \leq x \leq b$, be in $L(a,b)$, and let α, β be real. Suppose that the boundary-value problem*

$$y''(x) + q(x)y(x) = 0, \quad a \leq x \leq b, \tag{10.2.1}$$
$$y(a)\cos\alpha = y'(a)\sin\alpha, \quad y(b)\cos\beta = y'(b)\sin\beta, \tag{10.2.2}$$

has only the trivial solution. Then there is a real and continuous function $G(x,t)$, $a \leq x$, $t \leq b$, symmetric in the sense that

$$G(x,t) = G(t,x), \tag{10.2.3}$$

and such that for any continuous $f(x)$, $a \leq x \leq b$, the unique solution of

$$y''(x) + q(x)y(x) + f(x) = 0, \quad a \leq x \leq b, \tag{10.2.4}$$

satisfying the boundary conditions (10.2.2) is given by

$$y(x) = \int_a^b G(x,t)f(t)\,dt. \tag{10.2.5}$$

For the proof, we choose solutions u, v of (10.2.1) satisfying the conditions

$$u(a) = \sin\alpha,\, u'(a) = \cos\alpha, \quad v(b) = \sin\beta,\, v'(b) = \cos\beta. \tag{10.2.6}$$

These are linearly independent, since otherwise (10.2.1)–(10.2.2) would have a non-trivial solution. Thus their (constant) Wronskian

$$W = u'v - uv' \tag{10.2.7}$$

is not zero. It is easy to see that the required Green's function is then given by

$$G(x,t) = \begin{cases} u(x)v(t), & \text{if } t \geq x, \\ v(x)u(t), & \text{if } t \leq x. \end{cases} \tag{10.2.8}$$

We next recall information concerning the parametric case

$$y'' + (\lambda + q(x))y = 0, \quad a \leq x \leq b, \tag{10.2.9}$$

where $q(x)$ is still $L(a,b)$. The above lemma states that $G(x,t)$ exists provided that 0 is not an eigenvalue of the Sturm-Liouville problem (10.2.9)–(10.2.2). In particular, it will exist if all these eigenvalues are positive, and this can be arranged by a shift in the λ-origin. Let the eigenvalues of (10.2.9)–(10.2.2) be $\lambda_0 < \lambda_1 < \ldots$, and let $y_0(x)$, $y_1(x), \ldots$ be the associated eigenfunctions; we may assume them real, and normalized so that

$$\int_a^b y_n^2(x)\, dx = 1, \quad n = 0, 1, \ldots. \tag{10.2.10}$$

We need the fundamental fact that if $g(x)$ is representable in the form

$$g(x) = \int_a^b G(x,t)f(t)dt, \quad a \leq x \leq b, \tag{10.2.11}$$

where $f(x)$ is continuous (or at least in $L^2(a,b)$), then the eigenfunction expansion

$$g(x) = \sum_0^\infty c_n y_n(x), \quad c_n = \int_a^b y_n(t)g(t)\, dt, \tag{10.2.12}$$

is true, in fact with uniform and absolute convergence, provided of course that 0 is not an eigenvalue. Equivalently, we have that this expansion holds if $g(x)$ satisfies

$$g'' + qg + f = 0, \quad a \leq x \leq b, \tag{10.2.13}$$

and the boundary conditions (10.2.2).

10.3   Transition to a set of integral equations[1]

We now return to the multiparameter Sturm-Liouville problem

$$y_r''(x_r) + \left\{ \sum_{s=1}^k \lambda_s p_{rs}(x_r) + q_r(x_r) \right\} y_r(x_r) = 0, \quad a_r \leq x_r \leq b_r \tag{10.3.1}$$

$$y_r(a_r)\cos\alpha_r = y_r'(a_r)\sin\alpha_r, \quad y_r(b_r)\cos\beta_r = y_r'(b_r)\sin\beta_r; \tag{10.3.2}$$

---

[1]Observe that we have changed the sign in front of the q_r-term without loss of generality.

the intervals $[a_r, b_r]$, $r = 1, \ldots, k$, are to be all finite and the q_r, p_{rs} all real and continuous. We consider the case of "condition (A)," that in which

$$\det p_{rs}(x_r) > 0, \quad a_r \leq x_r \leq b_r, \ r = 1, \ldots, k. \tag{10.3.3}$$

Then (10.3.1)–(10.3.2) has an infinite sequence of eigenvalues $(\lambda_1^{(n)}, \ldots, \lambda_k^{(n)})$, where $n = (n_1, \ldots, n_k)$ runs through all sets of non-negative integers, n_r being the number of zeros of $y_r(x_r)$ in the open interval (a_r, b_r) (cf., Theorem 5.5.1).

There is an infinite sequence of eigenfunctions of the form

$$y(x) = \prod_1^k y_r(x_r), \tag{10.3.4}$$

products of non-trivial solutions of (10.3.1)–(10.3.2) associated with $(\lambda_1^{(n)}, \ldots, \lambda_k^{(n)})$. These are mutually orthogonal in the sense

$$\int_a^b y_{n'}(x)y_n(x)p(x)\, dx = 0, \quad n \neq n'. \tag{10.3.5}$$

Here we have written

$$a = (a_1, \ldots, a_k), \quad b = (b_1, \ldots, b_k), \quad p(x) = \det p_{rs}(x_r), \quad dx = \prod_1^k dx_r. \tag{10.3.6}$$

In (10.3.5) n, n' are distinct k-tuples of non-negative integers, and so differ in at least one component.

There are various ways of fixing the constant factors in the specification of the eigenfunctions. At this point, it is most convenient to suppose them normalized, so that

$$\int_a^b |y_n(x)|^2 p(x)\, dx = 1. \tag{10.3.7}$$

In addition, we shall suppose them real-valued; we recall that the eigenvalues are real. This leaves $y_n(x)$ determinate except for a sign-factor ± 1, which will be left undetermined.

We denote by $G_r(x_r, t_r)$ the Green's functions of the problems given by the differential equations

$$y_r''(x_r) + q_r(x_r)y_r(x_r) + f_r(x_r) = 0, \tag{10.3.8}$$

together with the respective boundary conditions (10.3.2), provided that these Green's functions exist.

Actually, we shall need to ensure that these Green's functions not merely exist, but arise from boundary-value problems with positive spectrum. This may be ensured by a change of λ-origin. Let us write

$$\lambda_s = \lambda_s{}^\dagger - \tau\mu_s, \quad s = 1, \ldots, k, \tag{10.3.9}$$

where the real numbers $\tau, \mu_1, \ldots, \mu_k$ are to be determined. In view of (10.3.3) and Theorem 4.4.1, we can choose the μ_s so that

$$\sum_1^k \mu_s p_{rs}(x_r) > 0, \quad a_r \leq x_r \leq b_r, \ r = 1, \ldots, k. \tag{10.3.10}$$

The effect of the substitution (10.3.9) on the differential equations (10.3.1) is to replace the λ_s by the $\lambda_s{}^\dagger$ and the $q_r(x_r)$ by

$$q_r{}^\dagger(x_r) = q_r(x_r) - \tau \sum \mu_s p_{rs}(x_r). \tag{10.3.11}$$

We can thus choose $\tau > 0$ so that the $q_r{}^\dagger(x_r)$ should be all negative, and whose absolute values are arbitrarily large. We can thus arrange that the boundary-value problems, with ρ_r as parameters,

$$y_r''(x_r) + \{q_r{}^\dagger(x_r) + \rho_r\}y_r(x_r) = 0, \tag{10.3.12}$$

together with the boundary conditions (10.3.2), should have all positive spectra, all eigenvalues being positive.

In what follows, we shall suppose such a transformation made, if necessary, so that the boundary-value problem

$$y_r''(x_r) + \{q_r(x_r) + \rho_r\}y_r(x_r) = 0, \quad r = 1, \ldots, k, \tag{10.3.13}$$

all have positive spectra, when taken with the respective boundary conditions (10.3.2).

Thus in particular the Green's functions $G_r(x_r, t_r)$ will all exist. The problems (10.3.8), (10.3.2) will have the unique solutions

$$y_r(x_r) = \int_{a_r}^{b_r} G_r(x_r, t_r) f_r(t_r) \, dt_r. \tag{10.3.14}$$

We now make the substitution (10.3.14) in (10.3.1), and get

$$f_r(x_r) - \int_{a_r}^{b_r} G_r(x_r, t_r) \left\{ \sum_1^k \lambda_s p_{rs}(t_r) f_r(t_r) \right\} dt_r = 0, \quad r = 1, \ldots, k \tag{10.3.15}$$

We are thus led to a second eigenvalue problem; an eigenvalue of this problem will be a k-tuple $\lambda_1, \ldots, \lambda_k$ such that all of the equations (10.3.15) have a non-trivial solution.

The relation between the two problems is described in

Theorem 10.3.1. *Let the $p_{rs}(x_r)$, $q_r(x_r)$ be continuous and satisfy (10.3.3). Assume that, by a change of λ-origin if necessary, the eigenvalue problems (10.3.13), (10.3.2) have purely positive eigenvalues, and let the $G_r(x_r, t_r)$ be the Green's functions of the problems (10.3.8), (10.3.2). Then the integral equation problem (10.3.15) and the Sturm-Liouville problem (10.3.1)–(10.3.2) have the same eigenvalues, all of which are simple. With the notation (10.3.4) for the eigenfunctions of (10.3.1)–(10.3.2) and the notation*

$$f_r^{(n)}(x_r) = -(d^2/dx_r^2 + q_r(x_r))\, y_r^{(n)}(x_r), \qquad (10.3.16)$$

the associated non-trivial solutions of (10.3.15) are given by (10.3.16), while the eigenfunctions of (10.3.15) are

$$f^{(n)}(x_r) = \prod_{1}^{k} f_r^{(n)}(x_r). \qquad (10.3.17)$$

For suppose that $\lambda_1, \ldots, \lambda_k$ is an eigenvalue of (10.3.1)–(10.3.2), with $y_r(x_r)$, $r = 1, \ldots, k$ being a set of non-trivial solutions. We then have (10.3.8) with

$$f_r(x_r) = \sum_{1}^{k} \lambda_s\, p_{rs}(x_r) y_r(x_r), \qquad (10.3.18)$$

and so (10.3.14) with this expression for f_r. Substituting (10.3.14) on the right of (10.3.18) we obtain (10.3.15). Also, since y_r, and so also f_r, is not identically equal to zero, we have that $\lambda_1, \ldots, \lambda_k$ must be an eigenvalue of (10.3.15) also.

Conversely, suppose that $\lambda_1, \ldots, \lambda_k$ is an eigenvalue of (10.3.15), the f_r being a set of non-trivial solutions. Then defining y_r by (10.3.14), we have that y_r satisfies (10.3.8) and the boundary conditions (10.3.2). Substituting (10.3.14) on the left of (10.3.15), we find that (10.3.8) implies (10.3.1). Thus the two problems have the same spectra.

It is trivial that the problem (10.3.1)–(10.3.2) has eigenvalues with multiplicity one only. In the case of (10.3.15), it is plain that if $f_r \not\equiv 0$, then y_r as defined by (10.3.14) is not identically zero, for otherwise it would follow from (10.3.15) that f_r vanished identically. Thus the number of linearly independent solutions of (10.3.15), for any one r, cannot exceed that for (10.3.1)–(10.3.2) for the same r, and so must be one.

Thus the spectra coincide in respect of multiplicities also; in the course of the proof, we have dealt with the relation between the two sets of eigenfunctions.

10.4   Orthogonality relations

It is convenient to introduce the scalar product

$$(f,g)_r = \int_{a_r}^{b_r} \int_{a_r}^{b_r} G_r(x_r,t_r)f(t_r)\overline{g(x_r)}\,dt_r\,dx_r, \qquad (10.4.1)$$

where, in the first place, $f(x_r)$, $g(x_r) \in C[a_r, b_r]$. We write G_r for the linear operator defined by $G_r(x_r, t_r)$ as an integral kernel, so that (10.3.14), for example, could be written $y_r = G_r f_r$. Write also p_{rs} for the operation of multiplication by the function $p_{rs}(x_r)$, and V_{rs} for the composite operation $p_{rs}G_r$; thus, for example, (10.3.15) can be written as

$$f_r - \sum_1^k \lambda_s\, V_{rs}f_r = 0, \quad r = 1,\ldots,k. \qquad (10.4.2)$$

The operators V_{rs} are symmetric with respect to the scalar product (10.4.1); that is to say, for any continuous complex-valued functions $f_r(x_r)$, $g_r(x_r)$, we have

$$(V_{rs}f_r, g_r)_r = (f_r, V_{rs}g_r)_r. \qquad (10.4.3)$$

To see this, let us write $y_r = G_r f_r$ as in (10.3.14), and likewise $z_r = G_r g_r$. It then turns out that

$$(V_{rs}f_r, g_r)_r = \int_{a_r}^{b_r} p_{rs}(x_r)y_r(x_r)\overline{z_r(x_r)}\,dx_r. \qquad (10.4.4)$$

A simple calculation shows that the right of (10.4.3) is also equal to the right of (10.4.4).

By means of this, we translate the orthogonality properties of the Sturm-Liouville eigenfunctions into their corresponding results for the integral equations. The former, that is to say (10.3.5), may be written

$$\det \int_{a_r}^{b_r} p_{rs}(x_r) y_r^{(n)}(x_r) y_r^{(n')}(x_r) \, dx_r = 0, \quad n \neq n'; \qquad (10.4.5)$$

here we have taken it, as we may, that the $y_r^{(n)}(x_r)$ are real. In view of the identity (10.4.4), this is equivalent to

$$\det \left(V_{rs} f_r^{(n)}, f_r^{(n')} \right) = 0, \quad n \neq n'. \qquad (10.4.6)$$

If we suppose the constant factors implicit in the choice of the $y_r^{(n)}(x_r)$ chosen so as to ensure the normalization (10.3.7), we shall also have

$$\det \left(V_{rs} f_r^{(n)}, f_r^{(n)} \right) = 1. \qquad (10.4.7)$$

10.5   Discussion of the integral equations

With a view to applying the result of [Section 11.10 of Atkinson (1972)] we introduce for each equation (10.3.15), or in abstract form (10.4.2), a Hilbert space H_r; we also need their tensor product X.

We define the space H_r as the completion, with respect to the norm associated with the scalar product (11.4.1)

$$(f, f)_r = \int_{a_r}^{b_r} \int_{a_r}^{b_r} G_r(x_r, t_r) \, f_r(t_r) \, \overline{f_r(x_r)} \, dt_r \, dx_r. \qquad (10.5.1)$$

Thus H_r will contain all continuous functions on $[a_r, b_r]$; it will contain certain "generalized functions" as well, and furthermore all elements of $L^2(a_r, b_r)$.

To be more specific, let us associate with a function $f_r(x_r)$ its Fourier expansion

$$f_r(x_r) = \sum c_m z_{rm}(x_r), \qquad (10.5.2)$$

where the $z_{rm}(x_r)$ are the eigenfunctions of the kernel $G_r(x_r, t_r)$, so that

$$z_{rm}(x_r) = \rho_{rm} \int_{a_r}^{b_r} G_r(x_r, t_r) \, z_{rm}(t_r) \, dt_r, \qquad (10.5.3)$$

normalized in the standard sense that

$$\int_{a_r}^{b_r} |z_{rm}(x_r)|^2 \, dx_r = 1. \qquad (10.5.4)$$

In other words, the $z_{rm}(x_r)$, ρ_{rm} are the normalized eigenfunctions and eigen-values of the Sturm-Liouville problem (10.3.13), with boundary conditions (10.3.2). We then have

$$(f_r, f_r)_r = \sum_0^\infty |c_m|^2 \rho_{rm}^{-1}. \tag{10.5.5}$$

Elements of H_r may thus be identified with expressions (10.5.2) for which

$$\sum_0^\infty |c_m|^2 \rho_{rm}^{-1} < \infty. \tag{10.5.6}$$

In the present Sturm-Liouville case, (10.5.6) will be equivalent to

$$\sum_0^\infty |c_m|^2 (1 + m^2)^{-1} < \infty. \tag{10.5.7}$$

It is clear that the continuous functions form a dense set in H_r.

The operator V_{rs} is defined and symmetric on the set of continuous functions, with respect to the scalar product in H_r. It can thus be extended uniquely to a symmetric operator on H_r, if we know that is bounded, in the norm of H_r, on the set of continuous functions. Writing such a function $f_r(x_r)$, so that the expansion (10.5.2) will be valid, at least in the L^2-sense, we have

$$\int_{a_r}^{b_r} G_r(x_r, t_r) f_r(t_r)\, dt_r = \sum_0^\infty c_m z_{rm}(x_r) \rho_{rm}^{-1}, \tag{10.5.8}$$

and so

$$
\begin{aligned}
(V_{rs} f_r)(x_r) &= p_{rs}(x_r) \sum_0^\infty c_m z_{rm}(x_r) \rho_{rm}^{-1} \\
&= \sum_0^\infty d_n z_{rn}(x_r),
\end{aligned}
\tag{10.5.9}
$$

say, where

$$d_n = \sum c_m \rho_{rm}^{-1} \int_{a_r}^{b_r} p_{rs}(x_r) z_{rm}(x_r) z_{rn}(x_r)\, dx_r; \tag{10.5.10}$$

here we take it, as we may, that the z_{rn} are real-valued.

We have to show that the ratio

$$\sum_0^\infty |d_n|^2 \rho_{rn}^{-1} \; / \; \sum_0^\infty |c_m|^2 \rho_{rm}^{-1} \tag{10.5.11}$$

admits a fixed upper bound, for all sets of z_{rm}, not all zero. Applying the Cauchy inequality to (10.5.10) we have

$$|d_n|^2 \leq \left\{ \sum_0^\infty |c_m|^2 \rho_{rm}^{-1} \right\} \left\{ \sum_0^\infty \rho_{rm}^{-1} \left(\int_{a_r}^{b_r} p_{rs} z_{rm} \, z_{rn} \, dx_r \right)^2 \right\};$$

Thus the ratio in (10.5.11) does not exceed

$$\sum_0^\infty \sum_0^\infty \rho_{rm}^{-1} \rho_{rn}^{-1} \left(\int_{a_r}^{b_r} p_{rs} z_{rm} \, z_{rn} \, dx_r \right)^2. \qquad (10.5.12)$$

Now

$$\left| \int_{a_r}^{b_r} p_{rs} z_{rm} \, z_{rn} \, dx_r \right| \leq \sup |p_{rs}(x_r)| \int_{a_r}^{b_r} |z_{rm} \, z_{rn}| \, dx_r,$$

and the last integral does not exceed 1, since the z_{rm} are normalised, as in (10.5.4). Using this fact in (10.5.12), we find that the norm of V_{rs} in the space H_r satisfies

$$\|V_{rs}\|_r \leq \sup |p_{rs}(x_r)| \sum_0^\infty \rho_{rm}^{-1}; \qquad (10.5.13)$$

this last sum is finite, since the eigenvalues ρ_{rm} of (10.3.13) behave, roughly, as m^2 as $m \to \infty$.

Rather similar arguments show that V_{rs} is a compact operator on H_r. We suppose we have a sequence of elements of H_r,

$$h^{(j)} = \sum_0^\infty c_m^{(j)} z_{rm}, \quad j = 1, 2, \ldots. \qquad (10.5.14)$$

which is bounded, in the sense that

$$\sum_0^\infty |c_m^{(j)}|^2 \rho_{rm}^{-1} \leq M < \infty, \quad j = 1, 2, \ldots. \qquad (10.5.15)$$

Without loss of generality, we can take it that the Fourier coefficients $c_m^{(j)}$ converge, so that, say,

$$c_m^{(j)} \to c_m, \quad j \to \infty, \qquad (10.5.16)$$

where

$$\sum |c_m|^2 \rho_{rm}^{-1} \leq M. \qquad (10.5.17)$$

We take these c_m as defining $h \in H_r$, as in (10.5.2) and then claim that

$$V_{rs} h^{(j)} \to V_{rs} h, \quad j \to \infty. \qquad (10.5.18)$$

To see this, we choose some (large) integer N, and write

$$h_1^{(j)} = \sum_{m \leq N} c_m^{(j)} z_{rm}, \quad h_2^{(j)} = \sum_{m > N} c_m^{(j)} z_{rm}, \tag{10.5.19}$$

with a similar notation for $h = h_1 + h_2$. We then have

$$V_{rs}h - V_{rs}h^{(j)} = V_{rs}h_2 + (V_{rs}h_1 - V_{rs}h_1^{(j)}) - V_{rs}h_2^{(j)}. \tag{10.5.20}$$

In the case of the first and last terms on the right of (10.5.20), we are applying V_{rs} to the subspace of H_r, generated by the z_{rm} for $m > N$. A slight modification of the argument leading to (10.5.13) shows that the norm of the restriction of V_{rs} to this subspace is not greater than

$$\sup |p_{rs}(x_r)| \sqrt{\left\{ \sum_{m > N} \rho_{rm}^{-1} \sum_n \rho_{rn}^{-1} \right\}}$$

and this can be made as small as we please by making N large. Since h_2, $h_2^{(j)}$ are uniformly bounded, we can ensure by taking N large that

$$||V_{rs}h_2||_r < \varepsilon, \quad ||V_{rs}h_2^{(j)}||_r < \varepsilon,$$

for any chosen $\varepsilon > 0$. Having chosen N in this way, we can then ensure by choosing j large, that

$$||V_{rs}(h_1 - h_1^{(j)})||_r < \varepsilon.$$

Thus, for large j, we have $||V_{rs}(h - h^{(j)})||_r < 3\varepsilon$, for any chosen $\varepsilon > 0$. This proves (10.5.18), and establishes the compactness of the V_{rs}.

To complete the requirements of [Section 11.10 of Atkinson (1972)], we need to verify a certain positive definiteness property, and to introduce a tensor product space X. The first of these amounts to the condition that

$$\det (V_{rs}h_r, h_r)_r > 0, \tag{10.5.21}$$

for any set of non-zero

$$h_r \in H_r, \quad r = 1, \dots, k. \tag{10.5.22}$$

If we write

$$u_r(x_r) = \int_{a_r}^{b_r} G_r(x_r, t_r) h_r(t_r) \, dt_r, \tag{10.5.23}$$

then for $f_r \in H_r$, $u_r \in L^2(a_r, b_r)$ and so the required property is that (cf. (10.4.4))

$$\det \int_{a_r}^{b_r} p_{rs}(x_r) |u_r(x_r)|^2 \, dx_r > 0, \tag{10.5.24}$$

and this is certainly the case (4.2.6) if the $u_r(x_r)$, $r = 1, \ldots, k$, are all not identically zero. Thus we need to know that if $h_r \neq 0$, as an element of H_r, then $u_r(x_r) \neq 0$. This is evident from (10.5.2), (10.5.8). Finally, we define the tensor product X of the spaces H_r as follows. Let X_0 be the space formed by finite linear combinations of products $h_1(x_1) \cdots h_k(x_k)$ of continuous functions on $[a_r,\, b_r]$, $r = 1,\, \ldots, k$. Such products will be decomposable elements. The scalar product

$$(f, g) = \prod_1^k \left\{ \int_{a_r}^{b_r} dx_r \int_{a_r}^{b_r} dt_r\, G_r(x_r, t_r) \right\} f(t_1, \ldots, t_k)\, \overline{g(x_1, \ldots, x_k)}$$

$$(10.5.25)$$

is then defined for all pairs of functions

$$f(x_1, \ldots, x_k), \quad g(x_1, \ldots, x_k) \tag{10.5.26}$$

continuous on the closed box

$$a_r \leq x_r \leq b_r, \quad r = 1, \ldots, k. \tag{10.5.27}$$

In particular, it is defined for pairs of elements of X_0; we specify that X is the completion of X_0 in the norm associated with the scalar product (10.5.25).

We need the observation that X includes all continuous functions on (10.5.27). For a continuous function can be approximated by elements of X_0 in the L^2-sense; for example, partial sums of the expansion of a continuous function in a multiple Fourier series will serve this purpose. A continuous function can therefore be approximated by elements of X_0 in the weaker sense given by (10.5.25). The induced operations $V_{rs}^\dagger$ on X, corresponding to V_{rs} on H_r, are given by applying these operations to the r-th arguments. For example, if $f(x_1, \ldots, x_k)$ is any continuous function, then

$$(V_{rs}\dagger f)(x_1, \ldots, x_k) = p_{rs}(x_r) \int_{a_r}^{b_r} G_r(x_r, t_r) f(x_1, \ldots, t_r, \ldots, x_k)\, dt_r.$$

$$(10.5.28)$$

Similarly,

$$(G_r\dagger f)(x_1, \ldots, x_k) = \int_{a_r}^{b_r} G_r(x_r, t_r) f(x_1, \ldots, t_r, \ldots, x_k)\, dt_r. \tag{10.5.29}$$

10.6   Completeness of eigenfunctions

We can now deal with this, under slightly restrictive conditions on the coefficients.

Theorem 10.6.1. *In the differential equations* (10.3.1), *let* $p_{rs}(x_r)$, $q_r(x_r)$, *be real and continuously twice differentiable in the finite intervals* $[a_r, b_r]$, $r = 1,\ldots,k$, *and let*

$$\det p_{rs}(x_r) > 0, \quad a_r \leq x_r \leq b_r, \quad r = 1,\ldots,k.$$

Let α_r, β_r, $r = 1,\ldots,k$, *be real. Then the eigenfunctions* (10.3.4) *are complete in the space of functions, which are in* L^2 *over the region* $a_r \leq x_r \leq b_r$, $r = 1,\ldots,k$.

As previously remarked, the completeness to be proved is in the mean-square sense, with the positive weight-function $\det p_{rs}(x_r)$; since the latter is continuous and positive, it will have positive upper and lower bounds, so that approximation in mean-square with this weight-function is equivalent to approximation in mean-square with a constant weight-function.

To prove the result, it will be sufficient to establish the Parseval equality for the eigenfunction expansion of a class of functions whose linear combinations are dense in the space concerned. A suitable class of functions for our purpose will be those of the form

$$u(x_1,\ldots,x_k) = \prod_1^k u_r(x_r), \tag{10.6.1}$$

where $u_r(x_r)$ is four times continuously differentiable and vanishes in neighborhoods of a_r, b_r. For each k-tuple $n = (n_1,\ldots,n_k)$ of non-negative integers, we form the Fourier coefficient

$$\begin{aligned}
c_n &= \det \int_{a_r}^{b_r} p_{rs}(x_r) u_r(x_r) y_r^{(n)}(x_r)\, dx_r \\
&= \int_a^b p(x)\, u(x)\, y^{(n)}(x)\, dx \tag{10.6.2}
\end{aligned}$$

with respect to the normalized eigenfunction (10.3.4). Here we have used the notation (10.3.6) in (11.6.2). What we have to show is that

$$\det \int_{a_r}^{b_r} p_{rs}(x_r)|u_r(x_r)|^2\, dx_r = \sum |c_n|^2. \tag{10.6.3}$$

For this purpose, we define functions

$$h_r(x_r) = -\{d^2/dx_r^2 + q_r(x_r)\}u_r(x_r), \tag{10.6.4}$$

and

$$h(x_1,\ldots,x_k) = \prod_1^k h_r(x_r), \tag{10.6.5}$$

We also define k functions

$$v_s(x_1, \ldots, x_k), \quad s = 1, \ldots, k, \tag{10.6.6}$$

as solutions of

$$\partial^2 u/\partial x_r^2 + q_r(x_r)u + \sum_1^k p_{rs}(x_r)v_s = 0, \quad r = 1, \ldots, k. \tag{10.6.7}$$

The v_s are uniquely determined by these equations since $\det p_{rs}(x_r)$ is to be positive.

In addition, the determinantal solution of (10.6.7) exhibits the v_s as continuously twice differentiable in each of the $x_1, \ldots, x_k$; it is for this reason that we have assumed the u_r to be four times differentiable, and the p_{rs}, q_r twice differentiable. Moreover, the operators

$$\partial^2/\partial x_r^2 + q_r(x_r)$$

may all be applied to the v_s. We note furthermore that the v_s will vanish whenever one of their arguments is in a neighborhood of the end-points of the associated interval. Thus they all satisfy the boundary conditions in all of the variables.

We write now

$$w_s(x_1, \ldots, x_k) = \prod_1^k (-\partial^2/\partial x_r^2 - q_r(x_r))v_s(x_1, \ldots, x_k). \tag{10.6.8}$$

We apply to the r-th equation in (10.6.7) the operator

$$D_r = \prod_{j \neq r} (-\partial^2/\partial x_j^2 - q_j(x_j)). \tag{10.6.9}$$

In view of (10.6.4)–(10.6.5) this gives

$$-h + \sum_1^k p_{rs}D_r\, v_s = 0, \quad r = 1, \ldots, k. \tag{10.6.10}$$

However, since the differential operator $-d^2/dx_r^2 - q_r(x_r)$ is inverted by the Green's function $G_r(x_r, t_r)$, acting as the kernel in an integral operator, we have

$$(D_r v_s)(x_1, \ldots, x_k) = \int_{a_r}^{b_r} G_r(x_r, t_r)w_s(x_1, \ldots, t_r, \ldots, x_k)dt_r.$$

Hence (10.6.10) gives

$$h(x_1, \ldots, x_k) = \sum_1^k p_{rs}(x_r) \int_{a_r}^{b_r} G_r(x_r, t_r) w_s(x_1, \ldots, t_r, \ldots, x_k) \, dt_r,$$

for $r = 1, \ldots, k$.

In other words, we have

$$h = \sum_1^k V_{rs}^\dagger w_s, \quad r = 1, \ldots, k, \tag{10.6.11}$$

where $V_{rs}^\dagger$ is the induced operator given by (10.5.28). We are now in a position to apply Theorem 11.10.1 of Atkinson (1972). In (10.6.11), we have the situation called for in *ibid.* (11.10.8) of the theorem in question, since h is decomposable and the w_s, being continuous, are in the tensor product space X concerned. The discrepancy in sign, as between the A_{rs} in *ibid.* (11.10.3) and the V_{rs} in (10.4.2), is of no importance; it may be removed by reversing the signs of the parameters.

We therefore have that the Parseval equality holds for the expansion of h in terms of eigenvectors of (10.4.2). Here the result (11.10.9) of Atkinson (1972) gives

$$\det(V_{rs} h_r, h_r)_r = \sum |\gamma_n|^2, \tag{10.6.12}$$

where

$$\gamma_n = \det(V_{rs} h_r, f_r^{(n)})_r. \tag{10.6.13}$$

However, it follows from (10.6.4), and the fact that u_r vanishes near a_r, b_r and so satisfies the boundary conditions, that

$$u_r(x_r) = \int_{a_r}^{b_r} G_r(x_r, t_r) h_r(t_r) \, dt_r.$$

Thus, by (10.4.4), the left of (10.6.12) is the same as the left of (10.6.3). In a similar way, the right of (10.6.12) is equal to the right of (10.6.13). This completes the proof of the completeness of eigenfunctions.

10.7   Completeness via partial differential equations

As we have seen in the preceding section, the coefficients in an eigenfunction expansion are easily found, given the function to be expanded. The difficulty is, of course, to show that the function being "expanded" is indeed represented by the series of eigenfunctions, even when the latter series is known to be convergent. Here we discuss an approach that places the difficulty elsewhere, in the theory of the completeness of eigenfunctions, for boundary-value problems for partial differential equations with one parameter.

We shall do this for the classical multiparameter Sturm-Liouville problem, and start by assembling information found in the intervening chapters. We take the opportunity to recall a more compact notation. We write $x = (x_1, \ldots, x_k)$ for a real k-tuple satisfying

$$a_r \leq x_r \leq b_r, \quad r = 1, \ldots, k, \tag{10.7.1}$$

and $n = (n_1, \ldots, n_k)$ for a k-tuple of non-negative integers. We have then

Theorem 10.7.1. *Let the functions $p_{rs}(x_r)$, r, $s = 1, \ldots, k$, and $q_r(x_r)$, $r = 1, \ldots, k$, be real and continuous in the finite real intervals (10.7.1). Let α_r, β_r, $r = 1, \ldots, k$, be real. We define p by*

$$p(x) = \det p_{rs}(x_r) > 0, \tag{10.7.2}$$

in the domain (10.7.1). Then the eigenvalue problem

$$y_r''(x_r) + \left\{ \sum \lambda_s p_{rs}(x_r) + q_r(x_r) \right\} y_r(x_r) = 0, \quad a_r \leq x_r \leq b_r, \tag{10.7.3}$$

$$y_r(a_r) \cos \alpha_r = y_r'(a_r) \sin \alpha_r, \quad y_r(b_r) \cos \beta_r = y_r'(b_r) \sin \beta_r, \tag{10.7.4}$$

$r = 1, \ldots, k$, *has an infinite sequence of eigenvalues*

$$(\lambda_1^{(n)}, \ldots, \lambda_k^{(n)}), \tag{10.7.5}$$

where $n = (n_1, \ldots, n_k)$ runs through all sets of non-negative integers, n_r being the number of zeros of $y_r(x_r)$ in the open interval (a_r, b_r). There is an infinite sequence of eigenfunctions of the form

$$y(x) = \prod_{r=1}^{k} y_r(x_r), \tag{10.7.6}$$

products of non-trivial solutions of (10.7.3)–(10.7.4) associated with (10.7.5).

As in § S123, these are mutually orthogonal in the sense

$$\int_a^b y_{n'}(x)\, y_n(x) p(x)\, dx = 0, \quad n \neq n' \tag{10.7.7}$$

where we have used the notation in (10.3.6). Furthermore, we can arrange that (see (10.3.7))

$$\int_a^b |y_n(x)|^2 \, p(x)\, dx = 1. \tag{10.7.8}$$

Let now

$$f(x) = f(x_1, \ldots, x_k) \tag{10.7.9}$$

be any continuous function in the "box" (10.7.1). We form the Fourier expansion

$$f(x) \sim \sum c_n \, y_n(x), \tag{10.7.10}$$

where

$$c_n = \int_a^b f(x) y_n(x) p(x)\, dx. \tag{10.7.11}$$

As noted earlier, we are concerned with whether equality holds in (10.7.10), in the pointwise, mean-square, uniform or any other sense. Allied with this is the question of whether there holds the Parseval equality

$$\int_a^b |f(x)|^2 p(x)\, dx = \sum |c_n|^2. \tag{10.7.12}$$

It should be noted that the weight-function (10.7.2) plays an essential part in the orthogonality (10.7.7), and so also in the eventual expansion theorem, by way of (11.6.2), (11.6.3). In particular, the most natural function-space for the expansion theorem is the space of measurable and square-integrable functions, with respect to the weight-function $p(x)$. Of course, if $p(x)$ is positive and continuous in (10.7.1), this space contains the same functions as the ordinary space of L^2-functions over (10.7.1) .

10.8   Preliminaries on the case $k = 2$

We consider the system (10.7.3)–(10.7.4) in the case $k = 2$, using the possibility of making transformations so that certain coefficients $p_{rs}(x_r)$ have fixed assigned signs. The system will take the form

$$y_r'' + \left\{ \sum_1^2 \lambda_s p_{rs} + q_r \right\} y_r = 0, \quad a_r \le x_r \le b_r, \quad r = 1, 2, \qquad (10.8.1)$$

together with boundary conditions

$$y_r(a_r) \cos \alpha_r = y_r'(a_r) \sin \alpha_r, \quad y_r(b_r) \cos \beta_r = y_r'(b_r) \sin \beta_r, \quad r = 1, 2.$$
$$(10.8.2)$$

We shall assume a transformation made so that

$$p_{12}(x_1) < 0, \ a_1 \le x_1 \le b_1, \quad p_{22}(x_2) > 0, \ a_2 \le x_2 \le b_2. \qquad (10.8.3)$$

The possibility of doing this, subject to condition (A), or (10.7.2) was established in Section 4.11; it was also shown there that the same can be arranged under condition (B), which is weaker than condition (A) if $k = 2$.

We now form the partial differential equations satisfied by products

$$u(x_1, x_2) = y_1(x_1)y_2(x_2),$$

namely,

$$\partial^2 u / \partial x_r^2 + \{ \sum_1^2 \lambda_s p_{rs} + q_r \} u = 0, \quad r = 1, 2, \qquad (10.8.4)$$

and eliminate the parameter λ_2; this gives

$$p_{22}\partial^2 u / \partial x_1^2 - p_{12}\partial^2 u / \partial x_2^2 + (p_{22}q_1 - p_{12}q_2)\, u + \lambda_1\, (p_{11}p_{22} - p_{12}p_{21}) = 0. \quad (10.8.5)$$

In addition, u will satisfy the boundary conditions

$$u \cos \alpha_r = (\partial u / \partial x_r) \sin \alpha_r, \quad x_r = a_r, \ r = 1, 2, \qquad (10.8.6)$$
$$u \cos \beta_r = (\partial u / \partial x_r) \sin \beta_r, \quad x_r = b_r, \ r = 1, 2. \qquad (10.8.7)$$

Thus the eigenfunctions of (10.8.1)–(10.8.2), that is to say, products of nontrivial solutions of (10.8.1)–(10.8.2), are eigenfunctions in the standard sense of (10.8.5)–(10.8.7); in particular, any eigenvalue (2-tuple) (λ_1, λ_2) of (10.8.1)–(10.8.2) in the multiparameter sense, yields an eigenvalue λ_1 of (10.8.5)-(10.8.7) in the usual sense. Our next task is to reverse the argument, and to show that all eigenfunctions and eigenvalues of (10.8.5)–(10.8.7) are to be obtained in this manner.

10.9   Decomposition of an eigensubspace

We now suppose that, conversely, λ_1 is an eigenvalue of the problem (10.8.5)–(10.8.7) and discuss whether for some λ_2 the pair (λ_1, λ_2) is an eigenvalue of the multiparameter problem (10.8.1)–(10.8.2); furthermore, we ask whether all non-trivial solutions of (10.8.5)–(10.8.7) for this value of λ_1 can be obtained by taking linear combinations of products of solutions of (10.8.1)–(10.8.2), with the various admissible values λ_2 and fixed λ_1.

We prove first a result involving a sort of Fourier coefficient.

Theorem 10.9.1. *Let* $u(x_1, x_2)$ *be a solution of* (10.8.5)–(10.8.7). *Let* $y_2(x_2)$ *be a solution of* (10.8.1)–(10.8.2) *with* $r = 2$. *Define*

$$y_1(x_1) = \int_{a_2}^{b_2} u(x_1, x_2)\, p_{22}(x_2)\, y_2(x_2)\, dx_2. \tag{10.9.1}$$

Then $y_1(x_1)$ *satisfies* (10.8.1)–(10.8.2) *with* $r = 1$.

We first check that $y_1(x_1)$ satisfies the boundary conditions (10.8.2). At $x_1 = a_1$, we have

$$y_1(a_1) \cos \alpha_1 - y_1'(a_1) \sin \alpha_1$$

$$= \int_{a_2}^{b_2} p_{22} y_2 \{ u(a_1, x_2) \cos \alpha_1 - (\partial/\partial\, x_2) u(a_1, x_2) \sin \alpha_1 \}\, dx_2.$$

Here the integrand on the right is zero by (10.8.6), and so we have the first of (10.8.2) with $r = 1$. The case of the boundary condition at $x_1 = b_1$ is similar. We must now verify that y_1 satisfies the differential equation in (10.8.1). We have

$$y_1'' + q_1 y_1 = \int_{a_2}^{b_2} p_{22} y_2 \{ \partial^2 u/\partial x_1^2 + q_1 u \}\, dx_2. \tag{10.9.2}$$

From (10.8.5), we have

$$p_{22} \{ \partial^2 u/\partial x_1^2 + q_1 u \} = p_{12} \{ \partial^2 u/\partial x_1^2 + q_2 u \} + \lambda_1 (p_{12} p_{21} - p_{11} p_{22}) u.$$

Substituting this in (10.9.2) we get

$$y_1'' + q_1 y_1 = p_{12} \int_{a_2}^{b_2} \{ \partial^2 u/\partial x_2^2 + q_2 u + \lambda_1 p_{21} u \}\, y_2\, dx_2 - \lambda_1 p_{11} \int_{a_2}^{b_2} p_{22} u y_2\, dx_2.$$

In view of (10.9.1), this may be re-written as

$$y_1'' + q_1 y_1 + \lambda_1 p_{11} y_1 = p_{12} \int_{a_2}^{b_2} \{\partial^2 u / \partial x_2^2 + q_2 u + \lambda_1 p_{21} u\}\, y_2 \, dx_2. \qquad (10.9.3)$$

We now use the fact that $y_2(x_2)$, $u(x_1, x_2)$ satisfy the same homogeneous boundary conditions in x_2 at a_2, b_2. From, this it follows that

$$\int_{a_2}^{b_2} (\partial^2 u / \partial x_2^2) y_2 \, dx_2 = \int_{a_2}^{b_2} u \, y_2'' \, dx_2.$$

Thus (10.9.3) may be replaced by

$$y_1'' + q_1 y_1 + \lambda_1 p_{11} y_1 = p_{12} \int_{a_2}^{b_2} (y_2'' + q_2 y_2 + \lambda_1 p_{21} y_2)\, u \; dx_2.$$

In view of (10.8.1), the right is the same as

$$-p_{12} \int_{a_2}^{b_2} \lambda_2 p_{22} y_2 u \, dx_2 = -\lambda_2 p_{12} y_1.$$

Thus

$$y_1'' + (\lambda_1 p_{11} + \lambda_2 p_{12} + q_1) y_1 = 0,$$

as was to be proved.

The above argument was of a formal nature; we do not need for the p_{rs} to have any specified signs, or even that they be real-valued. To complete the discussion, however, we must impose restrictions of this nature. We then have

Theorem 10.9.2. *Let the $p_{rs}(x_r)$, $q_r(x_r)$ be real and continuous, and let $p_{22}(x_2) > 0$, $p_{12}(x_1) < 0$, as in (10.8.3). Then for any real λ_1, the space of solutions of (10.8.5)–(10.8.7) has finite dimension, and is generated by products of solutions of (10.8.1)–(10.8.2), for a finite set of values of λ_2.*

We may assume that λ_1 is an eigenvalue of (10.8.5)–(10.8.7), since otherwise the space of solutions would have dimension 0. For such λ_1, we may consider (10.8.1)–(10.8.2) as providing two distinct eigenvalue problems in which λ_1 is fixed and λ_2 is the parameter to be determined. In the problem given by (10.8.1)–(10.8.2) with $r = 2$, the coefficient of $\lambda_2 y_2$ is $p_{22}(x_2)$, which we take to be positive. This is therefore a Sturm-Liouville problem of standard type, in which the eigenvalues for λ_2 form an increasing sequence, with no finite limit-point. In the problem given by (10.8.1)–(10.8.2) with $r = 1$, the coefficient of $\lambda_2 y_1$ is $p_{12}(x_1)$, which we take to be negative. In this case, therefore, the eigenvalues λ_2 form a decreasing sequence, with no finite limit-point. Thus the two sequences of eigenvalues λ_2 can have at most a finite number of members in common.

Let now
$$y_2^{(m)}(x_2), \quad m = 0, 1, \ldots$$
be a complete system of eigenfunctions of the problem (10.8.1)–(10.8.2) with
$r = 2$, and let $\lambda_2^{(m)}$, $m = 0, 1, \ldots$ be the associated eigenvalues; here we continue
to treat λ_1 as fixed. We form the Fourier coefficients

$$c_m(x_1) = \int_{a_2}^{b_2} u(x_1, x_2) y_2^{(m)}(x_2) p_{22}(x_2) \, dx_2. \qquad (10.9.4)$$

By Theorem 10.9.1, $c_m(x_1)$ must satisfy the eigenvalue problem (10.8.1)–(10.8.2)
with $r = 1$, and with $\lambda_2 = \lambda_2^{(m)}$. Hence either $c_m(x_1)$ is identically zero, or else
$\lambda_2^{(m)}$ is a common eigenvalue of the problem (10.8.1)–(10.8.2) given by fixing λ_1.
As we have shown, this is possible for at most a finite number of values of λ_2.
The eigenfunction expansion of one-parameter Sturm-Liouville thus ensures a
representation

$$u(x_1, x_2) = \sum c_m(x_1) y_2^{(m)}(x_2) \left\{ \int_{a_2}^{b_2} p_{22} |y_2^{(m)}|^2 d\xi_2 \right\}^{-1},$$

as a sum of products of solutions of (10.8.1)–(10.8.2) in which the sum is taken
over a certain finite set only. This proves the result.

We sum up our result for the completeness of eigenfunctions as

Theorem 10.9.3. *Let the $p_{rs}(x_r)$, $q_r(x)$, $a_r \le x_r \le b_r$, $r, s = 1, 2$ be real and
continuous, with $p_{22}(x_2) > 0$, $p_{12}(x_1) < 0$. If the eigenfunctions of the partial
differential equation problem (10.8.5)–(10.8.7) are complete in a space, then the
eigenfunctions of the two-parameter problem (10.8.1)–(10.8.2) are complete in
the same space.*

We have reduced the problem of the completeness of eigenfunctions in the
two-parameter case to that of the completeness of the eigenfunctions for a
one-parameter problem for a partial differential equation. One may consider
this last problem as solved classically; however, for completeness we sketch a
treatment here. Our treatment will rely rather heavily on the positivity of the
coefficients $p_{22}(x_2)$, $-p_{12}(x_1)$ of the second-order terms in (10.8.5), and to a
lesser extent on the positivity of the coefficient $(p_{11}p_{22} - p_{12}p_{21})$. We recall
that if the latter holds, that is to say, "condition (A)," then the former can
be arranged, though we can also arrange that $p_{22} > 0$, $p_{12} < 0$ under the less
demanding (if $k = 2$) condition (B). We first assume that condition (A) holds.

Theorem 10.9.4. *Let $p_{rs}(x_r)$, $q_r(x_r)$, $a_r \le x_r \le b_r$, $r, s = 1, 2$, be real and
continuous, and let $p(x) = p_{11}(x_1)p_{22}(x_2) - p_{12}(x_1)p_{21}(x_2) > 0$ in this domain.
Then the eigenfunctions of (10.8.1)–(10.8.2) form a complete set in the space
of functions $f(x_1, x_2)$, $a_r \le x_r \le b_r$, $r = 1, 2$, which are Lebesgue-measurable
and square-integrable over this rectangle.*

The proof of Theorem 10.9.4 proceeds as follows: On the basis of Section 4.11, we can arrange, by means of a linear transformation (4.11.2), that the $p_{rs}(x_r)$ have the fixed signs indicated in (4.11.3–4.11.4). Supposing this done, we note that we can choose a real γ, large and negative if necessary, so that the problems

$$z_r{}''(x_r) + \{q_r(x_r) + \gamma p_{r1}(x_r) + \lambda\} z_r(x_r) = 0, \qquad (10.9.5)$$

$$z_r(a_r) \cos \alpha_r = z_r'(a_r) \sin \alpha_r, \, z_r(b_r) \cos \beta_r = z_r'(b_r) \sin \beta_r, \qquad (10.9.6)$$

where $r = 1, 2$, both have purely positive spectrum.

We then write (10.8.5) in the form

$$Au + \lambda_1' Bu = 0; \qquad (10.9.7)$$

here the operators A, B are defined by

$$
\begin{aligned}
-A \;=\; & p_{22}\partial^2/\,\partial x_1^2 - p_{12}\partial^2/\partial x_2^2 + (p_{22}q_1 - p_{12}q_2) \\
& +\gamma(p_{11}p_{22} - p_{12}p_{21}) \qquad\qquad (10.9.8) \\
B \;=\; & (p_{11}p_{22} - p_{12}p_{21}), \quad \lambda_1 = \gamma + \lambda_1' \qquad (10.9.9)
\end{aligned}
$$

in which B acts, of course, by multiplication.

We are then concerned with the completeness of the eigenfunctions of the problem formed by (10.9.7), with disposable λ_1', and the boundary conditions (10.8.6)–(10.8.7).[2]

10.10 Completeness via discrete approximations

It is well-known that finite difference approximations are not only basic numerical tools, but can also serve to establish theoretical results of a general nature. In the case of the completeness of Sturm-Liouville eigenfunctions, this argument goes back a long way; references to the work of Plancherel and Fort will be found along with some discussion in Atkinson (1964a), with application to a similar problem involving rational functions. In the ordinary Sturm-Liouville case, a concise version of the argument has been given by B. M. Levitan; we refer for this to Titchmarsh (1962). Here we adapt this argument to the multiparameter case.

---

[2]Editorial Remark: In all manuscripts the proof of Theorem 10.9.4 ends here. For a plausible reconstruction of the remaining part of the proof, see the Notes at the end of this Chapter.

The argument proceeds in terms of the various Parseval equalities. In the ordinary Sturm-Liouville case, one replaces the differential equation by a recurrence or finite-difference equation, based on a finite subdivision of the basic interval, thus replacing the original boundary-value problem by an eigenvalue problem for a real symmetric matrix. The known completeness of the eigenvectors of such a matrix yields the Parseval equality for the expansion of a suitably chosen general vector. The final steps are to make the size of the subdivisions tend to zero, and to ensure that the Parseval equality converges uniformly in this process; this involves applying the Bessel inequality to a suitable function.

In the multiparameter version of the argument, the completeness of the eigenvectors of the approximating problems is no longer a matter of the completeness of the eigenvectors for a finite-dimensional one-parameter problem. Two methods are available. We can appeal to the general theory of arrays of symmetric matrices. In the resulting set of multiparameter algebraic equations, we apply the completeness properties established in Chapter 7 of Atkinson (1972). Alternatively, we can use the "Klein oscillation theorem" for sets of recurrence relations, which ensures that there are just enough eigenvalues (k-tuples) for completeness to hold. To obtain the uniform convergence, we need to apply the Bessel inequality to several functions.

10.11   The one-parameter case

In this and the next section, we review briefly the proof by discrete approximation of the completeness of the eigenfunctions for the one-parameter Sturm-Liouville case[3]

$$y'' + (\lambda p(x) - q(x))y = 0, \quad a \le x \le b, \; p(x) > 0, \text{ in } (a,b), \qquad (10.11.1)$$

with suitable separated boundary conditions. We denote by $y(x, \lambda)$ the solution with initial data

$$y(a, \lambda) = \sin \alpha, \quad y'(a, \lambda) = \cos \alpha,$$

and the eigenfunctions by $y_n(x) = y(x, \lambda_n)$, where $\lambda_0, \lambda_1, \ldots,$ are the eigenvalues.

We consider the expansion of a function f such that

$$\int p(x)|f(x)|^2 \, dx < \infty, \qquad (10.11.2)$$

---

[3]Note the change of sign before the coefficient $q(x)$ in this and subsequent sections.

all integrals being over (a, b). This takes the form

$$f(x) \sim \sum c_n y_n(x)/\rho_n, \tag{10.11.3}$$

where

$$c_n = \int p(x)y_n(x)f(x)\,dx, \tag{10.11.4}$$

$$\rho_n = \int p(x)|y_n(x)|^2\,dx. \tag{10.11.5}$$

The validity of this expansion in the space of functions given by (10.11.2) is characterized by the "Parseval equality"

$$\int p(x)|f(x)|^2\,dx = \sum |c_n|^2/\rho_n. \tag{10.11.6}$$

The same result with " $\geq$ " instead of " $=$ " is the "Bessel inequality," which is trivially true. The statement (10.11.6) is independent of the way the eigenfunctions are normalized.

For the sequel, it will be relevant to comment on the rate of convergence of the series on the right. For this purpose, we suppose that f, f' are absolutely continuous, that both vanish at the endpoints, and that

$$g = (qf - f'')/p \tag{10.11.7}$$

lies in the space (10.11.2). We apply the Bessel inequality to the expansion of g. We write

$$d_n = \int pgy_n\,dx$$

so that

$$d_n = \int (qf - f'')y_n\,dx = \int f(qy_n - y_n'')\,dx = \lambda_n \int py_n f\,dx = \lambda_n c_n, \tag{10.11.8}$$

where we have used integration by parts and the boundary conditions. The Bessel inequality for g then implies that

$$\int pg^2\,dx \geq \sum d_n^2/\rho_n = \sum (\lambda_n c_n)^2/\rho_n. \tag{10.11.9}$$

We deduce that, for any $\Lambda > 0$,

$$\sum_{\lambda_n \geq \Lambda} c_n^2/\rho_n \leq \Lambda^{-2} \int p|g|^2\,dx. \tag{10.11.10}$$

We proceed to discretize this argument.

10.12   The finite-difference approximation

We pass now to finite-dimensional approximations. For this we consider a sequence of integers m, tending to ∞, for each of which we divide the interval (a, b) into m equal parts, of length $h = (b - a)/m$, by nodes $x_j = x_j^{(m)}$ such that

$$a = x_0 < x_1 < \cdots < x_m = b. \qquad (10.12.1)$$

For an eigenfunction expansion, we replace $y(x)$ by a piecewise linear function $z(x) = z^{(m)}(x, \lambda)$, determined at these nodes by initial data

$$z(x_0) = \sin \alpha, \quad z(x_1) = \sin \alpha + h \cos \alpha, \qquad (10.12.2)$$

together with the three-term recurrence relation

$$h^{-2}\Delta^2 z(x_j) + z(x_j)\{\lambda p(x_j) - q(x_j)\} = 0, \quad j = 1, \ldots, m - 1 \qquad (10.12.3)$$

where the second-order difference operator Δ^2 is given by

$$\Delta^2 z(x_j) = z(x_{j+1}) - 2z(x_j) + z(x_{j-1}). \qquad (10.12.4)$$

We then set up the algebraic eigenvalue problem

$$z(x_m) \cos \beta = h^{-1}\{z(x_m) - z(x_{m-1})\} \sin \beta. \qquad (10.12.5)$$

For small $h > 0$, we shall have $\sin \alpha + h \cos \alpha \neq 0$ so that $z(x_m)$ is a polynomial of degree $m - 1$ in λ. Likewise, for small $h > 0$ we have $\cos \beta - h^{-1} \sin \beta \neq 0$ so that (10.12.5) is a polynomial equation of degree $m - 1$, and there will be just $m - 1$ eigenvalues all distinct and real, say,

$$\mu_n = \mu_n^{(m)}, \quad n = 0, \ldots, m - 2, \qquad (10.12.6)$$

which could be characterized by oscillatory arguments. The corresponding eigenfunctions $z(x, \mu_j)$ are orthogonal in the sense

$$h \sum p(x_j) z(x_j, \mu_n) z(x_j, \mu'_n) = 0, \quad n \neq n' \qquad (10.12.7)$$

with all summations over $1, \ldots, m - 1$. For any function $f(x)$, we have an expansion, valid at the x_j,

$$f(x_j) = \sum \gamma_n z(x_j, \mu_n)/\sigma_n, \qquad (10.12.8)$$

where

$$\gamma_n = \sum p(x_j) z(x_j, \mu_n) f(x_j), \quad \sigma_n = \sum p(x_j)|z(x_j, \mu_n)|^2. \qquad (10.12.9)$$

The Parseval equality is

$$\sum p(x_j)|f(x_j)|^2 = \sum |\gamma_n|^2/\sigma_n, \tag{10.12.10}$$

which is certainly valid in this finite-dimensional situation; this will need more attention in the multiparameter case.

In (10.12.9), we have suppressed the dependence on m, but will now re-introduce it and make $m \to \infty$, with a view to obtaining (10.11.6) in the limit, for suitable f. Let us assume that $f \in C''[a, b]$, and that $f \equiv 0$ in neighborhoods of a and b. It is clear that the left of (10.12.10) tends to the left of (10.11.6), if p is Riemann-integrable on $[a, b]$. It is also clear, by standard results [Faierman (1969)] on discrete approximation and initial-value problems, that for any fixed μ, we have

$$z^{(m)}(x, \mu) \to y(x, \mu), \quad z^{(m)'}(x, \mu) \to y'(x, \mu) \tag{10.12.11}$$

as $m \to \infty$, uniformly on $[a, b]$; in the latter case, $z^{(m)'}$ may be interpreted as a one-sided derivative when taken at a node. From this we see that if $\mu_n^{(m)}$ tends to a limit λ as $m \to \infty$, then λ must be an eigenvalue of the original problem, and that

$$h \sum_j p(x_j^{(m)}) z(x_j^{(m)}, \mu) f(x_j^{(m)}) \to \int p(x) y(x, \mu) f(x)\, dx, \tag{10.12.12}$$

$$h \sum_j p(x_j^{(m)}) z^2(x_j^{(m)}, \mu) \to \int p y^2(x, \mu)\, dx. \tag{10.12.13}$$

We now have to consider the convergence properties as $m \to \infty$ of the $m - 1$ eigenvalues of the discrete problem. For this purpose, we choose some (large) $\Lambda > 0$, and for each m divide the eigenvalues $\mu_n^{(m)}$ into two classes, those for which

$$|\mu_n^{(m)}| \leq \Lambda, \tag{10.12.14}$$

or

$$|\mu_n^{(m)}| > \Lambda. \tag{10.12.15}$$

Our first remark is that the number of eigenvalues in the class (10.12.14) with fixed Λ remains bounded as $m \to \infty$. In the contrary event, there would be an unbounded m-sequence with two distinct eigenvalues $\mu_n^{(m)}, \mu_n^{(m')}$, both tending to the same limit μ as $m \to \infty$. In view of (10.12.7), this would imply that

$$\int p(x) y^2(x, \mu)\, dx = 0,$$

which is impossible. It follows that by selection of a subsequence of m-values, we may arrange that the eigenvalues in class (10.12.14) all converge, say,

$$\mu_n^{(m)} \to \lambda_n, \quad n = 0, \ldots, N,$$

where the λ_n are distinct eigenvalues of the full problem, and

$$\gamma_n^2/\sigma_n \to c_n^2/\rho_n. \tag{10.12.16}$$

It remains to consider eigenvalues of the class (10.12.15). For this we need the analogue of the argument (10.11.7)–(10.11.10). We have, dropping the superscripts "(m)" and using the boundary conditions on f for large m,

$$\sum\{q(x_j)f(x_j) - h^{-2}\Delta^2 f(x_j)\}z(x_j, \mu_n)$$
$$= \sum\{q(x_j)z(x_j, \mu_n) - h^{-2}\Delta^2 z(x_j, \mu_n)\}f(x_j) = \mu_n \sum p(x_j)f(x_j)z(x_j, \mu_n).$$

We introduce, in analogy with (10.11.7), the auxiliary function

$$w_j = \{q(x_j)f(x_j) - h^{-2}\Delta^2 f(x_j)\}/p(x_j)$$

and write

$$\delta_n = \sum p(x_j)w_j z(x_j),$$

so that the above gives, in the notation (10.12.9),

$$\delta_n = \mu_n \gamma_n.$$

Again we have by the Bessel inequality (though exact in this case)

$$h \sum p(x_j)w_j^2 \geq \sum \delta_n^2/\sigma_n = \sum (\mu_n \gamma_n)^2/\sigma_n.$$

Hence

$$\sum_{|\mu_n| \geq \Lambda} \gamma_n{}^2/\sigma_n \leq \Lambda^{-2} h \sum p(x_j)w_j^2. \tag{10.12.17}$$

It might be noted that we did not prove that every eigenvalue of the continuous problem will appear by way of a limit from the discrete approximation. However, it is sufficient to know that the Parseval equality is true for a subset of the expansion coefficients.[4]

10.13   The multiparameter case

We now go through the analogue of the argument of §10.11 in the case of the system (5.2.1–5.2.2) in the original continuous case, leaving the discrete approximation to the next section. We now start with a given function $f(x_1, \ldots, x_k)$,

---

[4]The construction ends here in all manuscripts. For a possible reconstruction, see the Notes.

assumed to have continuous second-order partial derivatives, and to vanish in a neighborhood of the boundary of the box I. In analogy with (10.11.7), we now define k functions

$$g_s(x_1, ..., x_k), \quad s = 1, ..., k, \tag{10.13.1}$$

by

$$\partial^2 f / \partial x_r^2 - q_r f + \sum_1^k p_{rs}(x_r) g_s = 0, \tag{10.13.2}$$

for $r = 1, \ldots, k$, $a_r \leq x_r \leq b_r$. These functions are well-defined since we assume that $\det p_{rs} \neq 0$. We need the analogue of the results (10.11.8), connecting the Fourier coefficients of f and the g_s.

Let $(\lambda_1, \ldots, \lambda_k)$ be any eigenvalue, and $\prod y_r(x_r)$ the corresponding eigenfunction. We multiply the r-th equation (10.13.2) by $y_r(x_r)$ and integrate over (a_r, b_r). Integrating by parts as in (10.11.8) we get

$$\int_{a_r}^{b_r} (q_r f - \partial^2 f / \partial x_r^2) y_r \, dx_r = \int_{a_r}^{b_r} f(q_r y_r - y_r'') \, dx_r,$$

and hence, using (5.2.1),

$$\sum_{s=1}^k \lambda_s \int_{a_r}^{b_r} f p_{rs} y_r \, dx_r = \sum_{s=1}^k \int_{a_r}^{b_r} g_s p_{rs} y_r \, dx_r, \tag{10.13.3}$$

for $r = 1, \ldots, k$. Here, for any such r, the expressions on each side are functions of the x_t, where $t \neq r$.

Choosing now j, r, with $1 \leq j, r \leq k$, we multiply (10.13.3) by

$$P_{rj} \prod_{t \neq r} y_t(x_t),$$

where P_{rj} is the co-factor of p_{rj}, and integrate with respect to the $x_t, t \neq r$, and finally sum over r. This gives

$$\lambda_j \int f p \prod y_r \, dx = \int g_j p \prod y_r \, dx, \tag{10.13.4}$$

where $p = \det p_{rs}$, $1 \leq j \leq k$ and the product extends over all indices $r = 1, \ldots, k$. This is the desired relationship.

We may now deduce a Bessel inequality for each of the g_j, or for that matter for any linear combination of them. Going over to a compact notation, we define

$$c_n = \int p y^{(n)} f \, dx, \quad d_{jn} = \int p y^{(n)} g_j \, dx, \quad \rho_n = \int p |y^{(n)}|^2 \, dx.^5 \tag{10.13.5}$$

---

[5] The construction ends here in all manuscripts. For further details, see the Notes.

10.14   Finite difference approximations

With the y_r defined as in (10.3.1)–(10.3.2), for an unbounded sequence of integers $m \to \infty$, we form finite-difference approximations to these functions, denoted by

$$z_r(x_r) = z_r^{(m)}(x_r), \quad a_r \leq x_r \leq b_r, \ r = 1, \ldots, k \tag{10.14.1}$$

the superscript "(m)" will often be omitted. To form these, we first divide the intervals $[a_r, b_r]$ into m equal parts, of length

$$h_r = h_r^{(m)} = (b_r - a_r)/m \tag{10.14.2}$$

at nodes $x_{rj} = x_{rj}^{(m)}$ such that

$$a_r = x_{r0} < \cdots < x_{rm} = b_r. \tag{10.14.3}$$

For simplicity, we take m to be independent of r.

As in the one-parameter case, the z_r are initially defined at these nodes, starting with the initial data

$$z_r(x_{r0}) = \sin \alpha_r, \quad z_r(x_{r1}) = \sin \alpha_r + h_r \cos \alpha_r. \tag{10.14.4}$$

We then continue the definition at the nodes with the aid of recurrence relations

$$h_r^{-2} \Delta^2 z_r(x_{rj}) + z_r(x_{rj}) \left\{ \sum_s \lambda_s p_{rs}(x_{rj}) - q_r(x_{rj}) \right\} = 0, \tag{10.14.5}$$

where $j = 1, \ldots, m - 1$.

Finally, we complete the definition of z_r between the nodes by linear interpolation; z_r will be continuous and piecewise linear, z_r' being constant except at the nodes.

The functional dependence is expressed in full by

$$z_r^{(m)}(x_r) = z_r^{(m)}(x_r; \lambda_1, \ldots, \lambda_k).$$

At the node x_{rj}, this will be a polynomial in $\lambda_1, \ldots, \lambda_k$ of degree at most $j - 1$.

Corresponding to the boundary condition at b_r, we impose the condition

$$z_r(x_{rm}) \cos \beta_r - h_r^{-1}\{z_r(x_{rm}) - z_r(x_{r,m-1})\} \sin \beta_r = 0, \quad r = 1, \ldots, k. \tag{10.14.6}$$

Here the left is a polynomial in the λ_s, of degree at most $m - 1$.

The eigenvalues, k-tuples $\lambda_1, \ldots, \lambda_k$, will be common zeros of the equations (10.14.6). We denote them by

$$\mu_n^{(m)} = \mu_{1n}^{(m)}, \ldots, \mu_{kn}^{(m)}, \tag{10.14.7}$$

or more simply by $\mu_n = \mu_{1n}, \ldots, \mu_{kn}$, where n runs through some index set. In place of (10.12.7), we have

$$\sum_{s=1}^{k} \sum_{j=1}^{m-1} (\mu_{sn} - \mu_{sn'}) p_{rs}(x_{rj}) z_r(x_{rj}, \mu_n) z_r(x_{rj}, \mu_{n'}) = 0, \quad n \neq n',$$

where $r = 1, \ldots, k$, and so, if the eigenvalues are distinct, we have the orthogonality relation

$$\det \sum_{j=1}^{m-1} p_{rs}(x_{rj}) z_r(x_{rj}, \mu_n) z_r(x_{rj}, \mu_{n'}) = 0,$$

and so, by (4.2.4),

$$\sum_{u_1=1}^{m-1} \cdots \sum_{u_k=1}^{m-1} \det p_{rs}(x_{ru_r}) \prod_{r=1}^{k} z_r(x_{ru_r}, \mu_n) \prod_{r=1}^{k} z_r(x_{ru_r}, \mu_{n'}) = 0, \quad n \neq n'.$$

These constitute orthogonality relations between the eigenfunctions, which are now products

$$\prod_{r=1}^{k} z_r(x_{ru_r}, \mu_n) \tag{10.14.8}$$

defined on the k-dimensional grid

$$x_{1u_1}, \ldots, x_{ku_k}, \quad u_j = 1, \ldots, m - 1, \quad j = 1, \ldots, k. \tag{10.14.9}$$

At this point, a new element enters the argument, in that we need to use the completeness of the discrete eigenfunctions (10.14.8) on the set (10.14.9). Assuming this, we can set up the finite-dimensional Parseval equality. We define for the Fourier coefficients of f,

$$\gamma_n = \sum_{u_1} \cdots \sum_{u_k} \det p_{rs}(x_{ru_r}) f(x_{1u_1}, \ldots, x_{ku_k}) \prod z_r(x_{ru_r}, \mu_n)$$

and the normalization constants

$$\sigma_n = \sum_{u_1} \cdots \sum_{u_k} \det p_{rs}(x_{ru_r}) \prod z_r^2(x_{ru_r}, \mu_n).$$

The Parseval equality then takes the form

$$\sum_{u_1} \cdots \sum_{u_k} \det p_{rs}(x_{ru_r})|f(x_{1u_1},\ldots,x_{ku_k})|^2 = \sum |\gamma_n|^2/\sigma_n, \qquad (10.14.10)$$

and as before we have to make $m \to \infty$, and to show that the series on the right is uniformly convergent with respect to m.

For this we need to introduce k auxiliary functions w_s on the grid (10.14.9) by

$$h_r^{-2}\Delta_r^2 f - q_r f + \sum_{s=1}^{k} p_{rs}w_s = 0, \qquad r = 1,\ldots,k. \qquad (10.14.11)$$

This we may do since $\det p_{rs} \neq 0$. Following the steps of Section 10.12 we multiply by z_r and sum over the grid-values of x_r, the other x_t, $t \neq r$, remaining constant. We have

$$\sum (\Delta_r^2 f)\, z_r = \sum f(\Delta_r^2 z_r), \qquad (10.14.12)$$

and so we get

$$\sum \sum \lambda_s p_{rs} z_r f = \sum \sum p_{rs} z_r w_s, \qquad r = 1,\ldots,k, \qquad (10.14.13)$$

where one summation is over $s = 1,\ldots,k$, and the other over the x_{ru_r}. Both expressions are functions of the x_{vr}, $v \neq r$.

For some t, $1 \leq t \leq k$, we multiply (10.14.13) by

$$P_{rt} \prod_{v\neq r} z_v,$$

formed with arguments x_{vu_v}, $v \neq r$, sum over these variables, and finally sum over r. This gives

$$\mu_{sn}\gamma_n = \delta_{sn}{}^6$$

where the γ_{sn} are defined by the right-hand side of (10.14.13).

Notes for Chapter 10

For basic results on eigenfunction expansions in the one-parameter case Section 10.2, the reader may consult Atkinson (1964a) and Atkinson (1972).

An expansion theorem for periodic multiparameter Sturm-Liouville equations on a finite interval with boundary conditions of the form $y_r(T_r) = y_r(0) \exp(it_r)$,

---

[6]Curiously enough, this construction ends here in all manuscripts as well. For further discussion, see the Notes.

$y'_r(T_r) = y'_r(0) \exp(it_r)$, $r = 1, 2, \ldots, k$ was given by Guseĭnov (1980) with a similar result for finite difference equations (a topic to be considered in a later chapter).

The point of view that treats the question of an expansion theorem for functions in $C^2(0,1)$ via partial differential equations has been taken up by Faierman (1978) in the two-parameter case; see also many further references by this author in the bibliography regarding this question. The construction of a spectral measure in the three-parameter case is accomplished by Almamedov and Aslanov (1986a), (1986b). In particular, Faierman and Roach (1988b) consider the question of full and half-range expansions in the general multiparameter case.

An abstract expansion theorem is presented by Konstantinov (1994); see also Konstantinov (1995) and Konstantinov and Stadnyuk (1993). The uniform convergence of the expansion is studied by Rynne (1990) in the case of sufficiently differentiable coefficients. A nice though now dated survey on the eigenfunction expansion question is given by Volkmer (1984b).

Comments on the proof of Theorem 10.9.4: The theorem itself may be proved using discrete approximations as was done originally by Faierman (1966), (1969). The idea is that the transformation described in §4.11 that leads to (4.11.3-4.11.4) shows that we may assume without loss of generality that $p_{12}(x) < 0$. Hence the partial differential operator $-A$ defined in (10.9.8) is elliptic. Now the operator B is invertible, by assumption, so it follows that $B^{-1}A$ is symmetric in the rigged $(L^2, ())$ space with inner product defined by $[f,g] := (Bf, g)$. It is then a matter of showing that the operator $B^{-1}A$ is self-adjoint and the result follows. One possible continuation of the argument presented here can be found in either Faierman (1991b, Chapter 2 and pp. 57–59), Volkmer (1988, §6.8) or even Berezans'kiĭ (1968).

The technique being described, that is, of rewriting the problem as a generalized (symmetric) eigenvalue problem for a pair of operators (i.e., a partial differential operator and a positive multiplication operator) is one that has been exploited successfully in a number of subsequent papers by Faierman (1981a, 1981b, 1983a, 1985, 1986).

Comments on the construction in Section 10.12:

Note that the right-hand side of (10.12.17) is $O(\Lambda^{-2})$ since the product of the remaining terms are $O(1)$, as this is an approximation of the integral

$$\int p(x)g(x)^2 \, dx,$$

where $g(x_j) = w_j$ and this integral is finite by virtue of (10.11.7) and the assumptions on g at the outset. Since $|\mu_n| \geq \Lambda$ we get from (10.12.17) the lower bound

$$\sum_{|\mu_n|\geq\Lambda} \gamma_n^2/\sigma_n = O(\Lambda^{-2}),\qquad\qquad (10.14.14)$$

as $\Lambda \to \infty$, where the O-term is uniform in m. Thus,

$$\lim_{\Lambda\to\infty}\sup_m \sum_{|\mu_n|\geq\Lambda} \gamma_n^2/\sigma_n = 0.\qquad\qquad (10.14.15)$$

Combining (10.14.15) with (10.12.16) we deduce that

$$\sum_{n=0}^{m-2} \gamma_n^2/\sigma_n \to \sum_{n=0}^{\infty} c_n^2/\rho_n,\quad m\to\infty.$$

Hence, by (10.12.10), (10.11.6) and the fact that if the Parseval equality holds for a subset of Fourier coefficients of f then by the Bessel inequality it holds for all Fourier coefficients, we see that the Parseval equality holds for all functions that are $C^2[a,b]$ and vanish along with their first derivatives at both a,b. Since these are dense in $L_p^2[a,b]$ the completeness follows.

Comments on the construction in Section 10.13: We proceed as in the previous paragraph noting that Atkinson's "compact notation" helps to bring about a major simplification in the derivation of the completeness question. This notation is enforced in the sequel.

First we observe that by the Klein oscillation theorem the eigenvalues can be uniquely indexed according to the oscillation numbers of the non-trivial solutions of (5.2.1–5.2.2). Thus we can suppose that they are indexed serially and denote them by $\lambda_{1n},\ldots,\lambda_{kn}$, for $n=0,1,2,\ldots$, with their associated (real) solutions being denoted by $y_{1n}(x_1),\ldots,y_{kn}(x_k)$. The quantities c_n,d_{jn},ρ_n defined in (10.13.5) have

$$y^{(n)} = \prod_{r=1}^{k} y_{rn}(x_r).$$

Now the Parseval equality takes the form

$$\int P|f|^2\,dx = \sum |c_n|^2/\rho_n,\qquad\qquad (10.14.16)$$

because of the presence of the ρ_n-term in the general form (10.11.3). On the other hand, the Bessel inequality (which is clear) takes the form of the preceding equation with equality replaced by $\geq$. The equation that corresponds to (10.11.8) is (10.13.4), or

$$d_{jn} = \lambda_{jn}c_n.$$

The latter, when used in conjunction with the Bessel inequality, gives

$$\int p|g_j|^2\,dx \geq \sum |d_{jn}|^2/\rho_n = \sum (\lambda_{jn}c_n)^2/\rho_n.$$

As before, for $\Lambda > 0$ and $|\lambda_{jn}| \geq \Lambda$ we derive that (cf., (10.11.10))

$$\sum_{|\lambda_{jn}| \geq \Lambda} |c_n|^2/\rho_n \leq \Lambda^{-2} \int P|g_j|^2 \, dx.$$

We now move on to the last remaining gap, the one to be found at the end of Section 10.14.

Comments on the construction at the end of Section 10.14: The quantities μ_{sn}, δ_{sn} are defined, respectively, by (10.14.7) and

$$\delta_{sn} = \sum_{u_1} \cdots \sum_{u_k} \{\det p_{rs}(x_{ru_r})\} w_s(x_{1u_1}, \ldots, x_{ku_k}) \prod z_r(x_{ru_r}, \mu_n), \quad (10.14.17)$$

that is, the Fourier coefficient of w_s (of course this depends on m). We follow closely the argument of Section 10.12 leading to (10.12.16), which, when applied to this case, shows that

$$\gamma_n^2/\sigma_n \to c_n^2/\rho_n \text{ as } n \to \infty$$

(even if for a subsequence only). In order to prove the continuous Parseval equality (10.14.16), it is sufficient to show that

$$\lim_{N \to \infty} \sup_m \sum_{n \geq N} |\gamma_n^2/\sigma_n| = 0, \quad (10.14.18)$$

since this and the previous display together give that

$$\lim_{N \to \infty} \sum_{n=0}^{N-2} \gamma_n^2/\sigma_n = \sum_{n=0}^{\infty} c_n^2/\rho_n$$

which on account of (10.14.10) gives the desired result. Note that (10.14.18) holds if and only if

$$\lim_{\Lambda \to \infty} \sup_m \sum_{\sum_{s=1}^k |\mu_{sn}| > \Lambda} |\gamma_n^2/\sigma_n| = 0. \quad (10.14.19)$$

Let $\Lambda > 0$. If $\sum_{s=1}^k |\mu_{sn}| > \Lambda$, then there exists at least one s, $1 \leq s \leq k$, such that $|\mu_{sn}| > \Lambda/k$. Applying the Bessel inequality to w_s, we get

$$\sum_{u_1} \cdots \sum_{u_k} \{\det p_{rs}(x_{ru_r})\} |w_s(x_{1u_1}, \ldots, x_{ku_k})|^2 \geq \sum |\delta_{sn}|^2/\sigma_n$$

$$= \sum |\mu_{sn}\gamma_n|^2/\sigma_n$$

$$\geq \Lambda^2 k^{-2} \sum |\gamma_n^2/\sigma_n|.$$

But the left side is bounded uniformly as $m \to \infty$, by our assumptions on f and so on w_s, hence so is the right side upon taking the supremum and then the limit. This yields (10.14.19) and so (10.14.18). The Parseval equality now holds for functions $f(x_1, \ldots, x_k)$ assumed to have continuous second-order partial derivatives, and that vanish in a neighborhood of the boundary of the box I. The result for L^2 follows from a density argument.

10.15   Research problems and open questions

1. Is the dimension of the solution space referred to in Theorem 10.9.2 finite dimensional regardless of the sign conditions on the coefficients?

2. Determine the extent of the validity of Theorem 11.6.1 in the case where for some r we have $\det p_{rs}(x_r) = 0$ identically on a subinterval of $a_r \leq x_r \leq b_r$.

10.13   Research problems and open questions

Chapter 11

Limit-Circle, Limit-Point Theory

11.1   Introduction

We first recall a standard formulation for the one-parameter case. Let $q(x)$, $a \le x < \infty$, be real and continuous. For the differential equation

$$y'' + (\lambda - q(x))y = 0, \quad a \le x < \infty, \tag{11.1.1}$$

we introduce the following basic classification:

(i) the equation is in the "limit-circle" condition at ∞ if all solutions are in $L^2(a, \infty)$,

(ii) the equation is in the limit-point condition at ∞ if there is a solution not in $L^2(a, \infty)$. It is known, from the fundamental work of Weyl, that the classification does not depend on the choice of the parameter λ; if all solutions are in $L^2(a, \infty)$ for some λ, real or complex, then this is the case for all λ.

The classification thus pertains to the differential operator $-d^2/dx^2 + q(x)$, acting on a space of suitably differentiable functions on (a, ∞). It leads to a classification of this operator as being "essentially self-adjoint," or not when acting on a space of suitably differentiable functions satisfying a Sturmian initial condition at $x = a$, and to the notion of the "Deficiency-indices" of such an operator.

The classification has a crucial effect on the eigenfunction expansions associated with (11.1.1) and an initial condition

$$y(a) \cos \alpha = y'(a) \sin \alpha. \tag{11.1.2}$$

In the limit-point case, there is just one such expansion; it need not take the form of a discrete series expansion. In the limit-circle case, there is a whole family of series expansions. These may be obtained by imposing a boundary condition at ∞, in addition to (11.1.2). To formulate this, we choose a family of solutions $y(x, \lambda)$ of (11.1.1) with $y(a, \lambda) = \sin \alpha$, $y'(a, \lambda) = \cos \alpha$, choose some real μ to be an eigenvalue, and then determine the remaining eigenvalues by asking that

$$y'(x, \lambda)y(x, \mu) - y(x, \lambda)y'(x, \mu) \to 0. \tag{11.1.3}$$

In an alternative procedure, we choose any $b > a$, any real β, and impose together with (11.1.2) the boundary condition

$$y(b) \cos \beta = y'(b) \sin \beta. \tag{11.1.4}$$

The idea is then to set up the associated eigenfunction expansion over (a, b), and then to proceed to the limit in some sense as $b \to \infty$, with β possibly varying with b. It turns out that under certain conditions we obtain a unique eigenfunction expansion in this way in the limit-point case, and a family of eigenfunction expansions, of series type, in the limit-circle case.

In the one-parameter case (11.1.1), we have followed the standard practice in taking the coefficient of λ to be unity. This covers many cases of practical importance, either directly or by way of a change of variables, and is convenient from the point of view of classical operator theory. However, this cannot be done in the multiparameter situation. For the generalization of (11.1.1) given by

$$y'' + (\lambda p(x) - q(x))y = 0, \quad a \le x < \infty, \tag{11.1.5}$$

where $p(x)$ is continuous and positive, the corresponding classification deals with whether for all solutions the integral

$$\int_a^\infty p(x)|y(x)|^2 \, dx \tag{11.1.6}$$

converges, to a finite limit. If so, we speak of the limit-circle case, and if not of the limit-point case. Again, the classification is independent of the choice of λ.

In the multiparameter case, an additional complication presents itself, in that the coefficients of the spectral parameters need not have fixed signs. However, the manner of dealing with this must be deferred to a later section.

11.2   Fundamentals of the Weyl theory

We outline here proofs of certain basic propositions. The first of these covers the property, already referred to in §11.1, that the limit-point, limit-circle classification of (11.1.1), or indeed (11.1.5), is independent of the choice of λ.

Theorem 11.2.1. *Let $g_1(x)$, $g_2(x)$, $a \leq x < \infty$, be continuous, with $h(x)$ real and non-negative. Let all solutions of*

$$y'' + g_1 y = 0, \quad a \leq x < \infty, \tag{11.2.1}$$

satisfy

$$\int_a^\infty h|y|^2 \, dx < \infty. \tag{11.2.2}$$

Then if

$$|g_2(x) - g_1(x)| \leq K h(x), \quad a \leq x < \infty, \tag{11.2.3}$$

for some constant K, all solutions of

$$y'' + g_2 y = 0, \quad a \leq x < \infty, \tag{11.2.4}$$

also satisfy (11.2.2).

For the proof, we use the method of variation of parameters. Let y_1, y_2 be a pair of linearly independent solutions of (11.2.1); for convenience, we choose them so that

$$y_1' y_2 - y_2' y_1 = 1. \tag{11.2.5}$$

For any solution y of (11.2.4), we set

$$y = u_1 y_1 + u_2 y_2, \quad y' = u_1 y_1' + u_2 y_2', \tag{11.2.6}$$

so that

$$0 = u_1' y_1 + u_2' y_2, \quad (g_1 - g_2)y = u_1' y_1' + u_2' y_2'.$$

Hence, using (11.2.5),

$$u_1' = (g_1 - g_2)yy_2, \quad u_2' = (g_2 - g_1)yy_1.$$

Using (11.2.6), we get the system of linear differential equations

$$\begin{aligned} u_1' &= (g_1 - g_2)y_1 y_2 u_1 + (g_1 - g_2)y_2^2 u_2, & (11.2.7) \\ u_2' &= (g_2 - g_1)y_1^2 u_1 + (g_2 - g_1)y_1 y_2 u_2. & (11.2.8) \end{aligned}$$

Here the coefficients of u_1, u_2 on the right are in $L(a, \infty)$, in view of (11.2.3), the Cauchy inequality and our assumption that y_1, y_2 satisfy (11.2.2). Hence, by a known theorem on linear differential systems u_1, u_2 tend to finite limits as $x \to \infty$, and in particular are bounded. It thus follows from the first of (11.2.6) and the boundedness of u_1, u_2 that y itself satisfies (11.2.2), as was to be proved.

Weyl's second main theorem on this topic has a different character.

Theorem 11.2.2. *Let $g(x)$, $a \le x < \infty$, be continuous, and let*

$$\operatorname{Im} g(x) \ge 0, \quad a \le x < \infty. \tag{11.2.9}$$

Then

$$y'' + gy = 0, \quad a \le x < \infty, \tag{11.2.10}$$

has a solution, not identically zero, such that

$$\int_a^\infty \operatorname{Im} g(x)|y(x)|^2 \, dx < \infty. \tag{11.2.11}$$

Let y_1, y_2 be linearly independent solutions of (11.2.10), with real initial data, say, with

$$y_1(a) = 1, \ y_1'(a) = 0, \ y_2(a) = 0, \ y_2'(a) = 1. \tag{11.2.12}$$

Let $b > a$ be such that

$$\int_a^b \operatorname{Im} g(x) \, dx > 0: \tag{11.2.13}$$

if there is no such b, that is to say, if $\operatorname{Im} g(x)$ vanishes identically, the proposition is trivial, and so we suppose that (11.2.13) holds for suitably large b. For such b, we denote by $C(b)$ the collection of m-values such that if

$$y = y_1 + my_2, \tag{11.2.14}$$

then

$$\operatorname{Im} \{y'(b)/y(b)\} = 0, \text{ or } y(b) = 0. \tag{11.2.15}$$

It will turn out that $C(b)$ is a circle in the complex plane, and that if $D(b)$ denotes the closed disc bounded by this circle, then $D(b)$ includes $D(b')$ for any $b' > b$.

We note first that if y satisfies (11.2.10) and is given by (11.2.14), then

$$
\begin{aligned}
y'(b)\overline{y(b)} - \overline{y'(b)}y(b) &= \{y'(a)\overline{y(a)} - \overline{y'(a)}y(a)\} + \\
&\quad + \int_a^b \{y''(x)\overline{y(x)} - \overline{y''(x)}y(x)\} \, dx \\
&= (m - \overline{m}) + \int_a^b \{\overline{g} - g\}|y|^2 \, dx. \tag{11.2.16}
\end{aligned}
$$

The set $C(b)$ may equally be described by (11.2.15) or by

$$y'(b)\overline{y(b)} - \overline{y'(b)}y(b) = 0 \qquad (11.2.17)$$

and so, in view of (11.2.16) by

$$\int_a^b \operatorname{Im} g(x)|y(x)|^2\, dx = \operatorname{Im} m. \qquad (11.2.18)$$

We denote by $D(b)$ the set of m satisfying

$$\int_a^b \operatorname{Im} g(x)|y(x)|^2\, dx \le \operatorname{Im} m. \qquad (11.2.19)$$

We have at once the "nesting" property that

$$D(b) \supset D(b') \text{ if } b < b'; \qquad (11.2.20)$$

the effect of increasing b in (11.2.19) is to sharpen the inequality, so that the admissible set of m is certainly not increased.

We claim next that the sets $D(b)$ are, for b satisfying (11.2.13), bounded and non-empty. For the first statement, we expand the left of (11.2.19), using (11.2.14), as

$$\int_a^b \operatorname{Im} g\{|y_1|^2 + \overline{m}y_1\overline{y_2} + m\overline{y_1}y_2 + |m|^2|y_2|^2\}\, dx.$$

Here the coefficient of $|m|^2$ is positive by (11.2.13), and so if we make $m \to \infty$ in any manner, the left of (11.2.19) tends to ∞ as $|m|^2$, while the right is of order m. Thus $D(b)$ is bounded, subject to (11.2.13).

To see that $D(b)$, indeed $C(b)$, is non-empty we note that it contains the point

$$m = -y_1(b)/y_2(b). \qquad (11.2.21)$$

Here $y_2(b) \ne 0$, again subject to (11.2.13); if $y_2(b) = 0$, we would obtain a contradiction on using (11.2.16) with y_2 replacing y.

Thus, assuming (11.2.13) to hold for some b and so for all large b, we have that the sets $D(b)$ constitute a family of bounded nesting closed non-empty sets, which must have a point in common. Let m_0 be one such point; it need not be unique. Then (11.2.19), with $y = y_1 + m_0 y_2$, is true for all b, and so is true if we make $b \to \infty$ on the left. Thus (11.2.11) holds for this solution, which is not identically zero. This completes the proof.

We give the geometrical refinements that justify the terms "limit-circle," "limit-point."

Theorem 11.2.3. *Under the assumptions of Theorem 11.2.2, and assuming (11.2.13) to hold for large b, there are two possible cases;*

(i) all solutions of (11.2.10) satisfy (11.2.11), and the circles $C(b)$ tend to a limit-circle, or

(ii) there is a solution of (11.2.10), which does not satisfy (11.2.11), and the circles $C(b)$ tend to a point.

We prove that $C(b)$ is indeed a circle, subject to (11.2.13), and will obtain the result by calculating its radius. Using (11.2.15) we have that $C(b)$ is the inverse image of the closed real axis under the fractional-linear map

$$w = \{y_1'(b) + zy_2'(b)\}/\{y_1(b) + zy_2(b)\}; \qquad (11.2.22)$$

this is a non-degenerate conformal map in view of (11.2.12) and has as its inverse

$$z = \{wy_1(b) - y_1'(b)\}/\{y_2'(b) - wy_2(b)\}. \qquad (11.2.23)$$

Here $C(b)$ is obtained by letting w describe the real axis; it is necessarily either a circle or a straight line, and cannot be the latter since, as we showed earlier, it is bounded.

Its radius will be half the greatest distance between any point (11.2.21), with real w, and the particular point (11.2.23) of $C(b)$. Using (11.2.12), we find that this radius is

$$(1/2) \sup |y_2(b)\{wy_2(b) - y_2'(b)\}|^{-1},$$

the "sup" being over all real w. Since, subject to (11.2.13), which ensures that $y_2(b) \neq 0$,

$$\min |wy_2(b) - y_2'(b)| = |y_2(b)||\mathrm{Im}\, \{y_2'(b)/y_2(b)\}|,$$

we have that the radius of $C(b)$ is

$$|2y_2^2(b)\,\mathrm{Im}\, \{y_2'(b)/y_2(b)\}|^{-1} = |2\,\mathrm{Im}\, \{y_2'(b)\overline{y_2(b)}\}|^{-1}.$$

Using (11.2.16) with y_2 replacing y, we have that the radius is

$$\left\{ 2 \int_a^b \mathrm{Im}\, g(x)|y_2(x)|^2\, dx \right\}^{-1}. \qquad (11.2.24)$$

We now discuss the two possibilities

$$\int_a^\infty \mathrm{Im}\, g(x)|y_2(x)|^2\, dx \; < \; \infty, \qquad (11.2.25)$$

$$\int_a^\infty \mathrm{Im}\, g(x)|y_2(x)|^2\, dx \; = \; \infty. \qquad (11.2.26)$$

In the first of these, we have case (i) of the theorem. For then the circles $C(b)$ contract to a circle, and the discs $D(b)$, which they bound, have a non-trivial disc in common. In particular, these discs have at least two points in common, so that there are at least two values of m that satisfy (11.2.19) for all b. This gives us two linearly independent solutions satisfying (11.2.11), so that all solutions satisfy (11.2.11). If (11.2.26) holds, then the radius of the circles $C(b)$ tends to 0, so that they tend along with the discs $D(b)$ to a point; in addition, of course, we have a solution not satisfying (11.2.11). This completes the proof.

11.3   Dependence on a single parameter

In this section, we interpret and extend the results of the last section for the case of an equation of the form

$$y''(x) + \{\lambda p(x) - q(x)\}y(x) = 0, \quad a \le x < \infty. \tag{11.3.1}$$

It will be assumed throughout this section that $p(x)$, $q(x)$ are continuous and real-valued, and that $p(x)$ is non-negative and not identically zero. From Theorems 11.2.1–11.2.3 we have immediately

Theorem 11.3.1. *For the equation* (11.3.1), *there are two possibilities. Either* (i) *all solutions of* (11.3.1) *satisfy*

$$\int_a^\infty p(x)|y(x)|^2 \, dx < \infty, \tag{11.3.2}$$

for all values of λ, real or complex, or alternatively (ii) *for every value of λ, real or complex, there is at least one solution not satisfying* (11.3.2). *In either case, if* $\operatorname{Im} \lambda \ne 0$, *there is at least one non-trivial solution satisfying* (11.3.2).

The case that (11.3.2) holds for all y and all λ is naturally termed the "limit-circle case"; that that for all λ there is a solution not satisfying (11.3.2), is the "limit-point case".

For the former case, we can attach a certain uniformity to the result (11.3.2). The basis for this is

Theorem 11.3.2. *Let* (11.3.1) *be in the limit-circle case. Let, for some fixed α, $y(x, \lambda)$ denote the solution of* (11.3.1) *such that*

$$y(a, \lambda) = \sin \alpha, \quad y'(a, \lambda) = \cos \alpha. \tag{11.3.3}$$

Let $z_1(x)$, $z_2(x)$ be solutions of

$$z''(x) - q(x)z(x) = 0, \quad a \le x < \infty, \tag{11.3.4}$$

such that

$$z_1(a) = 0, \quad z_1'(a) = 1, \quad z_2(a) = 1, \quad z_2'(a) = 0. \tag{11.3.5}$$

Then there are entire functions $U_1(\lambda)$, $U_2(\lambda)$, which do not vanish together, such that, as $x \to \infty$,

$$y'(x, \lambda)z_2(x) - y(x, \lambda)z_2'(x) \to U_1(\lambda), \tag{11.3.6}$$

$$y(x, \lambda)z_1'(x) - y'(x, \lambda)z_1(x) \to U_2(\lambda). \tag{11.3.7}$$

This follows by a refinement of the proof of Theorem 11.2.1. We define functions $u_1(x, \lambda)$, $u_2(x, \lambda)$ by

$$y(x, \lambda) = \sum_1^2 u_s(x, \lambda)z_s(x), \quad y'(x, \lambda) = \sum_1^2 u_s(x, \lambda)z_s'(x), \tag{11.3.8}$$

and so have, as in (11.2.7)–(11.2.8),

$$u_1'(x, \lambda) = -\lambda p(x)\{z_1(x)z_2(x)u_1(x, \lambda) + z_2^2(x)u_2(x, \lambda)\},$$
$$u_2'(x, \lambda) = \lambda p(x)\{z_1^2(x)u_1(x, \lambda) + z_1(x)z_2(x)u_2(x, \lambda)\}. \tag{11.3.9}$$

The initial condition (11.3.3) gives

$$u_1(a, \lambda) = \cos\alpha, \quad u_2(a, \lambda) = \sin\alpha. \tag{11.3.10}$$

This has the form of a vector-matrix system

$$u'(x, \lambda) = \lambda\, A(x)u(x, \lambda), \quad u(a, \lambda) = u_0 \ne 0, \tag{11.3.11}$$

where $u(x, \lambda)$ is a two-vector, and $A(x)$ is a two-by-two matrix of functions in $L(a, \infty)$. It follows from known results in differential equation theory that $u_1(x, \lambda)$, $u_2(x, \lambda)$ tend as $x \to \infty$ to entire functions $U_1(\lambda)$, $U_2(\lambda)$, uniformly in any compact λ-region; these limiting functions cannot vanish together, since the initial values (11.3.10) are not both zero.

In particular, $u_1(x, \lambda)$, $u_2(x, \lambda)$ are uniformly bounded for

$$a \le x < \infty, \quad |\lambda| \le \Lambda,$$

for any fixed Λ. An easy application of the Gronwall inequality to (11.3.9)–(11.3.10) yields the bound

$$|u_1(x;\lambda)| + |u_2(x;\lambda)| \leq |u_0| \exp\left\{2|\lambda| \int_a^\infty p\{|z_1|^2 + |z_2|^2\}dt\right\}, \quad (11.3.12)$$

where $|u_0| = |\cos\alpha| + |\sin\alpha|$. From this, we derive

Theorem 11.3.3. *For the solution $y(x, \lambda)$ of (11.3.1) defined by (11.3.3), the integral*

$$\int_a^\infty p(x)|y(x;\lambda)|^2 \, dx \qquad\qquad (11.3.13)$$

converges uniformly in any bounded λ-region.

For it follows from (11.3.12) that, for any fixed $\Lambda > 0$ and all $|\lambda| \leq \Lambda$, there is a finite $C(\Lambda) > 0$ such that

$$|y(x;\lambda)| \leq C(\Lambda)\{|z_1(x)| + |z_2(x)|\}. \qquad\qquad (11.3.14)$$

We turn to a different aspect. We recall that a differential equation, in the real domain, on a half-axis (a, ∞), is said to be "oscillatory" if every solution has an infinity of zeros, that is to say, if the set of zeros of any solution has no upper bound. In the contrary event, when there exists a solution that is ultimately of fixed sign as $x \to \infty$, the differential equation is "non-oscillatory." In the case of (11.3.1), with $p(x) \geq 0$, the Sturm comparison theorem shows that if (11.3.1) is oscillatory when $\lambda = \lambda_1$, and $\lambda_1 < \lambda_2$, then (11.3.1) is oscillatory when $\lambda = \lambda_2$. Thus the real λ-axis may be divided into at most two intervals, according to whether (11.3.1) is oscillatory or non-oscillatory. In the limit-circle case there is only one such interval.

Theorem 11.3.4. *In the limit-circle case, (11.3.1) is either oscillatory for all real λ, or non-oscillatory for all real λ.*

We show that if $\lambda_1 \neq \lambda_2$, and (11.3.1) is oscillatory when $\lambda = \lambda_1$, then it is oscillatory when $\lambda = \lambda_2$. By a change of λ-origin, we can arrange that $\lambda_1 = 0$, and so it will be sufficient to show that if (11.3.4) is oscillatory, then so is (11.3.1), for any real λ.

Supposing then that (11.3.4) is oscillatory, we have that $z_1(x)$, $z_2(x)$ have an infinity of zeros; these interlace, and each takes alternating signs at the zeros of the other. For the λ-value in question, we have that $U_1(\lambda)$, $U_2(\lambda)$ are not both zero. Suppose, for example, that $U_1(\lambda) \neq 0$, so that $u_1(x, \lambda) \neq 0$ for large x. It then follows from (11.3.8) that, for large x, $y(x, \lambda)$ takes alternating signs at the successive zeros of $z_2(x)$. Thus, $y(x, \lambda)$ has an infinity of zeros, as was to be proved.

11.4   Boundary conditions at infinity

In the limit-circle case, the formulation in terms of eigenvalues and eigenfunctions can be retained, with slight modifications. There are, at least, three ways of specifying the behavior at ∞, to take the place of the boundary condition (11.1.4). These proceed

(i) in terms of the functions $U_1(\lambda)$, $U_2(\lambda)$ of (11.3.6)–(11.3.7). For some real λ, we ask for λ such that

$$U_1(\lambda) \sin \gamma = U_2(\lambda) \cos \gamma. \tag{11.4.1}$$

(ii) in terms of the orthogonality. We designate some real μ to be an eigenvalue, and specify that λ will be an eigenvalue if

$$(\lambda - \mu) \int_a^\infty p(x)y(x, \lambda)y(x, \mu) \, dx = 0. \tag{11.4.2}$$

(iii) in terms of a Wronskian. Again fixing some real μ as an eigenvalue, we say that λ is an eigenvalue if

$$y(x, \lambda)y'(x, \mu) - y(x, \mu)y'(x, \lambda) \to 0, \quad x \to \infty. \tag{11.4.3}$$

In the limit-circle case, all these formulations make sense. In (11.4.1) the eigenvalues are given as the zeros of an entire function, in view of Theorem 11.3.2; they thus form a discrete set, without finite limit-points. As to (ii), we note that the integral in (11.4.2) exists, in view of (11.3.2). Passing to (iii), we note that the limit in (11.4.3) exists, since it is given by the integral in (11.4.2).

Furthermore, these formulations are equivalent. As we have just noted, (11.4.2)–(11.4.3) are mutually equivalent, and so we have to consider the relation of one of them to (11.4.1), for suitable γ. Using (11.3.8) and the fact that $z_1'z_2 - z_1z_2' = 1$, we have

$$y(x, \lambda)y'(x, \mu) - y(x, \mu)y'(x, \lambda) = u_1(x, \mu)u_2(x, \lambda) - u_1(x, \lambda)u_2(x, \mu). \tag{11.4.4}$$

Thus (11.4.3) is equivalent to

$$U_1(\mu)U_2(\lambda) = U_1(\lambda)U_2(\mu), \tag{11.4.5}$$

which is equivalent to a condition of the form (11.4.1).

Conversely, a condition of the form (11.4.1) is equivalent to (11.4.2)–(11.4.3), for suitable μ. To see this, we use a substitution of polar coordinate type, defining a continuous $\varphi(x, \lambda)$ by

$$\varphi(x; \lambda) = \arg\{u_1(x, \lambda) + iu_2(x, \lambda)\}, \quad u(a, \lambda) = \alpha. \tag{11.4.6}$$

Write also
$$\psi(\lambda) = \lim \varphi(x, \lambda) = \arg\{U_1(\lambda) + iU_2(\lambda)\}. \tag{11.4.7}$$

In particular, we have
$$\varphi(x, 0) = \psi(0) = \alpha. \tag{11.4.8}$$

What we have to prove is that $\psi(\lambda)$ takes every value, $\mathrm{mod}\,\pi$, as λ increases on the real axis. In fact, $\psi(\lambda)$ is monotonic increasing, and tends to ∞ with λ.

This may be seen by considering the differential equation satisfied by $\varphi(x, \lambda)$, as a function of x; using (11.3.9), we obtain

$$\varphi'(x, \lambda) = \lambda\, p(x)(u_1z_1 + u_2z_2)^2/(u_1^2 + u_2^2). \tag{11.4.9}$$

This shows that if $\lambda > 0$, $\varphi(x, \lambda)$ is non-decreasing as a function of x, and that it is increasing at zeros of $p(x)y(x, \lambda)$; furthermore, it is increasing as a function of λ, if x is so large that

$$\int_a^x p(t)dt > 0. \tag{11.4.10}$$

Choosing $b > a$ so that (11.4.10) holds when $x = b$, let λ^* denote the first positive eigenvalue of (11.3.1) with initial conditions (11.3.3) and terminal boundary condition

$$y(b; \lambda)\{z_1'(b)\cos\alpha + z_2'(b)\sin\alpha\} = y'(b, \lambda)\{z_1(b)\cos\alpha + z_2(b)\sin\alpha\};$$

this has been chosen so that we must have

$$m(b, \lambda^*) \equiv \alpha \quad \mathrm{mod}\,\pi.$$

Since $\varphi(b, \lambda)$ is increasing in λ, this implies in view of (11.4.7) that $\varphi(b, \lambda) \geq \pi + \alpha$, and here in fact equality must hold. Since $\varphi(x, \lambda)$ is non-decreasing in x, we have $\psi(\lambda) \geq \pi + \alpha$. By (11.4.7), we have that $\psi(\lambda)$ increases over an interval of length at least π, as λ increases over $(0, \lambda^*)$; this proves that (11.4.1) gives an eigenvalue problem of the type (11.4.2)–(11.4.3).

In (11.4.2)–(11.4.3), we use one eigenvalue μ to determine the remaining eigenvalues. The question arises of whether we get precisely the same set of eigenvalues if this role of μ is taken over by another member of the set determined by μ. That this is so may be seen from the version (11.4.5). Thus, in the limit-circle case, if two functions of the form $y(x,\lambda)$, for various λ, are orthogonal to a third, in the sense (11.4.2), then they are orthogonal to one another.

The eigenvalue condition (11.4.5) may be put in the form

$$\psi(\lambda) = \psi(\mu) + n\pi, \qquad (11.4.11)$$

where n is an integer, not necessarily positive. This may be used to number the eigenvalues according to the oscillatory behaviour of the eigenfunctions even when (11.3.1) is oscillatory.

11.5   Linear combinations of functions

The foregoing depended essentially on the positivity, or at least non-negativity, of the coefficient $p(x)$ of the spectral parameter λ in the differential equation (11.3.10). When extending the investigation to k differential equations in k parameters, we have to consider both the positivity of a determinant of coefficients, as in earlier chapters and also the positivity of linear combinations of the coefficients appearing in a single equation. We start by taking up this last topic in respect of positivity, and without reference to differential equations.

Let

$$p_s(x), \quad s = 1,\ldots,k, \quad a \le x < \infty, \qquad (11.5.1)$$

be a set of real-valued functions. We can then define a set $U^+ \subset \mathbf{R}^k$ as the collection of real k-tuples $(\mu_1,\ldots,\mu_k)$ such that

$$\sum_1^k \mu_s p_s(x) \ge 0, \quad a \le x < \infty. \qquad (11.5.2)$$

This will be closed as a point-set in $\mathbf{R}^k$. It will also be closed under the algebraic operations of addition and of multiplication by non-negative scalars. It may therefore be described as a cone; U^+ will in any case contain the origin $(0,\ldots,0)$. We shall be concerned with the richer situation described in

Theorem 11.5.1. *Let the functions* (11.5.2) *not all vanish together for any* x *in* $[a,\infty)$. *Then the following situations are equivalent:*

(i) U^+ *contains interior points,*

(ii) *there exist k linearly independent linear combinations of the $p_s(x)$, which satisfy (11.5.2),*

(iii) *there is a positive linear combination*

$$p(x) = \sum_{1}^{k} \rho_s p_s(x) \qquad (11.5.3)$$

such that, for certain constants $c_s > 0$,

$$|p_s(x)| \le c_s p(x), \quad a \le x < \infty, \qquad (11.5.4)$$

(iv) *there exist k linearly independent linear combinations of the $p_s(x)$, which are positive on $[a, \infty)$.*

If the above situations hold, then the interior of U^+ consists of those points $(\mu_1, \dots, \mu_k)$ such that for some positive A, B we have

$$Ap(x) \le \sum \mu_s p_s(x) \le Bp(x), \quad a \le x < \infty. \qquad (11.5.5)$$

Suppose first that (i) holds. Then U^+ contains an open set in $\mathbf{R}^k$, and so contains a set of k points

$$(\mu_{1j}, \dots, \mu_{kj}), \quad j = 1, \dots, k, \qquad (11.5.6)$$

which do not lie in any proper subspace, so that

$$\det \mu_{sj} \ne 0. \qquad (11.5.7)$$

Since these points are in U^+, we have

$$\sum_{1}^{k} \mu_{sj} p_s(x) \ge 0, \quad j = 1, \dots, k, \quad a \le x < \infty. \qquad (11.5.8)$$

This is the situation of (ii), and so we have (i) $\Rightarrow$ (ii).

Suppose next that (ii) holds. We take

$$p(x) = \sum_{1}^{k} \sum_{1}^{k} \mu_{sj} p_s(x). \qquad (11.5.9)$$

We then claim that

$$p(x) > 0, \quad a \leq x < \infty. \tag{11.5.10}$$

For it follows from (11.5.8) that $p(x) \geq 0$, and that equality can hold here only if we have equality in all of (11.5.8). This is excluded by (11.5.7), and the hypothesis that the $p_s(x)$ do not all vanish together. Using (11.5.7) again we have that there are numbers ν_{sj} such that

$$p(x) = \sum_{j=1}^{k} \nu_{sj} \sum_{t=1}^{k} \mu_{tj} p_s(x).$$

Since

$$0 \leq \sum \mu_{sj} p_s(x) \leq p(x),$$

by (11.5.8), (11.5.9), we deduce (11.5.4), with $c_s = \sum |\nu_{sj}|$. Thus (ii) $\Rightarrow$ (iii).

We next remark that (iii) $\Rightarrow$ (i). It follows from (11.5.4) that $(\rho_1, \ldots, \rho_k)$ is an interior point of U^+.

Finally, we consider (iv). It is trivial that (iv) $\Rightarrow$ (ii). In the converse direction, we know that (ii) $\Rightarrow$ (iii), so that it will be sufficient to show that (iii) $\Rightarrow$ (iv). This is proved by the remark that $\sum \mu_s p_s(x)$ will be positive if $\sum c_s |\rho_s - \mu_s| < 1$. In view of (11.5.3)–(11.5.4), the set of such $(\mu_1, \ldots, \mu_k)$, being open, certainly includes k linearly independent members.

Passing to the last assertion of the theorem, we suppose that the conditions (i)–(iv) all hold. We denote by $int\ U^+$ the interior of U^+. For every set $(\mu_1, \ldots, \mu_k)$, there will be a $B > 0$ satisfying the second of (11.5.5), in view of (11.5.4). Suppose first that there exists an $A > 0$ satisfying the first of (11.5.5). It then follows from (11.5.4) that $\sum \mu'_s p_s(x) > 0$ if $\sum |\mu'_s - \mu_s|$ is suitably small; we thus can conclude that $(\mu'_1, .., \mu'_k) \in int\ U^+$. Suppose conversely that the latter holds; we have to show that there is an $A > 0$ satisfying (11.5.5). Since $(\mu_1, \ldots, \mu_k) \in int\ U^+$, it can be expressed as a convex linear combination of k linearly independent elements of U^+, or for that matter as a sum of k such elements. Writing

$$p^{\dagger}(x) = \sum \mu_s p_s(x), \tag{11.5.11}$$

there will be a representation

$$p^{\dagger}(x) = \sum \sum \sigma_{sj} p_s(x),$$

where the σ_{sj} satisfy the same conditions (11.5.7)–(11.5.8) as the μ_{sj}. It then follows, as in (iii) above, that

$$|p_s(x)| \leq c_s^{\dagger} p^{\dagger}(x), \quad s = 1, \ldots, k. \tag{11.5.12}$$

From (11.5.3), we then have

$$p(x) \leq \sum |\rho_s| c_s^{\dagger} p^{\dagger}(x), \quad a \leq x < \infty,$$

and this is equivalent to the first to the first of (11.5.5). This completes the proof of Theorem 11.5.1.

A positive-valued linear combination of the $p_s(x)$, for which there exist c_s satisfying (11.5.4), may be termed "dominating." The ratio of two such linear combinations is bounded from above and away from zero.

11.6   A single equation with several parameters

Let the functions $p_s(x)$, $s = 1, \ldots, k$, $q(x)$, be real and continuous on $[a, \infty)$. We consider the equation

$$y''(x) + \left\{ \sum_{s=1}^{k} \lambda_s p_s(x) + q(x) \right\} y(x) = 0, \quad a \leq x < \infty. \qquad (11.6.1)$$

We say that it is of "limit-circle type" if every solution satisfies

$$\int_a^\infty |p_s(x)||y(x)|^2 \, dx < \infty, \quad s = 1, \ldots, k. \qquad (11.6.2)$$

In the contrary event, that for some solution one of (11.6.2) fails, we say that it is of "limit-point type." The classification is not dependent on the choice of the parameters. We have, as a simple application of Theorem 11.2.1,

Theorem 11.6.1. *For the equation* (11.6.1), *there are just two possibilities. For all sets of values of* $\lambda_1, \ldots, \lambda_k$, *real or complex, all solutions satisfy* (11.6.2). *Alternatively, for every set of values of the* λ_s, *there is a solution not satisfying all of* (11.6.2).

The limit-circle, limit-point terminology may be shown to be appropriate under mild additional conditions on the $p_s(x)$. Let us assume that there is a linear combination of the $p_s(x)$, which is non-negative, and which does not vanish identically. We can then justify the limit-circle notion, in the case (11.6.2). Under slightly stronger conditions, we can justify the limit-point term, if (11.6.2) fails.

We need first

Theorem 11.6.2. *Let* $\lambda_1, \ldots, \lambda_k$ *be such that either*

$$\text{Im} \sum \lambda_s p_s(x) \geq 0, \quad a \leq x < \infty, \qquad (11.6.3)$$

or

$$\text{Im} \sum \lambda_s p_s(x) \leq 0, \quad a \leq x < \infty. \tag{11.6.4}$$

Then (11.6.1) *has a non-trivial solution such that*

$$\int_a^\infty \left|\text{Im}\left\{\sum \lambda_s p_s(x)\right\}\right| |y(x)|^2 \, dx < \infty. \tag{11.6.5}$$

Suppose further that the left of (11.6.3)–(11.6.4) *is not identically zero. Let* $Y_1(x)$, $Y_2(x)$ *be solutions of* (11.6.1), *which are linearly independent, and which have fixed real initial data (e.g., as in* (11.3.5)). *Write* $y(x) = Y_1(x) + mY_2(x)$, *and let* $C(b)$ *be the set* m-*values such that cf.,* (11.2.15), (11.2.17),

$$\text{Im}\left\{y'(b)\overline{y(b)}\right\} = 0. \tag{11.6.6}$$

Then if b *is such that*

$$\int_a^b \text{Im} \sum \lambda_s p_s(x) \, dx \neq 0, \tag{11.6.7}$$

the set $C(b)$ *is a circle. As* b *increases, the circles* $C(b)$ *nest.*

This comes from Theorem 11.2.2. The "nesting" of the circles $C(b)$ is strict, as b increases, only if the absolute value of the integral in (11.6.7) increases.

For the limit-circle case, we have then

Theorem 11.6.3. *Let one of* (11.6.3)–(11.6.4) *hold, without identical equality. Let all solutions of* (11.6.1) *satisfy* (11.6.2). *Then the circles* $C(b)$ *of the last theorem tend to a circle, of non-zero radius, as* $b \to \infty$.

This follows from case (i) of Theorem 11.2.3.

In connection with the limit-point case, we recall the definition of the set U^+ satisfying (11.5.2), and the conditions given in Theorem 11.5.1 for it to have an interior. If it does so, there will exist dominating linear combinations of the $p_s(x)$, characterized by (11.5.4); we denote anyone of these by $p(x)$. The conditions (11.6.2) are then equivalent to

$$\int_a^\infty p(x)|y(x)|^2 \, dx < \infty, \tag{11.6.8}$$

so that the limit-point case is characterized by the existence of a solution such that

$$\int_a^\infty p(x)|y(x)|^2 \, dx = \infty. \tag{11.6.9}$$

Theorem 11.6.4. *Let int U^+ be non-empty, and let there be, for some set of parameters and so for all, a solution satisfying (11.6.9). Let one of the points*

$$(\text{Im } \lambda_1, \ldots, \text{Im } \lambda_k), \quad (-\text{Im } \lambda_1, \ldots, -\text{Im } \lambda_k) \qquad (11.6.10)$$

be in int U^+. Then the circles $C(b)$ described in Theorem 11.6.2 tend to a point as $b \to \infty$.

The proof proceeds by showing that for at least one solution of (11.6.1),

$$\int |\text{Im } \sum \lambda_s p_s(x)||y(x)|^2 \, dx = \infty. \qquad (11.6.11)$$

To see this, we remark that (11.6.9) holds for at least one solution, and that

$$\text{Im } \sum \lambda_s p_s(x) \geq A p(x), \qquad (11.6.12)$$

by (11.5.5), for some $A > 0$. This proves (11.6.11), for the same $y(x)$ as in (11.6.9). The result now follows from Theorem 11.2.3.

We conclude this section by noting two straightforward adaptions of results for a single parameter.

Theorem 11.6.5. *In the limit-circle case, (11.6.1) is either oscillatory for all real sets $\lambda_1, \ldots, \lambda_k$, or non-oscillatory for all such sets.*

This follows from Theorem 11.3.4. From Theorem 11.3.3, we have

Theorem 11.6.6. *Let $y(x; \lambda) = y(x; \lambda_1, \ldots, \lambda_k)$ denote a solution of (11.6.1) with $y(a, \lambda)$, $y'(a, \lambda)$ fixed, independently of λ. Then, in the limit-circle case, the integrals*

$$\int_a^\infty |p_s(x)||y(x; \lambda)|^2 \, dx, \quad s = 1, \ldots, k, \qquad (11.6.13)$$

converge uniformly in any bounded λ-region.

11.7 Several equations with several parameters

After these preliminaries, we come to our main concern, namely, a set of equations

$$y_r''(x_r) + \left\{ \sum \lambda_s p_{rs}(x_r) + q_r(x_r) \right\} y_r(x_r) = 0, \qquad (11.7.1)$$

$$a_r \leq x_r < \infty, \quad r = 1, \ldots, k.$$

Here the $p_{rs}(x_r)$ are real and continuous. In this chapter, as in the next, we take up the case in which all k underlying intervals (a_r, ∞) are infinite; the case in which some only are infinite would also be of interest.

For the purposes of eigenfunction expansions, we are interested in spaces of functions defined on the product of the intervals

$$a_r \le x_r < \infty, \quad r = 1, \dots, k. \tag{11.7.2}$$

The relevant scalar product is given, for two such functions, f, g by

$$(f,g) = \int_{a_1}^{\infty} \cdots \int_{a_k}^{\infty} f(x_1, \dots, x_k)\overline{g(x_1, \dots, x_k)} \det p_{rs}(x_r)\, dx_1 \cdots dx_k. \tag{11.7.3}$$

Accordingly, we assume that

$$\det p_{rs}(x_r) > 0 \tag{11.7.4}$$

in the domain (11.7.2) and are concerned with functions $f(x_1, \dots, x_k)$ such that $(f, f) < \infty$.

In this section, we discuss the simultaneous limit-circle case, in which all k equations (11.7.1) are in the limit-circle case. This will mean that

$$\int_{a_r}^{\infty} |p_{rs}(x_r)||y_r(x_r)|^2\, dx_r < \infty, \quad r, s = 1, \dots, k, \tag{11.7.5}$$

for all solutions of (11.7.1). By Theorem 11.6.1, it is sufficient that this be the case for one set of values of the λ_s, since it will then be true for all sets, real or complex. Under this hypothesis, eigenfunctions will have finite norm. This will follow from

Theorem 11.7.1. *Let the $p_{rs}(x_r)$ satisfy (11.7.4) and let all equations (11.7.1) be of limit-circle type. Let*

$$y(x_1, \dots, x_k) = \prod_{1}^{k} y_r(x_r), \tag{11.7.6}$$

be a product of solutions of (11.7.1). Then

$$(y, y) < \infty, \tag{11.7.7}$$

where the scalar product on the left is given by (11.7.3).

This is evident since, by Theorem 4.2.3, adapted to infinite intervals,

$$(y, y) = \det \int_{a_r}^{\infty} p_{rs}(x)|y_r(x_r)|^2\, dx_r, \tag{11.7.8}$$

and all entries in this determinant are well-defined by (11.7.5).

The same result can be put somewhat differently.

Theorem 11.7.2. *With the assumptions of Theorem 11.7.1 let $y(x_1, \ldots, x_k)$ be a product of the form (11.7.6) where the y_r are solutions of the simultaneous partial differential equations*

$$\partial^2 y/\partial x_r^2 + \left\{ \sum \lambda_s p_{rs}(x_r) + q_r(x_r) \right\} y = 0, \quad r = 1, \ldots, k \qquad (11.7.9)$$

in the region (11.7.2). Then (11.7.7) holds.

This is so since y will now be a linear combinations of products of the form (11.7.6).

The continuous-dependence properties of the limit-circle case have an obvious consequence for sets of equations, which is needed for the discussion of limiting spectral functions.

Theorem 11.7.3. *Let (11.7.1) all be in the limit-circle case. In (11.7.6), let $y_r(x_r) = y_r(x_r, \lambda_1, \ldots, \lambda_k)$ be solutions of (11.7.1) with initial values $y_r(a_r)$, $y_r'(a_r)$ independent of $\lambda_1, \ldots, \lambda_k$. Then (y, y) is a continuous function of the λ_s.*

This follows from Theorem 11.6.6.

Theorem 11.7.1 gives conditions under which all products (11.7.6) satisfy $(y, y) < \infty$, so that we obtain, for each $\lambda_1, \ldots, \lambda_k$, 2^k linearly independent products with this property. Of course, we get at least one such product if we assume, for the $\lambda_1, \ldots, \lambda_k$, the existence of at least one non-trivial solution for each of (11.7.1), which satisfies (11.7.5). More fully, we have

Theorem 11.7.4. *For some given set $\lambda_1, \ldots, \lambda_k$ let the r-th equation (11.7.1) have just ν_r linearly independent solutions satisfying the corresponding k inequalities in (11.7.5). Then the number of linearly independent products $y(x_1, \ldots, x_k)$ of solutions of (11.7.1), such that $(y, y) < \infty$, is not less than*

$$\prod_1^k \nu_r. \qquad (11.7.10)$$

This follows at once from the representation (11.7.8). If $\nu_r = 2$, $r = 1, \ldots, k$, as in Theorem 11.7.1, the estimate (11.7.10) is of course precise, since 2^k is the maximum number of linearly independent products of solutions of (11.7.1). However, it is not apparent that it is precise in other cases.

11.8   More on positive linear combinations

The notions of Section 11.5 can be applied to assist in clarifying the point just raised. For the array of real continuous functions $p_{rs}(x_r)$, and for each $r = 1, \ldots, k$, we define the set U_r^+ of real k-tuples $(\mu_1, \ldots, \mu_k)$ such that

$$\sum_1^k \mu_s p_{rs}(x_r) \geq 0, \quad a_r \leq x_r < \infty. \tag{11.8.1}$$

We denote by U_r^- the reversed set, which give ≤ 0 in place of ≥ 0 here. We continue to assume that

$$\det p_{rs}(x) > 0. \tag{11.8.2}$$

It is immediate from the latter assumption that the sets U_r^+ contain non-zero elements. To see this, let P_{rs} denote the co-factor of $p_{rs}(x_r)$ in $\det p_{rs}(x_r)$; P_{rs} will be a function of all of the $x_1, \ldots, x_k$, except x_r. Taking, for definiteness, $r = 1$, and

$$\mu_s = P_{1s}(x_2, \ldots, x_k), \quad s = 1, \ldots, k, \tag{11.8.3}$$

we then have, by (11.8.2),

$$\sum_1^k \mu_s P_{1s}(x_1) > 0, \quad a_1 \leq x_1 < \infty, \tag{11.8.4}$$

for all choices

$$x_r \in [a_r, \infty), \quad r = 2, \ldots, k. \tag{11.8.5}$$

Generally, we have, for any $r = 1, \ldots, k$,

$$(P_{r1}, \ldots, P_{rk}) \in U_r^+, \tag{11.8.6}$$

for all sets of $x_t \geq a_t$, $t \neq r$.

Furthermore, any collection of k sets, chosen one from each of U_r^+, U_r^-, $r = 1, \ldots, k$, has a non-zero element in common. This may be seen from Theorem 4.4.1. For any set of $b_r > a_r$, we can find a set of μ_s so that the linear combinations $\sum \mu_s p_{rs}(x_r)$ have assigned fixed signs for all x_r, $a_r \leq x_r \leq b_r$; the μ_s may be supposed to satisfy $\sum \mu_s^2 = 1$. We then make the $b_r \to \infty$, choosing a subsequence so that the μ_s converge.

We need assumptions on the $p_{rs}(x)$, which ensure slightly stronger properties, in particular that the sets U_r^+ have non-empty interiors, and that expressions such as (11.8.6) can be used to give points of these interiors.

One such assumption concerns the expression

$$V(x_1,\ldots,x_k) = \{\det p_{rs}(x_r)\} \prod_{r=1}^{k} \left\{ \sum_{s=1}^{k} |p_{rs}(x_r)|^2 \right\}^{-1/2}, \qquad (11.8.7)$$

which may be viewed as a sort of measure of the solid angle between the k vectors represented by the rows of the matrix $p_{rs}(x_r)$. Write also

$$W(b_1,\ldots,b_k) = \min V(x_1,\ldots,x_k), \qquad (11.8.8)$$

the minimum being over

$$a_r \le x_r \le b_r, \quad r = 1,\ldots,k. \qquad (11.8.9)$$

Thus $W(b_1,\ldots,b_k)$ will be a positive non-increasing function of each of the b_r. We have then

Theorem 11.8.1. *Let* $W(b_1,\ldots,b_k)$ *have a positive lower bound if all but one of the* b_r *is bounded above. Then the point* (11.8.6) *is in the interior of* U_r^+.

For some fixed j, $1 \le j \le k$, and some fixed set of $x_t \ge a_t$, $t \ne j$, we form the $P_{j1},\ldots,P_{jk}$, and have to show that if, for some $\varepsilon > 0$,

$$|\mu_s - P_{js}| < \varepsilon, \quad s = 1,\ldots,k, \qquad (11.8.10)$$

then

$$\sum_{s=1}^{k} \mu_s P_{js}(x_j) > 0$$

for all $x_j \ge a_j$. In fact

$$\sum_{s=1}^{k} \mu_s P_{js}(x_j) \ge \det p_{rs}(x_r) - \varepsilon \sum_{s=1}^{k} |P_{js}(x_j)|$$

$$\ge V(x_1,\ldots,x_k) \prod_{1}^{k} \left\{ \sum_{s=1}^{k} |p_{rs}(x_r)|^2 \right\}^{1/2} - \varepsilon\sqrt{k} \left\{ \sum_{s=1}^{k} |p_{js}(x_j)|^2 \right\}^{1/2},$$

and this is positive if for all $x_r \le b_r$,

$$\varepsilon\sqrt{k} < W(b_1,\ldots,b_k) \prod_{r \ne j}^{k} \left\{ \sum_{s=1}^{k} |p_{rs}(x_r)|^2 \right\}^{1/2}, \qquad (11.8.11)$$

which is the case for sufficiently small ε. In particular, U_j^+ will have non-empty interior if there holds either (11.8.11) or

$$\varepsilon\sqrt{k} < \det p_{rs}(x_r) \left\{ \sum_{s=1}^{k} |p_{js}(x_j)|^2 \right\}^{-1/2} \qquad (11.8.12)$$

for some fixed $x_r > a_r$, $r \neq j$, and all $x_j \geq a_j$. This proves the result.

There is an alternative criterion of a less computational nature, not involving $\det p_{rs}(x_r)$ explicitly, apart from the standing requirement that it take only positive values.

Theorem 11.8.2. *For each r, $1 \leq r \leq k$, let the family of vectors*

$$p_{r1}(x_r), \ldots, p_{rk}(x_r), \quad a_r \leq x_r < \infty. \tag{11.8.13}$$

not all lie in any proper subspace of $\mathbf{R}^k$. Then the U_r^+ have non-empty interiors.

The proof consists in specifying elements of these interiors. We consider for each $b_r > a_r$ the dimension of the least subspace of $\mathbf{R}^k$ containing the vectors

$$p_{r1}(x_r), \ldots, p_{rk}(x_r), \quad a_r \leq x_r \leq b_r, \tag{11.8.14}$$

and note that this dimension is an non-decreasing function of b_r. If the dimension were less than k for all finite $b_r > a_r$, the same would be true for the whole family (11.8.13), which is contrary to hypothesis. We suppose the b_r so large that the families (11.8.14) of vectors do not lie in proper subspaces of $\mathbf{R}^k$.

To simplify notational aspects, we treat as typical the case of U_1^+, and show that it has non-empty interior. With b_r sufficiently large, as above let $h(x_2, \ldots, x_k)$ be any continuous positive function on the set $a_r \leq x_r \leq b_r$, $r = 2, \ldots, k$. We take

$$\rho_s = \int_{a_2}^{b_2} \cdots \int_{a_k}^{b_k} P_{1s}(x_2, \ldots, x_k) h(x_2, \ldots, x_k)\, dx_2 \cdots dx_k, \quad s = 1, \ldots, k,$$
$$\tag{11.8.15}$$

and claim that

$$(\rho_1, \ldots, \rho_k) \in \text{int } U_1^+. \tag{11.8.16}$$

In view of (11.8.3)–(11.8.4), we certainly have

$$(\rho_1, \ldots, \rho_k) \in U_1^+$$

since h is positive-valued. Thus if (11.8.16) were not true, $(\rho_1, \ldots, \rho_k)$ would be in the boundary of U_1^+. There would be a non-zero linear functional on $\mathbf{R}^k$, which took the value zero on $(\rho_1, \ldots, \rho_k)$, and which took non-negative values on U_1^+, and in particular on vectors of the form (11.8.6). If such a functional is given componentwise as $(\mu_1, \ldots, \mu_k)$, we would therefore have

$$\sum_1^k \mu_s P_{1s}(x_2, \ldots, x_k) \geq 0, \quad \sum \mu_s \rho_s = 0.$$

By (11.8.15), we would then have

$$\sum_{1}^{k} \mu_s P_{1s}(x_2, \ldots, x_k) = 0, \quad a_r \leq x_r \leq b_r, \ r = 2, \ldots, k, \qquad (11.8.17)$$

that is to say,

$$\begin{vmatrix} \mu_1 & \cdots & \mu_k \\ p_{21}(x_2) & \cdots & p_{2k}(x_2) \\ \vdots & & \vdots \\ p_{k1}(x_k) & \cdots & p_{kk}(x_k) \end{vmatrix} = 0, \qquad (11.8.18)$$

for the same set of $x_2, \ldots, x_k$.

If we fix all but one of the $x_2, \ldots, x_k$, allowing the remaining one, x_r, say, to vary over (a_r, b_r), this states that some linear combination of $p_{r1}(x_r), \ldots, p_{rk}(x_r)$ vanishes over $[a_r, b_r]$. Since (11.8.14) is not to lie in any proper subspace, we have that the linear combination concerned must be the trivial one. In other words, the co-factors of all elements in the determinant (11.8.18), except those in the top row, are zero.[1]

If $k = 2$, this says that μ_1, μ_2 are zero, which is contrary to the hypothesis that (μ_1, μ_2) is a non-zero functional. This contradiction proves the theorem in this case.

If $k > 2$, we argue that since the co-factors of the elements in the bottom row of (11.8.18) are all zero, the first $(k-1)$ rows are linearly dependent. It follows that the top row can be expressed as a linear combination of the next $(k-2)$ rows; these $(k-2)$ rows are themselves linearly independent, by the standing hypothesis (11.8.2). The same argument applies to any subset of $(k-2)$ from the last $(k-1)$ rows in (11.8.18). The top row in (11.8.18) is therefore expressible as a linear combination of each $(k-2)$ of the last $(k-1)$ rows. If the top row is not zero, this implies that the last $(k-1)$ rows are linearly dependent, contrary to (11.8.2). This completes the proof of Theorem 11.8.2.

---

[1]Alternately, we can choose x_2 such that the first two rows of the determinant on the left of (11.8.18) are linearly independent. This is possible otherwise all possible second rows would lie in a one-dimensional space. Then we can choose x_3 such that the first three rows are linearly independent (or else all possible third rows would lie in a two-dimensional space) and so on. We eventually obtain a contradiction to (11.8.18) without resorting to co-factors [Volkmer].

11.9   Further integrable-square properties

In Section 11.7, we used the rather obvious deduction that if the integrals in (11.7.5) all converge, then so does the determinant (11.7.8). We consider now the converse deduction. This is possible under either of the conditions of Theorem 11.8.1–11.8.2. In what follows, we assume the p_{rs}, q_r real and continuous and to satisfy the requirement that $\det p_{rs}(x_r) > 0$.

Theorem 11.9.1. *Let the $p_{rs}(x_r)$ satisfy one of the following two conditions:*

(i) *for each $r = 1, \ldots, k$, the function $W(b_1, \ldots, b_k)$, defined in (11.8.8), is bounded from zero, when the b_j, $j \neq r$, are bounded above,*

(ii) *for each r, the family of vectors (11.8.13) is not contained in any proper subspace of $\mathbf{R}^k$.*

Then, if y is a product of continuous functions $y_r(x_r)$, not identically zero, and there hold,

$$\int_{a_1}^{\infty} \cdots \int_{a_k}^{\infty} |y(x_1, \ldots, x_k)|^2 \det p_{rs}(x_r)\, dx_1 \cdots dx_k < \infty, \qquad (11.9.1)$$

we have (11.7.5).

For this purpose, we do not need that the $y_r(x_r)$ satisfy any differential equations.

We give the proof of (11.7.5) in the case $r = 1$; the general case can of course be reduced to this.

The result (11.9.1) remains in force if any of the upper limits ∞ in the integrals are replaced by finite $b_r > a_r$. Thus, if

$$\rho_s = \int_{a_2}^{b_2} \cdots \int_{a_k}^{b_k} \prod_2^k |y_r(x_r)|^2 P_{1s}(x_2, \ldots, x_k)\, dx_2 \cdots dx_k, \quad s = 1, \ldots, k,$$
$$(11.9.2)$$

we have

$$\int_a^{\infty} \left\{ \sum_1^k \rho_s P_{1s}(x_1) \right\} |y_1(x_1)|^2\, dx_1 < \infty; \qquad (11.9.3)$$

here P_{rs} denotes, as previously, the co-factor of p_{rs} in $\det p_{rs}(x_r)$. The conclusion, (11.7.5) with $r = 1$, will follow if we show that

$$(\rho_1, \ldots, \rho_k) \in \text{int } U_1^+. \qquad (11.9.4)$$

Suppose first that we have the situation of Theorem 11.8.1. We take the $b_r > a_r$, $r = 2, \ldots, k$, suitably large, and write

$$X(b_2, \ldots, b_k) = \inf W(b_1, b_2, \ldots, b_k) \tag{11.9.5}$$

the "inf" being over $b_1 \geq a_1$. We have then, in the notation of (11.8.7)–(11.8.8),

$$V(x_1, \ldots, x_k) \geq X(b_2, \ldots, b_k) \tag{11.9.6}$$

provided that $a_r \leq x_r \leq b_r$, $r = 2, \ldots, k$. We have to show that for some $\varepsilon > 0$, and any set μ_s satisfying

$$\sum |\mu_s - \rho_s| < \varepsilon$$

we have

$$\sum_1^k \mu_s p_{1s}(x) \geq 0, \quad a_1 \leq x_1 < \infty$$

so that $(\mu_1, \ldots, \mu_k) \in U_1^+$. Equivalently, we must show that for some $\delta > 0$, we have

$$\sum \rho_s p_{1s}(x_1) \geq \delta \sqrt{\sum |p_{1s}(x_1)|^2}, \quad a_1 \leq x_1 < \infty. \tag{11.9.7}$$

On reference to (11.9.2), we see that the left of (11.9.7) is equal to

$$\int_{a_2}^{b_2} \cdots \int_{a_k}^{b_k} \prod_1^k |y_r(x_r)|^2 \det p_{rs}(x_r)\, dx_1 \cdots dx_k, \tag{11.9.8}$$

where the determinant extends as usual over $r, s = 1, \ldots, k$. This yields (11.9.7) with δ given by

$$X(b_2, \ldots, b_k) \prod_{r=2}^k \left\{ \int_{a_r}^{b_r} |y_r(x_r)|^2 \sqrt{\sum_{s=1}^k |p_{rs}(x_r)|^2}\, dx_r \right\}.$$

This will be positive if the $b_2, \ldots, b_k$ are so large that the $y_r(x_r)$ do not vanish identically over $[a_r, b_r]$, $r = 2, \ldots, k$.

This completes the proof of Theorem 11.9.1, if the assumptions on the $p_{rs}(x_r)$ are those of Theorem 11.8.1. The case (ii) of the theorem is covered by (11.8.15)–(11.8.16).

We deduce that under the conditions of Theorem 11.9.1, the estimate (11.7.10) for the number of linearly independent products of solutions of (11.7.1), satisfying $(y, y) < \infty$, is precise. In particular, all such products will satisfy $(y, y) < \infty$ if and only if all the equations (11.7.1) are of limit-circle type.

Notes for Chapter 11

This exposition differs from all the classical presentations of this subject (Coddington and Levinson ([955], Atkinson [1964a], etc.) Indeed, this construction leads immediately to the limit-point/limit-circle theory in the multiparameter case and so seems preferable from a pedagogical point of view.

The Weyl theory in the one-parameter case is now extremely well developed with a literature that is both extensive and important; it has seen applications to a multitude of problems in mathematical physics. The limit-point/limit-circle classification for multiparameter Sturm-Liouville problems was developed by Sleeman (1973b) and reconsidered by Atkinson (1977) who studied the deficiency index theory as well. A basic question is on how to best interpret the L^2-theory associated with a given differential equation (determination of the weight-function that induces the measure, etc.). Explicit criteria on the coefficients for limit-point/limit-circle appear to be lacking in the multiparameter case.

A corresponding Weyl theory for Volterra-Stieltjes integral equations is developed in Mingarelli (1983) with the aim of unifying the discrete and continuous versions of Sturm-Liouville arising in the one-parameter case. However, the multiparameter case does not seem to have been considered at all for such equations. The same considerations apply to the case of dynamic equations on time scales.

11.10   Research problems and open questions

1. Give a detailed analysis of the cases where some of the intevals in (11.7.1) are infinite while others are finite.

2. Is the estimate in Theorem 11.7.4 precise in cases where $\nu_r \neq 2$ and $r = 1, 2, \ldots, k$?

Chapter 12

Spectral Functions

12.1   Introduction

In this chapter, we discuss the eigenfunction expansions associated with a set
of Sturm-Liouville equations

$$y_r''(x_r) + \{\sum_1^k \lambda_s p_{rs}(x_r) + q_r(x_r)\}y_r(x_r) = 0, \qquad (12.1.1)$$

$$a_r \leq x_r < \infty, \quad r = 1, \ldots, k,$$

in the singular case, when the intervals are semi-infinite, and boundary conditions are given at the initial points a_r, only. We take these conditions in the
form

$$y_r(a_r)\cos\alpha_r = y_r'(a_r)\sin\alpha_r, \quad r = 1, \ldots, k, \qquad (12.1.2)$$

where the α_r are fixed and real, and without loss of generality, are subject to
$0 \leq \alpha_r < \pi$, $r = 1, \ldots, k$. As in the last chapter, we shall assume the $p_{rs}(x_r)$,
$q_r(x_r)$ real and continuous, and to satisfy

$$\det p_{rs}(x_r) > 0, \quad a_r \leq x_r < \infty, \quad r = 1, \ldots, k. \qquad (12.1.3)$$

The above represents only one of a great variety of problems. Clearly, it would
be possible to discuss mixed problems, in which some of the intervals are infinite
and some not, or problems in which some equations have singularities at both
ends.

We start by following a method that is standard in the one-parameter case, in which the general idea is to deduce an expansion theorem by a limiting process from the finite-interval case. We have said "an" rather than "the" expansion theorem, since there may be more than one such theorem. Whether there is in fact a multiplicity of expansion theorems depends on the limit-point, limit-circle classification.

12.2   Spectral functions

We must start by re-phrasing the eigenfunction expansion theorem for the finite-interval case, so as to be able to carry out more readily the limiting processes $b_r \to \infty$. The α_r in (12.1.2) are taken as fixed, and we write $y_r(x_r; \lambda)$ for the solution of (12.1.1) such that

$$y_r(a_r; \lambda) = \sin \alpha_r, \quad y_r'(a_r; \lambda) = \cos \alpha_r; \qquad (12.2.1)$$

here λ stands for the k-tuple $(\lambda_1, \ldots, \lambda_k)$. We write

$$y(x; \lambda) = \prod y_r(x_r; \lambda); \qquad (12.2.2)$$

here x denotes the k-tuple $(x_1, \ldots, x_k)$.

For any $b = (b_1, \ldots, b_k)$, where $a_r < b_r < \infty$, $r = 1, \ldots, k$, we write

$$\rho(b; \lambda) = \int_a^b p(x)|y(x; \lambda)|^2 \, dx; \qquad (12.2.3)$$

here $p(x) = \det p_{rs}(x_r)$, $\int_a^b = \int_{a_1}^{b_1} \cdots \int_{a_k}^{b_k}$, and $dx = dx_1 \cdots dx_k$.

Suppose now that we have boundary conditions at the b_r, say,

$$y_r(b_r) \cos \beta_r = y_r'(b_r) \sin \beta_r, \quad r = 1, \ldots, k. \qquad (12.2.4)$$

We denote the eigenvalues of the problem given by (12.1.1)–(12.1.2) and (12.2.4) by $\lambda^{(n)} = (\lambda_1^{(n)}, \ldots, \lambda_k^{(n)})$, where, as previously, n runs through k-tuples of non-negative integers. The $\lambda^{(n)}$ will, of course, depend on the b_r, β_r; where appropriate, we write $\lambda^{(n)} = \lambda^{(n)}(b; \beta)$.

The completeness of eigenfunctions, for the finite interval case, is then expressed by the Parseval equality; this states that, for a rather general class of functions $f(x) = f(x_1, \ldots, x_k)$, that

$$\int_a^b p(x)|f(x)|^2 \, dx = \sum |g(\lambda^{(n)})|^2 / \rho(b; \lambda^{(n)}), \qquad (12.2.5)$$

where $g(\lambda)$ is a sort of Fourier transform of $f(x)$, given by

$$g(\lambda) = \int_a^b f(x)y(x;\lambda)p(x)\,dx. \tag{12.2.6}$$

We need to make $b \to \infty$ (i.e., $b_r \to \infty$, $r = 1,\ldots,k$) for some convenient, yet adequately extensive class of $f(x)$, for example, those that are suitably differentiable and have compact support.

For this purpose, we rewrite (12.2.5) as a multi-dimensional Stieltjes integral, with respect to a distribution given by a certain function $\tau(b,\beta,\lambda)$. To define this, we specify first that $\tau = 0$ if one or more of the components λ_r of λ is zero. Supposing then that $\lambda_r \neq 0$, $r = 1,\ldots,k$, we define

$$\tau(b,\beta,\lambda) = \prod(\lambda_r/|\lambda_r|)\sum\{\rho(b;\lambda^{(n)})\}^{-1}, \tag{12.2.7}$$

where the sum is over eigenvalues $\lambda^{(n)}$ whose components $\lambda_r^{(n)}$, $r = 1,\ldots,k$, satisfy one of inequalities

$$\lambda_r \le \lambda_r^{(n)} < 0, \text{ or } 0 < \lambda_r^{(n)} \le \lambda_r. \tag{12.2.8}$$

We can now re-state the Parseval equality (12.2.5) in the form

$$\int_a^b p(x)|f(x)|^2\,dx = \int_{-\infty}^{\infty} |g(\lambda)|^2 d\tau(b,\beta,\lambda). \tag{12.2.9}$$

Here the integral on the right is extended over the entire real λ-region. The spectral function τ defines a measure in the following manner. We consider the box

$$\sigma_r < \lambda_r < \tau_r, \quad r = 1,\ldots,k. \tag{12.2.10}$$

Then the measure associated with this set is to be

$$\sum \pm\tau(b,\beta,\lambda), \tag{12.2.11}$$

where λ runs through all corners of the box, that is to say, points whose r-th coordinates are either σ_r or τ_r, and the $+$ or $-$ sign is taken according to whether an even or an odd number of the coordinates have the lower possible values σ_r. It may be shown that in the case of uncoupled parameters, when $p_{rs}(x_r) \equiv 0$ for $r \neq s$, the spectral function is equal to the product of the spectral functions of the individual problems.

12.3   Rate of growth of the spectral function

The possible behavior of the spectral function in the one-parameter case may be illustrated by the simplest example, that of the differential equation $y'' + \lambda y = 0$ over (a, b), with initial data $y(a, \lambda) = \sin \alpha$, $y'(a, \lambda) = \cos \alpha$, where $0 \le \alpha < \pi$. Simple direct calculations show that if $0 < \alpha < \pi$, the spectral function $\tau(b, \beta, \lambda)$ grows like $\lambda^{1/2}$ as $\lambda \to +\infty$, while if $\alpha = 0$, this has to be replaced by $\lambda^{3/2}$. This phenomenon extends to the multiparameter case. We have

Theorem 12.3.1. *Let the $p_{rs}(x_r)$, $q_r(x_r)$ be continuous, and let (12.1.3) hold. Let $0 \le \alpha_r < \pi$, $r = 1, \dots, k$. Then for any fixed set $b_{r0} > a_r$, $r = 1, \dots, k$, there is a constant A such that, if $b_r \ge b_{r0}$, $r = 1, \dots, k$, then*

$$|\tau(b, \beta, \lambda)| \le A\{1 + \sum |\lambda_s|^\gamma\}, \tag{12.3.1}$$

for all choices of the β_r, $r = 1, \dots, k$, where

$$\gamma = k/2 + \sum_{\alpha_r = 0} 1. \tag{12.3.2}$$

In other words, in forming the index γ, we are to count $1/2$ for every r such that $\alpha_r \in (0, \pi)$, and $3/2$ for every r such that $\alpha_r = 0$.

The proof follows arguments used in the one-parameter case. For $\Lambda > 0$, we consider a region of the form

$$|\lambda_s| \le \Lambda, \quad s = 1, \dots, k. \tag{12.3.3}$$

We apply the Parseval equality (12.2.9) (or (12.2.5)) to a function $f(x)$ to be chosen later, with the integral on the right restricted to the region (12.3.3); actually, the Bessel inequality would suffice for this purpose, the completeness of eigenfunctions not being required at this point. We get

$$\int_a^b p(x)|f(x)|^2 \, dx \ge \int_{-\Lambda}^{\Lambda} \cdots \int_{-\Lambda}^{\Lambda} |g(\lambda)|^2 d\tau(b, \beta, \lambda)$$

$$\ge \min |g(\lambda)|^2 \int_{-\Lambda}^{\Lambda} \cdots \int_{-\Lambda}^{\Lambda} d\tau(b, \beta, \lambda), \tag{12.3.4}$$

where the minimum is taken over (12.3.3). This gives

$$|\tau(b, \beta, \lambda)| \le \{\min |g(\lambda)|^2\}^{-1} \int_a^b p(x)|f(x)|^2 \, dx \tag{12.3.5}$$

subject to (12.3.3).

We now proceed to the details. We choose

$$f(x) = \prod f_r(x_r), \qquad (12.3.6)$$

where, for $r = 1$ to k,

$$f_r(x_r) = 1, \quad a_r \leq x_r < a_r + \delta, \qquad (12.3.7)$$

and

$$f_r(x_r) = 0, \quad x_r \geq a_r + \delta. \qquad (12.3.8)$$

Here δ has the form

$$\delta = \eta(1 + \Lambda)^{-1/2}, \qquad (12.3.9)$$

where η, to be chosen later, satisfies in any case

$$0 < \eta < \min_r (b_{r0} - a_r). \qquad (12.3.10)$$

We then have, if $b_r \geq b_{r0}$, $r = 1, \ldots, k$,

$$\int_a^b p(x)|f(x)|^2 \, dx \leq M_1 \eta^k (1 + \Lambda)^{-k/2}, \qquad (12.3.11)$$

where

$$M_1 = \max \det p_{rs}(x_r), \quad a_r \leq x_r \leq b_{r0}, \; r = 1, \ldots, k. \qquad (12.3.12)$$

We next arrange that $g(\lambda)$ should have a positive lower bound, subject to (12.3.3). As a first step, we suppose η chosen so that

$$y_r(x_r; \lambda) \neq 0, \quad a_r < x_r < a_r + \delta, \; r = 1, \ldots, k, \qquad (12.3.13)$$

again subject to (12.3.3). It will then follow from the special form chosen for f that

$$|g(\lambda)| \geq M_2 \prod \int_{a_r}^{a_r + \delta} |y_r(x_r; \lambda)| \, dx_r, \qquad (12.3.14)$$

where

$$M_2 = \min \det p_{rs}(x_r), \quad a_r \leq x_r \leq a_r + \delta, \quad r = 1, \ldots, k. \qquad (12.3.15)$$

Summing up so far, we have (cf., (12.3.5) and (12.3.11))

$$|\tau(b, \beta, \lambda)| \leq (M_1/M_2^2)\eta^k (1 + \Lambda)^{-k/2} \prod \left\{ \int_{a_r}^{a_r + \delta} |y_r(x_r; \lambda)| \, dx_r \right\}^{-2}, \qquad (12.3.16)$$

subject to (12.3.3), to $b_r \geq b_{r0}$, $r = 1, \ldots, k$, and to η being chosen so that (12.3.13) holds.

We have from (12.1.1)–(12.1.2) that

$$y_r(x_r) = \sin \alpha_r + (x_r - a_r) \cos \alpha_r - \int_{a_r}^{x_r} (x_r - t_r) \sum \lambda_s p_{rs}(t_r) y_r(t_r) dt_r.$$
(12.3.17)

If

$$M = |\max p_{rs}(x_r)|, \quad a_r \leq x_r \leq b_{r0}, \ r, s = 1, \ldots, k,$$

we have from (12.3.17) that, for $a_r \leq x_r \leq a_r + \delta$,

$$|y_r(x_r)| \leq \sin \alpha_r + \delta |\cos \alpha_r| + k\Lambda \delta M \int_{a_r}^{x_r} |y_r(t_r)| dt_r,$$

and so, by the Gronwall inequality,

$$|y_r(x_r)| \leq \{\sin \alpha_r + \delta |\cos \alpha_r|\} \exp(k\Lambda \delta^2 M).$$
(12.3.18)

We apply these first in the case $\alpha_r \in (0, \pi)$. If we impose on δ the requirements

$$\delta |\cos \alpha_r| < \sin \alpha_r, \quad k\Lambda \delta^2 M < 1,$$
(12.3.19)

it will follow from (12.3.18) that

$$|y_r(x_r)| \leq 2e, \quad a_r \leq x_r \leq a_r + \delta,$$

and so, by (12.3.17), that

$$|y_r(x_r) - \sin \alpha_r - (x_r - a_r) \cos \alpha_r| \leq 2k\Lambda \delta^2 eM.$$

If we strengthen the requirements (12.3.19) to

$$\delta |\cos \alpha_r| \leq (1/4) \sin \alpha_r, \quad k\Lambda \delta^2 eM \leq (1/8) \sin \alpha_r,$$
(12.3.20)

it will follow that

$$y_r(x_r) \geq (1/2) \sin \alpha_r, \quad a_r \leq x_r \leq a_r + \delta,$$

so that (12.3.13) holds, and

$$\int_{a_r}^{a_r+\delta} |y_r(x_r)| dx_r \geq (\delta/2) \sin \alpha_r.$$
(12.3.21)

To satisfy (12.3.20), we can choose η, independently of Λ, so that

$$\eta |\cos \alpha_r| \leq (1/4) \sin \alpha_r, \quad k\eta^2 eM \leq (1/8) \sin \alpha_r,$$
(12.3.22)

and we shall then have from (12.3.21) that

$$\int_{a_r}^{a_r+\delta} |y_r(x_r)| \, dx_r \geq (\eta/2)(1+\Lambda)^{-1/2} \sin \alpha_r. \tag{12.3.23}$$

Take next the case $\alpha_r = 0$. Imposing the condition $k\Lambda\delta^2 M < 1$ we have from (12.3.18) that $|y_r(x_r)| \leq \delta e$, and so from (12.3.17) that

$$|y_r(x_r) - (x_r - a_r)| \leq (x_r - a_r) \, k\Lambda\delta^2 Me.$$

We impose the stronger condition $k\Lambda\delta^2 Me < 1/2$ and have

$$y_r(x_r) \geq (x_r - a_r)/2, \quad a_r \leq x_r \leq a_r + \delta,$$

so that (12.3.13) will hold, and then

$$\int_{a_r}^{a_r+\delta} |y_r(x_r)| \, dx_r \geq \delta^2/4. \tag{12.3.24}$$

The condition $k\Lambda\delta^2 Me < 1/2$ will be satisfied if $k\eta^2 Me < 1/2$, and we shall then have, from (12.3.24), that

$$\int_{a_r}^{a_r+\delta} |y_r(x_r)| \, dx_r \geq (1/4)\eta^2/(1+\Lambda). \tag{12.3.25}$$

Summing up, we choose η to satisfy (12.3.10), the inequalities (12.3.22) for all r such that $\alpha_r \in (0, \pi)$, and $k\eta^2 Me < 1/2$; this can be done with η independent of Λ, of the β_r, and of the b_r provided that

$$b_r \geq b_{r0} > a_r, \quad r = 1, \ldots, k. \tag{12.3.26}$$

Inserting the bounds (12.3.23) or (12.3.25), as the case may be, in (12.3.16) we have

$$|\tau(b, \beta, \lambda)| \leq M_3(1+\Lambda)^\gamma,$$

where M_3 is independent of Λ, β_r and b_r, subject to (12.3.26), and

$$\gamma = -k/2 - 2 \left\{ \sum_{\alpha_r > 0} (-1/2) + \sum_{\alpha_r = 0} (-1) \right\}.$$

This is equivalent to (12.3.2). This completes the proof of Theorem 12.3.1.

We need two particular consequences.

Theorem 12.3.2. *Subject to* (12.3.1), $\tau(b, \beta, \lambda)$ *is uniformly bounded in the region* (12.3.3), *for any choice of the* β_r, *whether constant or varying with* b.

Theorem 12.3.3. *Subject to* (12.3.26), *the sum*

$$\sum 1/\rho(b; \lambda^{(n)}), \tag{12.3.27}$$

extended over the eigenvalues $\lambda^{(n)}$ *of* (12.1.1)–(12.1.2), (12.2.4), *which satisfy*

$$|\lambda_r^{(n)}| \leq \Lambda, \quad r = 1, \ldots, k, \tag{12.3.28}$$

is uniformly bounded, for any fixed Λ, *regardless of the choice of the* β_r, *whether constant or varying with* b.

12.4   Limiting spectral functions

We now use the boundedness properties proved in the last section as an aid to discussing the convergence properties of sequences of spectral functions

$$\tau(b^{(m)}, \beta^{(m)}, \lambda), \quad m = 1, 2, \ldots, \tag{12.4.1}$$

where the sequence of k-tuples

$$b^{(m)} = (b_1^{(m)}, \ldots, b_k^{(m)}), \tag{12.4.2}$$

is subject to

$$b_r^{(m)} > a_r, \quad b_r^{(m)} \to \infty, \tag{12.4.3}$$

and the $\beta^{(m)}$ are subject only to

$$0 < \beta_r^{(m)} \leq \pi, \quad r = 1, \ldots, k. \tag{12.4.4}$$

A real-valued function $\tau(\lambda)$ of the k real arguments $\lambda_1, \ldots, \lambda_k$ will be termed a limiting spectral function if we have, with pointwise convergence,

$$\tau(b^{(m)}, \beta^{(m)}, \lambda) \to \tau(\lambda), \tag{12.4.5}$$

as $m \to \infty$, through some sequence of $b^{(m)}, \beta^{(m)}$, satisfying the above conditions (12.4.3)–(12.4.4).

Two results on limiting spectral functions can be stated immediately. The question of whether such functions exist at all is covered by

Theorem 12.4.1. *Any sequence* (12.4.1) *subject to* (12.4.3)–(12.4.4), *contains a pointwise convergent subsequence.*

This follows from the k-dimensional version of the Helly selection principle, on the basis of the uniform boundedness established in the last section, together with the "positively monotonic" property. We recall this definition herewith for completeness.

Let $a = (a_1, \ldots, a_k)$, $b = (b_1, \ldots, b_k)$, where $b_i > a_i$ for all $i = 1, \ldots, k$. The box $[a, b]$ is defined by $[a, b] = \prod_{r=1}^{k} [a_r, b_r]$. For such an $I = [c, d]$, we use the notation

$$M_r f(x_1, \ldots, x_{r-1}, x_{r+1}, \ldots, x_k)$$

$$= f(x_1, \ldots, x_{r-1}, d_r, x_{r+1}, \ldots, x_k) - f(x_1, \ldots, x_{r-1}, c_r, x_{r+1}, \ldots, x_k).$$

A function f of several variables $(x_1, \ldots, x_k)$ is said to be positively monotonic on $[a, b]$ if for all $I = [c, d] \subseteq [a, b]$ there holds

$$M_1 M_2 \cdots M_k f \geq 0.$$

Certain properties of a limiting spectral function are inherited directly from the members of the sequences of which they are the limits.

Theorem 12.4.2. *A limiting spectral function is positively monotonic, and satisfies a bound of the form* (12.3.1).

Here γ is given by (12.3.2); the requirement that $b_r > b_{r0}$ will be satisfied by any sequence $b^{(m)}$ for large m, in view of (12.4.3).

If we simply want to prove that there exists at least one limiting spectral function, we may as well keep the $\beta^{(m)}$ fixed independently of m. However, in some cases, something is lost by this procedure. We discuss more flexible procedures in what follows.

12.5 The full limit-circle case

We recall that for the equations (12.1.1), it is possible for all, or none, or some only to be of limit-circle type. The uniqueness, and, so to speak, the discreteness, of limiting spectral functions may be expected to depend greatly on which of these situations hold. The present section is devoted to the case in which all equations are of limit-circle type. We assume that for all equations

(12.1.1), and all solutions y_r, not only those satisfying the initial conditions (12.1.2), we have

$$\int_{a_r}^{\infty} |p_{rs}||y_r|^2 \, dx_r < \infty, \quad r, s = 1, \ldots, k. \tag{12.5.1}$$

We recall that, by Section 11.7, we can then classify the equations (12.1.1) as oscillatory or non-oscillatory, regardless of the choice of the real λ_s; of course, the equations need not be all oscillatory, or all non-oscillatory together.

We term the "full limit-circle case" that in which all of (12.1.1) are of limit-circle type. This resembles closely the finite-interval case, and in a suitable formulation includes it. We assume, as usual, that the $p_{rs}(x_r)$, $q_r(x_r)$ are real and continuous and that $\det p_{rs}(x_r)$ is always positive. We then extend to the infinite-interval case the notion of boundary conditions and eigenvalues. In the oscillatory case, however, we do not get the same characterization of eigen-values in terms of oscillation numbers of zeros. However, these "generalized eigenvalues" will still have the property that they have no finite limit-point. In default of an oscillatory characterization of these eigenvalues, we suppose them numbered serially, in order of non-decreasing magnitude, in some sense.

The situation in which all the equations (12.1.1) are of limit-circle type resembles the finite interval case in several respects:

(i) the use of multi-dimensional Stieltjes integrals can be avoided, integrals over λ being replaced by discrete sums over a λ-set without finite limit-points,

(ii) there is an infinity of limiting spectral functions, corresponding to a one-parameter choice for each equation of a boundary condition at infinity,

(iii) there is an orthogonal set of eigenfunctions, satisfying these boundary conditions.

For a positively monotonic function $\tau(\lambda)$, where $\lambda = (\lambda_1, \ldots, \lambda_k)$, let us say that τ is of discrete type if for all λ, except for a denumerable set without finite points of accumulation, the corresponding measure of the rectangular parallelopiped given by the set of $(\mu_1, \ldots, \mu_k)$ such that $|\mu_s - \lambda_s| < \epsilon$, $s = 1, \ldots, k$, is zero for some $\epsilon > 0$. The exceptional points, such that the measure of such a box has a positive lower bound, for all $\nu > 0$, may be called the jump-points of τ, the lower bounds being the associated weights. The effect will be that a Stieltjes integral $\int h(\lambda) d\tau(\lambda)$, for continuous $h(\lambda)$ may be replaced by a discrete sum, over the jump-points of $\tau(\lambda)$, of the product of the value of $h(\lambda)$ and the associated weight.

We prove that limiting spectral functions are of this type, under the limit-circle hypothesis.

Theorem 12.5.1. *Let the $p_{rs}(x_r)$, $q_r(x_r)$ be real and continuous, and let $\det p_{rs}(x_r)$ be positive. Let all the equations (12.1.1) be of limit-circle type, in the sense of Section 11.7. Then all limiting spectral functions of the system (12.1.1)–(12.1.2) are of discrete type.*

We assume that we have the situation (12.4.1)–(12.4.5), so that $\tau(\lambda)$ is a limiting spectral function. We write $\lambda^{(n)}(b, \beta)$ for the eigenvalues associated with the Sturm-Liouville problem given by (12.1.1)–(12.1.2) and (12.2.4) and use the notation (12.2.3).

It follows from Theorem 12.3.1 that for any fixed $\Lambda > 0$, there is a constant C such that

$$\sum 1/\rho(b^{(m)}; \lambda) < C, \qquad (12.5.2)$$

where the summation is extended over those eigenvalues

$$\lambda^{(n)}(b^{(m)}, \beta^{(m)}), \qquad (12.5.3)$$

which lie in the region

$$|\lambda_s| \leq \Lambda, \quad s = 1, \ldots, k. \qquad (12.5.4)$$

Now it follows from Section 11.6–11.7 that the terms on the left of (12.5.1) are uniformly bounded, for fixed Λ. We deduce that the number of terms on the left of (12.5.2) is bounded as $m \to \infty$, for fixed Λ. Thus the number of eigenvalues $\lambda^{(n)}(b^{(m)}, \beta^{(m)})$ lying in any bounded λ-region is bounded as $m \to \infty$.

We now suppose the eigenvalues (12.5.3), where n runs through k-tuples of nonnegative integers, re-numbered serially as

$$\mu^{(j)}(b^{(m)}, \beta^{(m)}), \quad j = 0, 1, \ldots \qquad (12.5.5)$$

in order of magnitude in some suitable sense; for example, they may be ordered in such a way that $\sum |\lambda_s|$ is a non-decreasing function of j, as Λ runs through the sequence (12.5.3).

The sequences

$$\mu^{(j)}(b^{(m)}, \beta^{(m)}), \quad m = 1, 2, \ldots \qquad (12.5.6)$$

will either all be bounded, or else will be bounded up to some point, say, for $0 \leq j \leq j_0$, and unbounded for $j > j_0$. In either event, we can choose a subsequence of $m = 1, 2, \ldots$, so that all the sequence (12.5.5) either converge to finite limits, or tend to ∞; if the latter occurs, it will occur from some point onward. Actually, the latter possibility will not occur, but we do not need to assume this now.[1]

---

[1] This is left as an exercise to the reader.

Notes for Chapter 12

Here we consider the multiparameter analogue of an explicit expansion formula associated with the differential equation

$$y'' + (\lambda p - q)y = 0, \quad a \le x \le b;$$

where p, q are continuous over $[a, b]$ and $y(x, \lambda)$ will be a solution with fixed initial data at $x = a$. To sketch the method of Atkinson (1964a), we apply at the end-point b a family of boundary conditions

$$Ky(b, \lambda) \cos \beta = y'(b, \lambda) \sin \beta,$$

the condition at $x = a$ remaining fixed throughout, and then average the resulting expansions over $0 \le \beta \le \pi$.

This expansion formula takes the form, in Parseval equality version,

$$\int pf^2 \, dx = \int g^2(x, \lambda) \, d\tau(\lambda),$$

where

$$\tau'(\lambda) = \pi^{-1} \{Ky^2(b, \lambda) + K^{-1}y'^2(b, \lambda)\}^{-1}.$$

(See Atkinson [1964a, pp. 240–243]. The method of Atkinson [1964a] proceeds in the real domain, whereas Clark [1987], Bennewitz (1989) use the Titchmarsh-Weyl function. For matrix extensions, see Clark [1987], Clark and Gesztesy [2004] and the references therein.)

We consider here the analogue of this argument for the k-parameter case of a system

$$y_r{''}(x_r) + \left\{ \sum_1^k \lambda_s p_{rs}(x_r) + q_r(x_r) \right\} y_r(x_r) = 0, \quad r = 1, \ldots, k,$$

Here we ask that each equation have a solution in its respective finite interval $[a_r, b_r]$, not identically zero, and satisfying suitable boundary conditions. We use fixed initial conditions, say,

$$y_r(a_r) \cos \alpha_r = y_r'(a_r) \sin \alpha_r, \quad r = 1, \ldots, k,$$

while at b_r we use a condition with a parameter,

$$K_r y_r(b_r) \cos \beta_r = y_r'(b_r) \sin \beta_r, \quad r = 1, \ldots, k,$$

We set up the eigenfunction expansion, in Parseval form, and average over all the β_r, from 0 to π. Formally, the result is

$$\int \cdots \int p|f|^2 \, dx = \pi^{-k} \int \cdots \int |g(\lambda)|^2 \prod_1^k \{K_r y_r^2 + K_r^{-1} y_r'^2\}^{-1} \, d\lambda,$$

where

$$dx = dx_1 \cdots dx_k, \quad d\lambda = d\lambda_1 \cdots d\lambda_k, \quad p = \det p_{rs}(x_r),$$

$$g(\lambda) = \int pfy(x, \lambda) \, dx, \quad y = \prod y_r(x_r, \lambda).$$

For the proof, we use the fact that

$$\partial(\theta_1, \ldots, \theta_k)/\partial(\lambda_1, \ldots, \lambda_k)$$

$$= \prod_1^k Rr^{-2} \int_{a_1}^{x_l} \cdots \int_{a_k}^{x_k} \det p_{rs}(u_r)\{(y_r(u_r))^2 \, du_r\}. \qquad (12.5.7)$$

Here the phase angles θ_r are given by

$$K_r y_r(b_r) \cos \theta_r = y_r'(b_r) \sin \theta_r, \quad r = 1, \ldots, k,$$

and

$$R_r^2 = K_r y_r^2(b_r, \lambda) + K_r^{-1} y_r'^2(b_r, \lambda).$$

The basic ideas of this chapter are summarized, to some extent, in Atkinson (1977). Gadzhiev (1982) considers the multiparameter eigenvalue problem for Dirac operators along with an extension of the limit-point, limit-circle classification, in the two-parameter case. Another presentation of the Weyl theory is given by Isaev (1986b).

The extension of (12.2.6)–(12.2.9) in the singular case by means of a limiting precedure and use of Helly's selection theorem was considered by Browne (1972b). We note with interest that drafts of this book were available and in circulation as far back as 1972 as noted by Browne (1972a, Acknowledgment). Atkinson (1977, p. 10) states "It would seem that little is known about the nature as regards continuity or rate of growth, of these multi-dimensional spectral functions."

12.6 Research problems and open questions

1. *Find general criteria for limit-point or limit-circle in the multiparameter case*, much as in the one equation, one-parameter (Sturm-Liouville) case.

Current work by Mirzoev (2008) contains criteria for the limit-circle classification of a single Sturm-Liouville equation in several parameters. For example, the development of a criterion in the spirit of Levinson's (1949) classical limit-point theorem would be of interest.

2. Undertake a study of the continuity and rate of growth of the multidimensional spectral functions considered in this chapter (cf., the basic estimates in § 12.3).

3. In his 1972 doctoral dissertation, Stephen S. P. Ma (one of Atkinson's former students) studied the spectral theory of the following system of equations (which we will call a *Ma system* for simplicity).

Consider the matrix differential system (see § 2.12)

$$(\lambda A + \mu B + C)u = 0$$

coupled with a single Sturm-Liouville equation in two parameters

$$-(s(t)y')' + r(t)y = (\lambda p(t) + \mu q(t))y$$

and satisfying a set of homogeneous separated boundary conditions on the finite closed interval $I = [a, b]$, *viz.*

$$\sin \alpha y(a) - \cos \alpha s(a)y'(a) = 0,$$
$$\cos \beta y(b) - \sin \beta s(b)y'(b) = 0.$$

Here A, B, C are hermitian $n \times n$ matrices with complex valued entries and α, β are as usual in $(0, \pi]$. It is assumed that s, s', r, p, q are real valued continuous functions in I and we retain $p(t)y(t)$ as the motation for the vector function

$$p(t)y(t) = \mathrm{col}(p(t)y_1(t), \ldots, p(t)y_n(t)), \quad etc.$$

An eigenvalue of a Ma system is then defined to be a pair λ, μ such that both (12.6)-(12.6) have a nontrivial solution satisfying the boundary conditions. The eigenfunction is then by definition the vector function $y(t)u$.

In this little known work, Ma (1972a) he develops a corresponding Weyl theory for such systems, gives a completeness theorem, and produces the existence and uniqueness of the spectral function along with some criteria for the limit-point, limit-circle classification.

One possible question is: *Can one undertake such a study in the multiparameter case?*

Appendix on Sturmian Lemmas

A.1 Introduction

We are concerned here with the zeros of solutions of

$$y'' + \{f(x) + g(x)\}y = 0, \quad a \leq x \leq b, \tag{A.1.1}$$

where f, g are real, and g is in some sense small compared with f. We will assume that $g \in L(a, b)$, as we have in mind the situation that f becomes large on account of the presence of one or more parameters, as in the case

$$y'' + \{\lambda f(x) + g(x)\}y = 0, \quad \lambda \to \infty, \tag{A.1.2}$$

Here if f is positive, or positive in part, the number of zeros becomes unbounded as in the well-known expression

$$\pi^{-1}\sqrt{\lambda} \int_a^b \sqrt{\max\{f(x), 0\}} \, dx, \tag{A.1.3}$$

while if $f(x) < 0$, one is led to non-oscillation, in a certain sense. We first obtain specific results on the oscillatory case under various hypotheses, and then pass to the non-oscillatory one. The methods will be a combination of the Sturm comparison theorem and various Prüfer-type substitutions. We note that not all the results listed in this appendix are used in the text, some are here simply for interest's sake.

A.2 The oscillatory case, continuous f

For a continuous function f, the notation

$$\varphi = \operatorname{Re} \sqrt{f}, \quad \varphi(x) \ge 0, \tag{A.2.1}$$

will be used occasionally; note that φ is continuous whenever f is continuous. This formulation provides a convenient way of excluding intervals where $f(x)$ is negative.

Theorem A.2.1. *Let $f(x)$, $a \le x \le b$, be real and continuous, and let $N(a,b,y)$ denote the number of zeros in $(a,b]$ of y, a non-trivial solution of*

$$y'' + f(x)y = 0, \quad a \le x \le b. \tag{A.2.2}$$

Then for any positive integer n

$$\left| N(a,b,y) - \pi^{-1} \int_a^b \varphi(x)\, dx \right| \le n + \pi^{-1}(b-a)\sqrt{\delta_n}, \tag{A.2.3}$$

where

$$\delta_n = \max |f(x_2) - f(x_1)|, \quad |x_2 - x_1| \le (b-a)/n. \tag{A.2.4}$$

We divide the interval $[a,b]$ into n equal parts, and denote the points of subdivision by

$$a = c_0 < c_1 < \cdots < c_n = b.$$

Write

$$f_{r1} = \min f(x), \qquad f_{r2} = \max f(x), \qquad c_r \le x \le c_{r+1}.$$

In the first place, we assume that for all r we have $f_{r2} \ge 0$. Let ν_r denote the number of zeros in $(c_r, c_{r+1}]$ of the solution $y(x)$ of (A.2.2). Let ν_{r1} denote the minimum number of zeros in the same interval of any solution of

$$y_{r1}'' + f_{r1}y_{r1} = 0,$$

and ν_{r2} the maximum number of zeros there of any solution of

$$y_{r2}'' + f_{r2}y_{r2} = 0.$$

From the Sturm comparison theorem, we have then that

$$\nu_{r1} \le \nu_r \le \nu_{r2},$$

and so

$$\nu_{r1} \geq \pi^{-1}(c_{r+1} - c_r)\mathrm{Re}\sqrt{f_{r1}}, \tag{A.2.5}$$

$$\nu_{r2} \leq \pi^{-1}(c_{r+1} - c_r)\mathrm{Re}\sqrt{f_{r2}} + 1. \tag{A.2.6}$$

Thus, adding the inequalities (A.2.5), we get

$$N(a, b, y) \geq \pi^{-1}\sum(c_{r+1} - c_r)\mathrm{Re}\sqrt{f_{r1}}, \tag{A.2.7}$$

and likewise from (A.2.6),

$$N(a, b, y) \leq \pi^{-1}\sum(c_{r+1} - c_r)\mathrm{Re}\sqrt{f_{r2}} + n. \tag{A.2.8}$$

We now compare these sums to the associated integrals. Suppose first that $0 \leq f_{r1} < f_{r2}$. Then

$$\mathrm{Re}\sqrt{f_{r2}} - \mathrm{Re}\sqrt{f_{r1}} = \sqrt{f_{r2}} - \sqrt{f_{r1}} \leq \sqrt{f_{r2} - f_{r1}}.$$

Since $f_{r2} - f_{r1} \leq \delta_n$ and in this case $f_{r2} \geq 0$, we have that

$$\mathrm{Re}\sqrt{f_{r2}} - \mathrm{Re}\sqrt{f_{r1}} \leq \sqrt{\delta_n}.$$

The same result is easily checked in the remaining cases, namely,

$$0 \leq f_{r1} = f_{r2}, \quad f_{r1} \leq 0 \leq f_{r2}$$

It then follows that the sums in (A.2.7)–(A.2.8) differ from

$$\pi^{-1}\int_a^b \varphi(x)\,dx$$

by not more than $\pi^{-1}(b - a)\sqrt{\delta_n}$, and this proves the result.

The case $f_{r2} < 0$ for some r is handled simply since this implies $f(x) < 0$ on $[c_r, c_{r+1}]$ and so every solution of (A.2.2) has at most one zero on that interval by Sturm theory. It follows that $|N(c_r, c_{r+1}, y) - \pi^{-1}\int_{c_r}^{c_{r+1}} \varphi(x)\,dx| \leq 1$ for such r (in the case where $f(x) < 0$ for every $r = 0, 1, \ldots, n-1$, we get an upper bound equal to n which is less than the right side of (A.2.3)). Thus, (A.2.3) is trivially satisfied. The general result is obtained by applying the first case to those intervals where $f(x) \geq 0$ and carefully adding up the individual contributions.

The error term

$$n + \pi^{-1}(b - a)\sqrt{\delta_n}$$

involves an arbitrary choice of the integer n. We can be more specific about an optimal choice of n if we assume that $f(x)$ satisfies a Lipschitz condition.

A.3 The Lipschitz case

Theorem A.3.1. *Let $f(x)$ in (A.2.2) satisfy*

$$|f(x_2) - f(x_1)| \le K(x_2 - x_1), \quad a \le x_1 \le x_2 \le b, \qquad (A.3.1)$$

for some fixed $K \ge 0$. Then the number of zeros in $(a, b]$ of a non-trivial solution of (A.2.2) differs from

$$\pi^{-1} \int_a^b \varphi(x)\, dx \qquad (A.3.2)$$

by not more than

$$n + \frac{(b-a)^{3/2}\sqrt{K}}{\sqrt{n}\pi}, \qquad (A.3.3)$$

where n is given by the choice

$$n = [K^{1/3}(2\pi)^{-2/3}(b-a) + 1], \qquad (A.3.4)$$

and $[\ ]$ denotes the integral-part function.

The situation often presents itself that in (A.2.2) the coefficient of y is the sum of two terms, cf., (A.1.1), of which one predominates. In the simplest case, we have

Theorem A.3.2. *Let $f(x)$, $g(x)$, $a \le x \le b$, be real and continuous. We write, for every positive integral n,*

$$\delta_n = \max |f(x_2) - f(x_1)|, \quad |x_2 - x_1| \le (b-a)/n, \qquad (A.3.5)$$

$$A = \sup_{x \in [a,b]} |g(x)|. \qquad (A.3.6)$$

Then the number of zeros in $(a, b]$ of a non-trivial solution of

$$y'' + \{f(x) + g(x)\}y = 0 \qquad (A.3.7)$$

differs from (A.3.2) by not more than

$$n + \pi^{-1}(b-a)(\sqrt{\delta_n} + 3\sqrt{A}). \qquad (A.3.8)$$

Proof. We apply Theorem A.2.1. If

$$|x_2 - x_1| \le (b-a)/n,$$

we have

$$|f(x_2) + g(x_2) - f(x_1) - g(x_1)| \leq \delta_n + 2A, \qquad \text{(A.3.9)}$$

and also

$$\text{Re } \sqrt{f} - \sqrt{A} \leq \text{Re } \sqrt{(f+g)} \leq \text{Re } \sqrt{f} + \sqrt{A}. \qquad \text{(A.3.10)}$$

Thus the difference referred to in Theorem A.2.1 does not exceed

$$n + \pi^{-1}(b-a)\sqrt{A} + \pi^{-1}(b-a)\sqrt{(\delta_n + 2A)},$$

and this in turn does not exceed (A.3.8).

A.4 Oscillations in the differentiable case

In the case that $f(x)$ is positive and suitably smooth, an effective method of carrying out estimations on the number of zeros is given by modified Prüfer substitutions, such as $\sqrt{f(x)}\, y(x) = r\sin\theta$, $y'(x) = r\cos\theta$, $r > 0$ or the more usual

$$\tan\theta = \sqrt{f}\, y/y'.$$

This leads in the ordinary Sturm-Liouville case to useful and rather precise estimates for eigenvalues; the smoother the f the more precise the eigenvalue estimates. However, in the multiparameter case, even under the relatively simple "condition A," one is inevitably led into the discussion of

$$y'' + f(x)y = 0, \quad a \leq x \leq b, \qquad \text{(A.4.1)}$$

where $f(x)$ changes sign, and so into "turning-point" problems. We shall not enter here into the rather difficult topic of the approximate integration of (A.4.1) when $f(x)$ changes sign at one or more points; instead, we use what is to be had by use of the Sturm comparison theorem.

Here, as in the previous version, y is a non-trivial solution of (A.1.1), f is to be positive but now continuously differentiable, while g is integrable. As before, we interpret $\theta(x)$ so as to be continuous, equal to a multiple of π if $y = 0$.

Lemma A.4.1. Let $f(x) > 0$ in (A.1.1) be continuously differentiable over (a,b) and $g \in L(a,b)$. Then the number $N(a,b,y)$ of zeros in $(a,b]$ of a non-trivial solution of (A.1.1) differs from

$$\pi^{-1} \int_a^b \sqrt{f}\, dx \qquad \text{(A.4.2)}$$

by not more than

$$1 + \frac{\pi^{-1}}{4} \int_a^b \frac{|f'|}{f} \, dx + \frac{\pi^{-1}}{\min \sqrt{f}} \int_a^b |g| \, dx,$$

that is,

$$\left| N(a,b,y)\pi - \int_a^b \sqrt{f} \right| \leq \pi + \frac{1}{4} \int_a^b \frac{|f'|}{f} \, dx + \frac{1}{\min \sqrt{f}} \int_a^b |g| \, dx. \qquad (A.4.3)$$

Defining θ as above we find that in fact, θ is absolutely continuous. Simple calculations yield

$$\theta' = \sqrt{f} + \{f'/(4f)\} \sin 2\theta + \{g/\sqrt{f}\} \sin^2 \theta. \qquad (A.4.4)$$

Integrating (A.4.4) over $[a,b]$ and using simple estimations we find that

$$\left| \theta(b) - \theta(a) - \int_a^b \sqrt{f} \right| \leq (1/4) \int_a^b \frac{|f'|}{f} \, dx + \frac{1}{\min \sqrt{f}} \int_a^b |g| \, dx.$$

On the other hand, since θ increases at every zero of y we get that

$$[\pi^{-1}(\theta(b) - \theta(a))] \leq N(a,b,y) \leq [\pi^{-1}(\theta(b) - \theta(a))] + 1.$$

Combining these two previous displays we get the stated result.

Note that the estimate (A.4.3) is not generally optimal; nevertheless, it is sufficient for the purposes of obtaining spectral asymptotics in this special case. Specifically, Lemma A.4.1 may be used to derive asymptotic estimates for the eigenvalues of (A.1.2) subject to a pair of homogeneous separated boundary conditions, at least when f is positive and continuously differentiable and g is integrable over (a,b). Indeed, from (A.4.3), we obtain the estimate

Theorem A.4.1. *Let $f(x)$ be positive over $[a,b]$ and continuously differentiable over (a,b). Let g be integrable over (a,b). Then as $\lambda \to \infty$ the number of zeros $N(a,b,y)$ of any non-trivial solution of (A.1.2) admits the estimate*

$$N(a,b,y) \sim \pi^{-1}\sqrt{\lambda} \int_a^b \sqrt{f} \, dx.$$

A.5   The Lebesgue integrable case

We consider now the case (A.1.2), where $f \in L(a,b)$, and as before $g \in L(a,b)$. We allow f to take both signs, and require it to be in part positive. Writing

$$f_1(x) = \max\{f(x), 0\}, \quad f_2(x) = -\min\{f(x), 0\}, \qquad (A.5.1)$$

we assume that

$$\int_a^b f_1(x)\,dx > 0. \tag{A.5.2}$$

We have then

Theorem A.5.1. *As* $\lambda \to \infty$,

$$N(a,b,y) \sim \pi^{-1}\sqrt{\lambda}\int_a^b \sqrt{f_1}\,dx. \tag{A.5.3}$$

Similar results may be proved allowing more general parameter-dependence and with a general second-order term in the differential equation. Reference may be made to Atkinson and Mingarelli (1987); the present special case permits slight simplifications in the proof.

One first proves the result in the case that $f_2 \equiv 0$, so that $f \geq 0$, and $\sqrt{f}$ is real and in $L^2(a,\,b)$. For any $\delta > 0$, we choose a non-negative step-function w_1, constant in a set of intervals I_j, $j = 1,\ldots,M$, approximating to $\sqrt{f_1}$ in the sense that

$$\int_a^b |\sqrt{f_1} - w_1|^2 dx < (b-a)\delta^2. \tag{A.5.4}$$

We replace this by the positive step-function

$$h_1 = w_1 + \delta, \tag{A.5.5}$$

so that

$$\int_a^b |\sqrt{f_1} - h_1|^2\,dx < 4(b-a)\delta^2. \tag{A.5.6}$$

It follows that

$$\int_a^b |\sqrt{f_1} - h_1|\,dx < 2(b-a)\delta. \tag{A.5.7}$$

We now apply a Prüfer transformation, in which the angle is given by

$$\tan\theta = y\sqrt{(\lambda h_1)}/y'. \tag{A.5.8}$$

This may be done in each of the intervals I_j, and leads to the differential equation

$$\theta' = \sqrt{(\lambda f_1)} + \sqrt{\lambda}\{\cos^2\theta(h_1 - \sqrt{f_1}) + \sin^2\theta(f_1/h_1 - \sqrt{f_1})\} + g\,\sin^2\theta/(h_1\sqrt{\lambda}). \tag{A.5.9}$$

We integrate these equations over their respective intervals, and combine the results. If $N_1(a, b, y)$ denotes the number of zeros for a solution of (A.1.2), we get

$$\pi N_1(a, b, y)/\sqrt{\lambda} \;=\; \int_a^b \sqrt{f_1}\, dx + O\left\{ \int_a^b |h_1 - \sqrt{f_1}|\, dx \right\}$$

$$+\;\; O\left\{ \int_a^b |f_1/h_1 - \sqrt{f_1}|\, dx \right\} + O(M/\sqrt{\lambda}) + O\{1/(\delta\lambda)\}.$$

Here the second term on the right is of order $O(\delta)$, by (A.5.7). For the third term on the right, we note that

$$f_1/h_1 - \sqrt{f_1} = (\sqrt{f_1} - h_1)^2/h_1 + (\sqrt{f_1} - h_1)$$

and this also yields $O(\delta)$, by (A.5.6)–(A.5.7), and the fact that $h_1 \geq \delta$. This completes the proof in the case when $f_2 \equiv 0$.

Passing to the general case we denote by $N(a, b, y)$ the number of zeros of a solution (A.1.2) in the case $f = f_1 - f_2$. By the Sturm comparison theorem, we have $N(a, b, y) \leq N_1(a, b, y) + 1$, and so we need to show that

$$\liminf_{\lambda \to \infty} \lambda^{-1/2} N(a, b, y) \geq \pi^{-1} \int_a^b \sqrt{f_1}\, dx. \qquad (A.5.10)$$

We define h_1, w_1 as before, and define h_2, w_2 similarly, corresponding to f_2. In each I_j, we define θ as before, and derive the differential equation (A.5.9), together with an extra term on the right, namely,

$$-\sqrt{\lambda} f_2 \sin^2 \theta/h_1,$$

which is of order

$$\sqrt{\lambda}\, h_1^{-1}\{h_2^2 + (\sqrt{f_2} - h_2)^2 + h_2|\sqrt{f_2} - h_2|\}. \qquad (A.5.11)$$

The method now is to integrate (A.5.9) over a subset of (a, b), so as to deal with this error-term, without loosing too much of the leading term, the integral in (A.5.3). The subset in question, S_1, say, is that described by

$$h_1 > \sqrt{\delta}, \quad h_2 < \sqrt{\delta}.$$

This consists of a finite number M_1 of intervals, since h_1, h_2 are step-functions. Using the analogues of (A.5.6)–(A.5.7), we find that the integral of (A.5.11) over S_1 is of order $O\{\sqrt{(\lambda\delta)}\}$. Thus, from integrating (A.5.9) over S_1, we get

$$\pi N(a, b, y)/\sqrt{\lambda} > \int_{S_1} \sqrt{f_1}\, dx + O(M_1/\sqrt{\lambda}) + O(\sqrt{\delta}) + O(1/(\delta\lambda)).$$

To complete the proof, we show that, if $S_2 = (a, b)\backslash S_1$,

$$\int_{S_2} \sqrt{f_1}\, dx = O(\sqrt{\delta}).$$ (A.5.12)

Now S_2 comprises sets S_3, S_4 described, respectively, by

$$h_1 \leq \sqrt{\delta},$$

and

$$h_1 > \sqrt{\delta}, \quad h_2 \geq \sqrt{\delta}.$$

In the case of S_3, we write $\sqrt{f_1} = h_1 + (\sqrt{f_1} - h_1)$ and use (A.5.7). For the case of S_4, it will be sufficient, by the Schwarz inequality, to prove that meas $(S_4) = O(\delta)$.

To see this, we note that in S_4 we have

$$|\sqrt{f_1} - h_1| + |\sqrt{f_2} - h_2| \geq \sqrt{\delta},$$

since $f_1 f_2 = 0$. We then get the required result by squaring both sides integrating and using (A.5.6) and its analogue. This completes the proof.

A.6   The nonoscillatory case

We present here an auxiliary result on the behavior of the Prüfer angle to be used in Chapters 5, 6, and 9.

Theorem A.6.1. *Let* $\theta(x, \mu)$, $a \leq x \leq b$, $0 \leq \mu < l$, *be the solution of*

$$\theta'(x, \mu) = \cos^2 \theta(x, \mu) + p(x, \mu) \sin^2 \theta(x, \mu),$$ (A.6.1)

with initial condition

$$\theta(a, \mu) = \alpha,$$ (A.6.2)

where $\alpha \in [0, \pi)$ *is fixed. Let* $p(x, \mu) \in L(a, b)$ *for each* $\mu \in [0, l]$ *and let*

$$p(x, \mu) \leq g(x) - h(x, \mu),$$ (A.6.3)

where $g(x) \geq 0$, $g(x) \in L(a, b)$, $h(x, \mu) \geq 0$, $h(x, \mu) \in L(a, b)$ *for each* μ, *and for each* $[a', b']$ *with* $a \leq a' < b' \leq b$ *let*

$$\int_{a'}^{b'} h(x, \mu)\, dx \to \infty,$$ (A.6.4)

as $\mu \to \infty$. *Then, as* $\mu \to \infty$,

$$\theta(b, \mu) \to 0.$$ (A.6.5)

As previously, it follows from (A.6.1) that θ is increasing in x when a multiple of π, so that, since $\theta(a, \mu) \geq 0$, $\theta(x, \mu) > 0$, $a < x \leq b$. We then show that, for large μ, $\theta(x, \mu)$ has an upper bound in $(\pi/2, \pi)$. For the proof, we choose some $\delta = \max(\alpha, \pi/2) + \varepsilon$, where $\varepsilon > 0$ is such that $\delta < \pi$, and claim that for large μ we have

$$\theta(x, \mu) < \delta, \tag{A.6.6}$$

for all $x \in [a, b]$.

Supposing the contrary, that is, for every M there is a $\mu > M$ such that (A.6.6) fails. We denote by x_0 the first x-value in $[a, b]$ such that $\theta(x_0, \mu) = \delta$. We choose a subdivision of (a, b), denoted by

$$a = \xi_0 < \xi_1 < \cdots < \xi_n = b \tag{A.6.7}$$

such that

$$\int_{\xi_r}^{\xi_{r+1}} (1 + g(x))dx < \varepsilon/2, \quad r = 0, \ldots, n - 1. \tag{A.6.8}$$

Since $\theta' \leq 1 + g(x)$, we have, for any x_1, x_2, with $a \leq x_1 < x_2 \leq b$,

$$\theta(x_2, \mu) - \theta(x_1, \mu) \leq \int_{x_1}^{x_2} (1 + g(x))dx, \tag{A.6.9}$$

and so, if $\xi_r \leq x_1 < x_2 \leq \xi_{r+1}$,

$$\theta(x_2, \mu) - \theta(x_1, \mu) < \varepsilon/2. \tag{A.6.10}$$

Let x_0 lie in the interval $[\xi_r, \xi_{r+1}]$ of the subdivision (A.6.7). By (A.6.10), we must have $r \geq 1$. Again by (A.6.10), we must have $\theta(x, \mu) > \pi/2$ in $[\xi_{r-1}, \xi_r]$, while also $\theta(x, \mu) \leq \delta$. Hence

$$\sin \theta(x, \mu) \geq \sin \delta, \quad \xi_{r-1} \leq x \leq \xi_r. \tag{A.6.11}$$

Integrating (A.6.1) over this interval we deduce that

$$\theta(\xi_r, \mu) - \theta(\xi_{r-1}, \mu) \leq \varepsilon/2 - \sin^2 \delta \int_{\xi_{r-1}}^{\xi_r} h(x, \mu)\, dx. \tag{A.6.12}$$

Here the left is bounded above and below, while the right tends to $-\infty$ as $\mu \to \infty$. This gives a contradiction, and so proves (A.6.6).

To complete the proof, we suppose that for some ε as above, we have $\theta(b, \mu) > \varepsilon$, for arbitrarily large μ. We now consider the behavior of θ in the interval $[\xi_{n-1}, b]$. The above arguments show that $\theta(x, \mu) > \varepsilon/2$, so that $\sin \theta(x, \mu) \geq \min\{\sin(\varepsilon/2), \sin \delta\}$. The argument of (A.6.12) then shows that

$$\theta(b, \mu) - \theta(\xi_{n-1}, \mu) \to -\infty$$

as $\mu \to \infty$, which is impossible. This completes the proof.

Notes for the Appendix

A simpler proof of the Atkinson-Mingarelli theorem, Theorem A.5.1, in the particular case where f is continuously differentiable and admits a finite number of simple zeros may be found in Constantin-McKean (1999).

Various developments in this direction include but are not restricted to results by Eberhard et al. (1994) dealing with connection formulas for Sturm-Liouville equations with an arbitrary number of turning points and the resulting spectral asymptotics; the problem of determining the spectral asymptotics of Sturm-Liouville equations with indefinite matrix coefficients as in Volkmer (2004) and a study of the asymptotics of the eigenvalues of general equations with indefinite weights having zeros and or poles undertaken by Eberhard et al. (2001).

Research in the area of Sturm-Liouville equations with indefinite weights and their extensions has been productive of late and we cite the papers by D'yachenko (2000), Volkmer (1996), Tumanov (2000), Cao et al. (2003), Binding et al. (2002b) as samples of the literature. Spectral asymptotics for Dirac equations were presented in Binding and Volkmer (2001a).

The inverse spectral problem for Sturm-Liouville equations with indefinite weights has also gathered much attention lately with, for example, papers by Binding et al. (2002a), Marasi and Akbarfam (2007), Akbarfam (1999), Akbarfam and Mingarelli (2002), (2006), Hosseinabady (2003), and the references within these.

On the general subject of oscillations of Sturm-Liouville equations with indefinite weights we cite, in particular, Kong et al. (2001), Binding and Volkmer (2001b), Mingarelli (1994), Allegretto and Mingarelli (1989), and Deift and Hempel (1986), where, in the latter, a rigorous proof of an oscillation theorem of Richardson (1918) and Haupt (1911) is given in the case of continuous coefficients. Another reference on this subject is the current monograph by Zettl (2005).

A.7 Research problems and open questions

1. Find a version of Theorem A.6.1 for Prüfer angles θ satisfying a differential equation of the form

$$\theta'(x, \mu) = r(x, \mu) + q(x, \mu) \cos^2 \theta(x, \mu) + p(x, \mu) \sin^2(x, \mu),$$

where, for all real μ,

$$\theta(a, \mu) = \alpha, \quad \alpha \in (0, \pi].$$

2. Observe that the constant on the right of (A.2.3) is sharp in the case when $n = 1$ and f is a constant positive function on $[a, b]$, and therefore, the same is true of (A.3.3).

The objective is then to *improve the bound on the right of* (A.4.3) in Lemma A.4.1.

3. Find matrix versions of the results of this Appendix in the spirit of Atkinson (1964a, Chapter 10) where zeros are now thought of as *conjugate points*.

4. Find three-term recurrence relation analogues of some/all the results in this section.

5. Determine corresponding Sturmian results for even more general equations in the unified framework of either Volterra-Stieltjes integral equations (Atkinson [1964a, Chapters 11–12], Mingarelli [1983]) or differential equations on time-scales (Agarwal *et al.* [2002] and the references therein).

6. Can one exploit (A.4.4) in the event that additional data is given, data such as $f \in C^m(a, b)$, $m > 1$?

7. Find analogues of any of the resuls of this section for *complex valued* coefficients $f(x)$, $g(x)$ and real x as usual, cf., Hille (1969).

Bibliography

[Abramov, 1994] Abramov, A. A. (1994). A method for solving multiparameter eigenvalue problems that arise in the application of the Fourier method. *Zh. Vychisl. Mat. i Mat. Fiz.*, 34(10):1524–1525.

[Agarwal et al., 2002] Agarwal, R. P., Bohner, M., and O'Regan, D. (2002). Time scale boundary value problems on infinite intervals. *J. Comput. Appl. Math.*, 141(1–2):27–34. Dynamic equations on time scales.

[Agranovich et al., 1999] Agranovich, M. S., Katsenelenbaum, B. Z., Sivov, A. N., and Voitovich, N. N. (1999). *Generalized method of eigenoscillations in diffraction theory*. WILEY-VCH Verlag Berlin GmbH, Berlin. Translated from the Russian manuscript by Vladimir Nazaikinskii.

[Akbarfam and Mingarelli, 2002] Akbarfam, A. J. and Mingarelli, A. B. (2002). Higher order asymptotic distribution of the eigenvalues of nondefinite Sturm-Liouville problems with one turning point. *J. Comput. Appl. Math.*, 149(2):423–437.

[Alama et al., 1989] Alama, S., Deift, P. A., and Hempel, R. (1989). Eigenvalue branches of the Schrödinger operator $H - \lambda W$ in a gap of $\sigma(H)$. *Comm. Math. Phys.*, 121(2):291–321.

[Albeverio et al., 2008] Albeverio, S., Hryniv, R., and Mykytyuk, Y. (2008). On spectra of non-self-adjoint Sturm-Liouville operators. *Selecta Math. (N.S.)*, 13(4):571–599.

[Aleskerov, 1990] Aleskerov, F. (1990). *Expansion in eigenfunctions of self-adjoint singular multiparameter differential operators with a matrix coefficient*. PhD thesis, Baku. (Abstract of dissertation).

[Al-Gwaiz, 2008] Al-Gwaiz, M. A. (2008). *Sturm-Liouville theory and its applications*. Springer Undergraduate Mathematics Series. Springer-Verlag London Ltd., London.

[Aliyev and Kerimov, 2008] Aliyev, Y. N. and Kerimov, N. B. (2008). The basis property of Sturm-Liouville problems with boundary conditions depending quadratically on the eigenparameter. *Arab. J. Sci. Eng. Sect. A Sci.*, 33(1):123–136.

[Allahverdiev and Džabarzade, 1979] Allahverdiev, D. È. and Džabarzade, R. M. (1979). Convergence of multiple expansions in eigenelements of a two-parameter operator system. *Akad. Nauk Azerbaĭdzhan. SSR Dokl.*, 35(6):16–20.

[Allahverdiev and Isaev, 1981] Allahverdiev, B. P. and Isaev, G. A. (1981). Oscillation theorems for multiparameter problems with boundary conditions that depend on spectral parameters. *Izv. Akad. Nauk Azerbaĭdzhan. SSR Ser. Fiz.-Tekhn. Mat. Nauk*, 2(6):17–23 (1982).

[Allegretto and Mingarelli, 1989] Allegretto, W. and Mingarelli, A. B. (1989). Boundary problems of the second order with an indefinite weight-function. *J. Reine Angew. Math.*, 398:1–24.

[Almamedov and Aslanov, 1986a] Almamedov, M. and Aslanov, A. (1986a). On the construction of a spectral measure for selfadjoint three-parameter problems. *Sov. Math., Dokl.*, 33:739–741.

[Almamedov and Aslanov, 1986b] Almamedov, M. S. and Aslanov, A. A. (1986b). Construction of a spectral measure for selfadjoint three-parameter problems. *Dokl. Akad. Nauk SSSR*, 288(4):780–782.

[Almamedov and Isaev, 1985] Almamedov, M. and Isaev, G. (1985). Solvability of nonselfadjoint linear operator systems, and the set of decomposability of multiparameter spectral problems. *Sov. Math., Dokl.*, 31:472–474.

[Almamedov et al., 1985] Almamedov, M. S., Aslanov, A. A., and Isaev, G. A. (1985). On the theory of two-parameter spectral problems. *Dokl. Akad. Nauk SSSR*, 283(5):1033–1035.

[Almamedov et al., 1987] Almamedov, M. S., Aslanov, A. A., and Isaev, G. A. (1987). Tensor determinants that are connected with multiparameter spectral problems. *Izv. Vyssh. Uchebn. Zaved. Mat.* English translation: Soviet Math. (Iz. VUZ) 31 (1987), no. 11, 1–8.

[Amer, 1998] Amer, M. A. (1998). Constructive solutions for nonlinear multiparameter eigenvalue problems. *Comput. Math. Appl.*, 35(11):83–90.

[Andrew, 1996] Andrew, A. L. (1996). Solution of nonlinear and multiparameter eigenvalue problems. In *Computational techniques and applications: CTAC95 (Melbourne, 1995)*, pages 97–104. World Sci. Publ., River Edge, NJ.

[Annaby, 2005] Annaby, M. H. (2005). Multivariate sampling theorems associated with multiparameter differential operators. *Proc. Edinb. Math. Soc. (2)*, 48(2):257–277.

[Arscott, 1964a] Arscott, F. M. (1964a). *Periodic differential equations. An introduction to Mathieu, Lamé, and allied functions.* International Series of Monographs in Pure and Applied Mathematics, Vol. 66. A Pergamon Press Book. The Macmillan Co., New York.

[Arscott, 1964b] Arscott, F. M. (1964b). Two-parameter eigenvalue problems in differential equations. *Proc. London Math. Soc. (3)*, 14:459–470.

[Arscott, 1974a] Arscott, F. M. (1974a). Integral-equation formulation of two-parameter eigenvalue problems. In *Spectral theory and asymptotics of differential equations (Proc. Conf., Scheveningen, 1973)*, pages 95–102. North-Holland Math. Studies, Vol. 13. North-Holland, Amsterdam.

[Arscott, 1974b] Arscott, F. M. (1974b). Transform theorems for two-parameter eigenvalue problems in Hilbert space. In *Ordinary and partial differential equations (Proc. Conf., Univ. Dundee, Dundee, 1974)*, pages 302–307. Lecture Notes in Math., Vol. 415. Springer, Berlin.

[Arscott, 1987] Arscott, F. M. (1987). Studies in multiplicative solutions to linear differential equations. *Proc. Roy. Soc. Edinburgh Sect. A*, 106(3–4):277–305.

[Arscott and Darai, 1981] Arscott, F. M. and Darai, A. (1981). Curvilinear coordinate systems in which the Helmholtz equation separates. *IMA J. Appl. Math.*, 27(1):33–70.

[Arscott and Sleeman, 1968] Arscott, F. M. and Sleeman, B. D. (1968). Multiplicative solutions of linear differential equations. *J. London Math. Soc.*, 43:263–270.

[Aslanov, 2004] Aslanov, A. (2004). Definiteness conditions in the multiparameter spectral theory. *WSEAS Trans. Math.*, 3(3):717–720.

[Aslanov and Isahanly, 1998] Aslanov, A. and Isahanly, H. (1998). A multidimensional complex analytical view on the multi-parameter spectrum and the construction of the spectral measure (english). *Khazar Math. J.*, 1(1):3–65.

[Atkinson, 1963] Atkinson, F. V. (1963). Boundary problems leading to orthogonal polynomials in several variables. *Bull. Amer. Math. Soc.*, 69:345–351.

[Atkinson, 1964a] Atkinson, F. V. (1964a). *Discrete and continuous boundary problems.* Academic Press, New York, London.

[Atkinson, 1964b] Atkinson, F. V. (1964b). Multivariate spectral theory: the linked eigenvalue problem for matrices. Technical report, U.S. Army Research Center, Madison, Wisconsin.

[Atkinson, 1968] Atkinson, F. V. (1968). Multiparameter spectral theory. *Bull. Amer. Math. Soc.*, 74:1–27.

[Atkinson, 1972] Atkinson, F. V. (1972). *Multiparameter eigenvalue problems: Matrices and compact operators.* Academic Press, New York, London.

[Atkinson, 1975] Atkinson, F. V. (1975). Limit-n criteria of integral type. *Proc. Roy. Soc. Edinburgh Sect. A*, 73:167–198.

[Atkinson, 1977] Atkinson, F. V. (1977). Deficiency-index theory in the multi-parameter Sturm-Liouville case. In *Differential equations (Proc. Internat. Conf., Uppsala, 1977)*, pages 1–10. Sympos. Univ. Upsaliensis Ann. Quingentesimum Celebrantis, No. 7. Almqvist & Wiksell, Stockholm.

[Atkinson, 1981] Atkinson, F. V. (1981). A class of limit-point criteria. In *Spectral theory of differential operators (Birmingham, Ala., 1981)*, volume 55 of *North-Holland Math. Stud.*, pages 13–35. North-Holland, Amsterdam.

[Atkinson and Mingarelli, 1987] Atkinson, F. V. and Mingarelli, A. B. (1987). Asymptotics of the number of zeros and of the eigenvalues of general weighted Sturm-Liouville problems. *J. Reine Angew. Math.*, 375/376:380–393.

[Bailey, 1981] Bailey, P. B. (1981). The automatic solution of two-parameter Sturm-Liouville eigenvalue problems in ordinary differential equations. *Appl. Math. Comput.*, 8(4):251–259.

[Battle, 2008] Battle, L. (2008). Stieltjes Sturm-Liouville equations: eigenvalue dependence on problem parameters. *J. Math. Anal. Appl.*, 338(1):23–38.

[Bennewitz, 1981] Bennewitz, C. (1981). Spectral theory for Hermitean differential systems. In *Spectral theory of differential operators (Birmingham, Ala., 1981)*, volume 55 of *North-Holland Math. Stud.*, pages 61–67. North-Holland, Amsterdam.

[Bennewitz, 1989] Bennewitz, C. (1989). Spectral asymptotics for Sturm-Liouville equations. *Proc. London Math. Soc. (3)*, 59(2):294–338.

[Berezans′kiĭ, 1968] Berezans′kiĭ, J. M. (1968). *Expansions in eigenfunctions of selfadjoint operators.* Translated from the Russian by R. Bolstein, J. M. Danskin, J. Rovnyak and L. Shulman. Translations of Mathematical Monographs, Vol. 17. American Mathematical Society, Providence, R.I.

[Berezanskiĭ and Konstantinov, 1992] Berezanskiĭ, Y. M. and Konstantinov, A. Y. (1992). Expansion in eigenvectors of multiparameter spectral problems. *Funktsional. Anal. i Prilozhen.*, 26(1):81–83. Translation in Funct. Anal. Appl. 26(1992), no. 1, 65 67.

[Berezansky and Konstantinov, 1992] Berezansky, Y. M. and Konstantinov, A. Y. (1992). Expansion in eigenvectors of multiparameter spectral problems. *Ukraïn. Mat. Zh.*, 44(7):901–913. *Translation in Ukrainian Math. J.* 44 (1992), no. 7, 813–823 (1993).

[Bhattacharyya et al., 2001] Bhattacharyya, T., Binding, P. A., and Seddighi, K. (2001). Multiparameter Sturm-Liouville problems with eigenparameter dependent boundary conditions. *J. Math. Anal. Appl.*, 264(2):560–576.

[Bhattacharyya et al., 2002] Bhattacharyya, T., Košir, T., and Plestenjak, B. (2002). Right-definite multiparameter Sturm-Liouville problems with eigenparameter-dependent boundary conditions. *Proc. Edinb. Math. Soc. (2)*, 45(3):565–578.

[Bhattacharyya and Mohandas, 2005] Bhattacharyya, T. and Mohandas, J. P. (2005). Two-parameter uniformly elliptic Sturm-Liouville problems with eigenparameter-dependent boundary conditions. *Proc. Edinb. Math. Soc. (2)*, 48(3):531–547.

[Bihari, 1976] Bihari, I. (1976). Notes on the eigenvalues and zeros of the solutions of half-linear second order ordinary differential equations. *Period. Math. Hungar.*, 7(2):117–125.

[Billigheimer, 1970] Billigheimer, C. E. (1970). Regular boundary problems for a five-term recurrence relation. *Pacific J. Math.*, 35:23–51.

[Binding, 1980a] Binding, P. (1980a). Another positivity result for determinantal operators. *Proc. Roy. Soc. Edinburgh Sect. A*, 86(3-4):333–337.

[Binding, 1980b] Binding, P. (1980b). On the use of degree theory for nonlinear multiparameter eigenvalue problems. *J. Math. Anal. Appl.*, 73(2):381–391.

[Binding, 1981a] Binding, P. (1981a). Multiparameter definiteness conditions. *Proc. Roy. Soc. Edinburgh Sect. A*, 89(3-4):319–332.

[Binding, 1981b] Binding, P. (1981b). Variational methods for one and several parameter nonlinear eigenvalue problems. *Canad. J. Math.*, 33(1):210–228.

[Binding, 1982a] Binding, P. (1982a). Left definite multiparameter eigenvalue problems. *Trans. Amer. Math. Soc.*, 272(2):475–486.

[Binding, 1982b] Binding, P. (1982b). Multiparameter variational principles. *SIAM J. Math. Anal.*, 13(5):842–855.

[Binding, 1982c] Binding, P. (1982c). On a problem of B. D. Sleeman. *J. Math. Anal. Appl.*, 85(2):291–307. (Erratum, Vol. 90, (1982), 270–271).

[Binding, 1983c] Binding, P. (1982–1983c). Multiparameter definiteness conditions. II. *Proc. Roy. Soc. Edinburgh Sect. A*, 93(1–2):47–61.

[Binding, 1983a] Binding, P. (1983a). Abstract oscillation theorems for multiparameter eigenvalue problems. *J. Differential Equations*, 49(3):331–343.

[Binding, 1983b] Binding, P. (1983b). Dual variational approaches to multiparameter eigenvalue problems. *J. Math. Anal. Appl.*, 92(1):96–113.

[Binding, 1984] Binding, P. (1984). Perturbation and bifurcation of nonsingular multiparametric eigenvalues. *Nonlinear Anal.*, 8(4):335–352.

[Binding, 1989] Binding, P. (1989). Multiparameter root vectors. *Proc. Edinburgh Math. Soc. (2)*, 32(1):19–29.

[Binding, 1991] Binding, P. (1991). Indefinite Sturm-Liouville theory via examples of two-parameter eigencurves. In *Ordinary and partial differential equations, Vol. III (Dundee, 1990)*, volume 254 of *Pitman Res. Notes Math. Ser.*, pages 38–49. Longman Sci. Tech., Harlow.

[Binding and Browne, 1977] Binding, P. and Browne, P. J. (1977). A variational approach to multiparameter eigenvalue problems for matrices. *SIAM J. Math. Anal.*, 8(5):763–777.

[Binding and Browne, 1978a] Binding, P. and Browne, P. J. (1978a). Positivity results for determinantal operators. *Proc. Roy. Soc. Edinburgh Sect. A*, 81(3–4):267–271.

[Binding and Browne, 1978b] Binding, P. and Browne, P. J. (1978b). A variational approach to multi-parameter eigenvalue problems in Hilbert space. *SIAM J. Math. Anal.*, 9(6):1054–1067.

[Binding and Browne, 1980] Binding, P. and Browne, P. J. (1980). Comparison cones for multiparameter eigenvalue problems. *J. Math. Anal. Appl.*, 77(1):132–149.

[Binding and Browne, 1981] Binding, P. and Browne, P. J. (1981). Spectral properties of two-parameter eigenvalue problems. *Proc. Roy. Soc. Edinburgh Sect. A*, 89(1–2):157–173.

[Binding and Browne, 1983] Binding, P. and Browne, P. J. (1983). A definiteness result for determinantal operators. In *Ordinary differential equations and operators (Dundee, 1982)*, volume 1032 of *Lecture Notes in Math.*, pages 17–30. Springer, Berlin.

[Binding and Browne, 1984] Binding, P. and Browne, P. J. (1984). Multiparameter Sturm theory. *Proc. Roy. Soc. Edinburgh Sect. A*, 99(1–2):173–184.

[Binding and Browne, 1988] Binding, P. and Browne, P. J. (1988). Applications of two parameter spectral theory to symmetric generalised eigenvalue problems. *Appl. Anal.*, 29(1–2):107–142.

[Binding and Browne, 1989a] Binding, P. and Browne, P. J. (1989a). Eigencurves for two-parameter selfadjoint ordinary differential equations of even order. *J. Differential Equations*, 79(2):289–303.

[Binding and Browne, 1989b] Binding, P. and Browne, P. J. (1989b). Two parameter eigenvalue problems for matrices. *Linear Algebra Appl.*, 113:139–157.

[Binding and Browne, 1991] Binding, P. and Browne, P. J. (1991). Asymptotics of eigencurves for second order ordinary differential equations. II. *J. Differential Equations*, 89(2):224–243.

[Binding et al., 1987] Binding, P., Browne, P. J., and Picard, R. H. (1987). Spectral properties of two-parameter eigenvalue problems, II. *Proc. Roy. Soc. Edinburgh Sect. A*, 106(1–2):39–51. (Erratum, Vol. 115 (1–2) (1990), 87–90).

[Binding et al., 1982a] Binding, P., Browne, P. J., and Turyn, L. (1981–1982a). Existence conditions for two-parameter eigenvalue problems. *Proc. Roy. Soc. Edinburgh Sect. A*, 91(1–2):15–30.

[Binding et al., 1984a] Binding, P., Browne, P. J., and Turyn, L. (1984a). Existence conditions for eigenvalue problems generated by compact multiparameter operators. *Proc. Roy. Soc. Edinburgh Sect. A*, 96(3–4):261–274.

[Binding et al., 1984b] Binding, P., Browne, P. J., and Turyn, L. (1984b). Spectral properties of compact multiparameter operators. *Proc. Roy. Soc. Edinburgh Sect. A*, 98(3–4):291–303.

[Binding et al., 1995] Binding, P., Farenick, D. R., and Li, C.-K. (1995). A dilation and norm in several variable operator theory. *Canad. J. Math.*, 47(3):449–461.

[Binding and Košir, 1996] Binding, P. and Košir, T. (1996). Second root vectors for multiparameter eigenvalue problems of Fredholm type. *Trans. Amer. Math. Soc.*, 348(1):229–249.

[Binding and Strauss, 2005] Binding, P. and Strauss, V. (2005). On operators with spectral square but without resolvent points. *Canad. J. Math.*, 57(1):61–81.

[Binding and Volkmer, 1986] Binding, P. and Volkmer, H. (1986). Existence and uniqueness of indexed multiparametric eigenvalues. *J. Math. Anal. Appl.*, 116(1):131–146.

[Binding and Volkmer, 1992] Binding, P. and Volkmer, H. (1992). On a problem of R. G. D. Richardson. *Proc. Edinburgh Math. Soc. (2)*, 35(3):337–348.

[Binding and Volkmer, 1996] Binding, P. and Volkmer, H. (1996). Eigencurves for two-parameter Sturm-Liouville equations. *SIAM Rev.*, 38(1):27–48.

[Binding and Volkmer, 2005] Binding, P. and Volkmer, H. (2005). Prüfer angle asymptotics for Atkinson's semi-definite Sturm-Liouville eigenvalue problem. *Math. Nachr.*, 278(12–13):1458–1475.

[Binding, 1981c] Binding, P. A. (1981c). On generalized and quadratic eigenvalue problems. *Applicable Analysis*, 12:27–45.

[Binding et al., 1992] Binding, P. A., Browne, P. J., and Seddighi, K. (1992). Two parameter asymptotic spectra. *Results Math.*, 21(1–2):12–23.

[Binding et al., 1994] Binding, P. A., Browne, P. J., and Seddighi, K. (1994). Sturm-Liouville problems with eigenparameter dependent boundary conditions. *Proc. Edinburgh Math. Soc. (2)*, 37(1):57–72.

[Binding et al., 2002a] Binding, P. A., Browne, P. J., and Watson, B. A. (2002a). Inverse spectral problems for left-definite Sturm-Liouville equations with indefinite weight. *J. Math. Anal. Appl.*, 271(2):383–408.

[Binding et al., 2002b] Binding, P. A., Browne, P. J., and Watson, B. A. (2002b). Spectral asymptotics for Sturm-Liouville equations with indefinite weight. *Trans. Amer. Math. Soc.*, 354(10):4043–4065 (electronic).

[Binding et al., 2003] Binding, P. A., Browne, P. J., and Watson, B. A. (2003). Decomposition of spectral asymptotics for Sturm-Liouville equations with a turning point. *Adv. Differential Equations*, 8(4):491–511.

[Binding et al., 1982b] Binding, P. A., Källström, A., and Sleeman, B. D. (1982b). An abstract multiparameter spectral theory. *Proc. Roy. Soc. Edinburgh Sect. A*, 92(3–4):193–204.

[Binding and Seddighi, 1987a] Binding, P. A. and Seddighi, K. (1987a). Elliptic multiparameter eigenvalue problems. *Proc. Edinburgh Math. Soc. (2)*, 30(2):215–228.

[Binding and Seddighi, 1987b] Binding, P. A. and Seddighi, K. (1987b). On root vectors of selfadjoint pencils. *J. Funct. Anal.*, 70(1):117–125.

[Binding and Sleeman, 1991] Binding, P. A. and Sleeman, B. D. (1991). Spectral decomposition of uniformly elliptic multiparameter eigenvalue problems. *J. Math. Anal. Appl.*, 154(1):100–115.

[Binding and Volkmer, 2001a] Binding, P. A. and Volkmer, H. (2001a). Existence and asymptotics of eigenvalues of indefinite systems of Sturm-Liouville and Dirac type. *J. Differential Equations*, 172(1):116–133.

[Binding and Volkmer, 2001b] Binding, P. A. and Volkmer, H. (2001b). Oscillation theory for Sturm-Liouville problems with indefinite coefficients. *Proc. Roy. Soc. Edinburgh Sect. A*, 131(5):989–1002.

[Blum and Chang, 1978] Blum, E. K. and Chang, A. F. (1978). A numerical method for the solution of the double eigenvalue problem. *J. Inst. Math. Appl.*, 22(1):29–42.

[Bôcher, M., 1894] Bôcher, M. (1894). *Über die Reihenentwickelungen der Potentialtheorie. Mit einem Vorwort von F. Klein.* Leipzig. B. G. Teubner. VIII + 258 S.

[Bôcher, M., 1898a] Bôcher, M. (1898a). The theorems of oscillation of Sturm and Klein (first paper). *Bull. Amer. Math. Soc.*, 4(7):295–313.

[Bôcher, M., 1898b] Bôcher, M. (1898b). The theorems of oscillation of Sturm and Klein (second paper). *Bull. Amer. Math. Soc.*, 4(8):365–376.

[Bôcher, M., 1899] Bôcher, M. (1899). The theorems of oscillation of Sturm and Klein (third paper). *Bull. Amer. Math. Soc.*, 5:22–43.

[Bôcher, M., 1900] Bôcher, M. (1900). Randwertaufgaben bei gewöhnlichen Differentialgleichungen. In *Enzyklopàdie der mathematischen Wissenschaften II A 7a*, pages 437–463. Tubner, Leipzig.

[Browne, 1972a] Browne, P. J. (1972a). A multi-parameter eigenvalue problem. *J. Math. Anal. Appl.*, 38:553–568.

[Browne, 1972b] Browne, P. J. (1972b). A singular multi-parameter eigenvalue problem in second order ordinary differential equations. *J. Differential Equations*, 12:81–94.

[Browne, 1974] Browne, P. J. (1974). Multi-parameter problems. In *Ordinary and partial differential equations (Proc. Conf., Univ. Dundee, Dundee, 1974)*, pages 78–84. Lecture Notes in Math., Vol. 415. Springer, Berlin.

[Browne, 1975] Browne, P. J. (1974-1975). Multi-parameter spectral theory. *Indiana Univ. Math. J.*, 24:249–257.

[Browne, 1977a] Browne, P. J. (1977a). Abstract multiparameter theory. I. *J. Math. Anal. Appl.*, 60(1):259–273.

[Browne, 1977b] Browne, P. J. (1977b). Abstract multiparameter theory. II. *J. Math. Anal. Appl.*, 60(1):274–279.

[Browne, 1977c] Browne, P. J. (1977c). A completeness theorem for a nonlinear multiparameter eigenvalue problem. *J. Differential Equations*, 23(2):285–292.

[Browne, 1978] Browne, P. J. (1978). The interlacing of eigenvalues for periodic multi-parameter problems. *Proc. Roy. Soc. Edinburgh Sect. A*, 80(3–4):357–362.

[Browne, 1980] Browne, P. J. (1980). Abstract multiparameter theory. III. *J. Math. Anal. Appl.*, 73(2):561–567.

[Browne, 1982] Browne, P. J. (1982). Multiparameter problems: the last decade. In *Ordinary and partial differential equations (Dundee, 1982)*, volume 964 of *Lecture Notes in Math.*, pages 95–109. Springer, Berlin.

[Browne, 1989] Browne, P. J. (1989). Two-parameter eigencurve theory. In *Ordinary and partial differential equations, Vol. II (Dundee, 1988)*, volume 216 of *Pitman Res. Notes Math. Ser.*, pages 52–59. Longman Sci. Tech., Harlow.

[Browne and Isaev, 1988] Browne, P. J. and Isaev, H. (1988). Symmetric multiparameter problems and deficiency index theory. *Proc. Edinburgh Math. Soc. (2)*, 31(3):481–488.

[Browne and Sleeman, 1979] Browne, P. J. and Sleeman, B. D. (1979). Regular multiparameter eigenvalue problems with several parameters in the boundary conditions. *J. Math. Anal. Appl.*, 72(1):29–33.

[Browne and Sleeman, 1985] Browne, P. J. and Sleeman, B. D. (1985). Inverse multiparameter eigenvalue problems for matrices. *Proc. Roy. Soc. Edinburgh Sect. A*, 100(1–2):29–38.

[Browne and Sleeman, 1986] Browne, P. J. and Sleeman, B. D. (1986). Inverse multiparameter eigenvalue problems for matrices. II. *Proc. Edinburgh Math. Soc. (2)*, 29(3):343–348. (Erratum: Vol. 30 (2) (1987), 323).

[Browne and Sleeman, 1988] Browne, P. J. and Sleeman, B. D. (1988). Inverse multiparameter eigenvalue problems for matrices. III. *Proc. Edinburgh Math. Soc. (2)*, 31(1):151–155.

[Bublik and Lyashenko, 1995] Bublik, B. N. and Lyashenko, B. N. (1995). The eigenvalue problem in multiparameter optimization problems of theoretical physics. *Kibernet. Sistem. Anal.*, 3:155–159, 192.

[Bujurke et al., 2008] Bujurke, N. M., Salimath, C. S., and Shiralashetti, S. C. (2008). Computation of eigenvalues and solutions of regular Sturm-Liouville problems using Haar wavelets. *J. Comput. Appl. Math.*, 219(1):90–101.

[Camp, 1928] Camp, C. C. (1928). An expansion involving p inseparable parameters associated with a partial differential equation. *Amer. J. Math.*, 50(2):259–268.

[Camp, 1938] Camp, C. C. (1938). On multiparameter expansions associated with a differential system and auxiliary conditions at several points in each variable. *Amer. J. Math.*, 60(2):447–452.

[Cao et al., 2003] Cao, X., Kong, Q., Wu, H., and Zettl, A. (2003). Sturm-Liouville problems whose leading coefficient function changes sign. *Canad. J. Math.*, 55(4):724–749.

[Carmichael, 1921a] Carmichael, R. D. (1921a). Boundary value and expansion problems: algebraic basis of the theory. *Amer. J. Math.*, 43(2):69–101.

[Carmichael, 1921b] Carmichael, R. D. (1921b). Boundary value and expansion problems; formulation of various transcendental problems. *Amer. J. Math.*, 43(4):232–270.

[Carmichael, 1922] Carmichael, R. D. (1922). Boundary value and expansion problems: oscillation, comparison, and expansion theorems. *Amer. J. Math.*, 44(2):129–152.

[Cernea, 2008] Cernea, A. (2008). Continuous version of Filippov's theorem for a Sturm-Liouville type differential inclusion. *Electron. J. Differential Equations*, pages No. 53, 7.

[Chanane, 2008] Chanane, B. (2008). Sturm-Liouville problems with parameter dependent potential and boundary conditions. *J. Comput. Appl. Math.*, 212(2):282–290.

[Clark, 1987] Clark, S. L. (1987). *Some qualitative properties of the spectral density function for Hamiltonian systems*. PhD thesis, University of Tennessee, Department of Mathematics, Knoxville, TN.

[Clark and Gesztesy, 2004] On Weyl-Jitchmarsh theory for singular finite difference Hamiltonian systems. *J. Comput. Appl. Math.*, 71(1–2):151–184.

[Coddington and Levinson, 1955] Coddington, E. A. and Levinson, N. (1955). *Theory of ordinary differential equations*. McGraw-Hill Book Company, Inc., New York–Toronto–London.

[Collatz, 1968] Collatz, L. (1968). Multiparametric eigenvalue problems in inner-product spaces. *J. Comput. System Sci.*, 2:333–341.

[Constantin, 1997] Constantin, A. (1997). A general-weighted Sturm-Liouville problem. *Ann. Scuola Norm. Sup. Pisa Cl. Sci. (4)*, 24(4):767–782 (1998).

[Constantin and McKean, 1999] Constantin, A. and McKean, H. P. (1999). A shallow water equation on the circle. *Comm. Pure Appl. Math.*, 52(8):949–982.

[Cordes, 1953] Cordes, H. O. (1953). Separation der Variablen in Hilbertschen Räumen. *Math. Ann.*, 125:401–434.

[Cordes, 1954a] Cordes, H. O. (1954a). Der Entwicklungssatz nach Produkten bei singulären Eigenwertproblemen partieller Differentialgleichungen, die durch Separation zerfallen. *Nachr. Akad. Wiss. Göttingen. Math.-Phys. Kl. Math.-Phys.-Chem. Abt.*, 1954:51–69.

[Cordes, 1954b] Cordes, H. O. (1954b). Über die Spektralzerlegung von hypermaximalen Operatoren, die durch Separation der Variablen zerfallen. I, II. *Math. Ann.*, 128:257–289; 373–411 (1955).

[Dai, 1999] Dai, H. (1999). Numerical methods for solving multiparameter eigenvalue problems. *Int. J. Comput. Math.*, 72(3):331–347.

[Deift and Hempel, 1986] Deift, P. A. and Hempel, R. (1986). On the existence of eigenvalues of the Schrödinger operator $H - \lambda W$ in a gap of $\sigma(H)$. *Comm. Math. Phys.*, 103(3):461–490.

[Dixon, 1907] Dixon, A. C. (1907). Harmonic expansions of functions of two variables. *Proc. London. Math. Soc. (2)*, 5:411–478.

[Doole, 1931] Doole, H. (1931). A certain multiple-parameter expansion. *Bull. Amer. Math. Soc.*, 37:439–446.

[D'yachenko, 2000] D'yachenko, A. V. (2000). Asymptotics of the eigenvalues of an indefinite Sturm-Liouville problem. *Mat. Zametki*, 68(1):139–143.

[Dymarskiĭ, 2006] Dymarskiĭ, Y. M. (2006). On the topological properties of manifolds of eigenfunctions generated by a family of periodic Sturm-Liouville problems. *Sovrem. Mat. Fundam. Napravl.*, 16:22–37.

[Dzhabar-Zadeh, 1999] Dzhabar-Zadeh, R. M. (1999). The multiparameter analogue of the resolvent operator. *Proc. Inst. Math. Mech. Acad. Sci. Azerb.*, 11:37–44, 215 (2000).

[Eberhard and Elbert, 2000] Eberhard, W. and Elbert, Á. (2000). On the eigenvalues of half-linear boundary value problems. *Math. Nachr.*, 213:57–76.

[Eberhard and Freiling, 1992] Eberhard, W. and Freiling, G. (1992). The distribution of the eigenvalues for second order eigenvalue problems in the presence of an arbitrary number of turning points. *Results Math.*, 21(1-2):24–41.

[Eberhard et al., 1994] Eberhard, W., Freiling, G., and Schneider, A. (1994). Connection formulas for second order differential equations with a complex parameter and having an arbitrary number of turning points. *Math. Nachr.*, 165:205–229.

[Eberhard et al., 2001] Eberhard, W., Freiling, G., and Wilcken-Stoeber, K. (2001). Indefinite eigenvalue problems with several singular points and turning points. *Math. Nachr.*, 229:51–71.

[Elbert, 1981] Elbert, Á. (1981). A half-linear second order differential equation. In *Qualitative theory of differential equations, Vol. I, II (Szeged, 1979)*, volume 30 of *Colloq. Math. Soc. János Bolyai*, pages 153–180. North-Holland, Amsterdam.

[Erdélyi et al., 1953] Erdélyi, A., Magnus, W., Oberhettinger, F., and Tricomi, F. G. (1953). *Higher transcendental functions. Vols. I, II*. McGraw-Hill Book Company, Inc., New York–Toronto–London. Based, in part, on notes left by Harry Bateman.

[Everitt, 1981] Everitt, W. N. (1981). On certain regular ordinary differential expressions and related differential operators. In *Spectral theory of differential operators (Birmingham, Ala., 1981)*, volume 55 of *North-Holland Math. Stud.*, pages 115–167. North-Holland, Amsterdam.

[Everitt et al., 1983] Everitt, W. N., Kwong, M. K., and Zettl, A. (1983). Oscillation of eigenfunctions of weighted regular Sturm-Liouville problems. *J. London Math. Soc. (2)*, 27(1):106–120.

[Faierman, 1966] Faierman, M. (1966). *Boundary value problems in differential equations*. PhD thesis, University of Toronto, Department of Mathematics, Toronto, Ontario, Canada, M5S 1A1.

[Faierman, 1969] Faierman, M. (1969). The completeness and expansion theorems associated with the multi-parameter eigenvalue problem in ordinary differential equations. *J. Differential Equations*, 5:197–213.

[Faierman, 1971] Faierman, M. (1971). On a perturbation in a two-parameter ordinary differential equation of the second order. *Canad. Math. Bull.*, 14:25–33.

[Faierman, 1972a] Faierman, M. (1972a). An oscillation theorem for a one-parameter ordinary differential equation of the second order. *J. Differ. Equations*, 11:10–37.

[Faierman, 1972b] Faierman, M. (1972b). Asymptotic formulae for the eigenvalues of a two-parameter ordinary differential equation of the second order. *Trans. Amer. Math. Soc.*, 168:1–52.

[Faierman, 1974] Faierman, M. (1974). The expansion theorem in multi-parameter Sturm-Liouville theory. In *Ordinary and partial differential equations (Proc. Conf., Univ. Dundee, Dundee, 1974)*, pages 137–142. Lecture Notes in Math., Vol. 415. Springer, Berlin.

[Faierman, 1975a] Faierman, M. (1974–1975a). Asymptotic formulae for the eigenvalues of a two parameter system of ordinary differential equations of the second order. *Canad. Math. Bull.*, 17(5):657–665.

[Faierman, 1975b] Faierman, M. (1974–1975b). A note on Klein's oscillation theorem for periodic boundary conditions. *Canad. Math. Bull.*, 17(5):749–755.

[Faierman, 1977] Faierman, M. (1977). On the distribution of the eigenvalues of a two-parameter system of ordinary differential equations of the second order. *SIAM J. Math. Anal.*, 8(5):854–870.

[Faierman, 1978] Faierman, M. (1978). Eigenfunction expansions associated with a two-parameter system of differential equations. *Proc. Roy. Soc. Edinburgh Sect. A*, 81(1–2):79–93.

[Faierman, 1979b] Faierman, M. (1978-1979b). An oscillation theorem for a two-parameter system of differential equations. *Quaestiones Math.*, 3(4):313–321.

[Faierman, 1979a] Faierman, M. (1979a). Distribution of eigenvalues of a two-parameter system of differential equations. *Trans. Amer. Math. Soc.*, 247:45–86.

[Faierman, 1980] Faierman, M. (1980). Bounds for the eigenfunctions of a two-parameter system of ordinary differential equations of the second order. *Pacific J. Math.*, 90(2):335–345.

[Faierman, 1981a] Faierman, M. (1981a). An eigenfunction expansion associated with a two-parameter system of differential equations. I. *Proc. Roy. Soc. Edinburgh Sect. A*, 89(1–2):143–155.

[Faierman, 1981b] Faierman, M. (1981b). An eigenfunction expansion associated with a two-parameter system of differential equations. In *Spectral theory of differential operators (Birmingham, Ala., 1981)*, volume 55 of *North-Holland Math. Stud.*, pages 169–172. North-Holland, Amsterdam.

[Faierman, 1982] Faierman, M. (1982). An eigenfunction expansion associated with a two-parameter system of differential equations. II. *Proc. Roy. Soc. Edinburgh Sect. A*, 92(1-2):87–93.

[Faierman, 1983a] Faierman, M. (1982–1983a). An eigenfunction expansion associated with a two-parameter system of differential equations. III. *Proc. Roy. Soc. Edinburgh Sect. A*, 93(3–4):189–195.

[Faierman, 1983b] Faierman, M. (1982–1983b). An oscillation theorem for a two-parameter system of differential equations with periodic boundary conditions. *Quaestiones Math.*, 5(2):107–118.

[Faierman, 1984] Faierman, M. (1984). A left definite two-parameter eigenvalue problem. In *Differential equations (Birmingham, Ala., 1983)*, volume 92 of *North-Holland Math. Stud.*, pages 205–211. North-Holland, Amsterdam.

[Faierman, 1985] Faierman, M. (1985). The eigenvalues of a multiparameter system of differential equations. *Applicable Anal.*, 19(4):275–290.

[Faierman, 1986] Faierman, M. (1986). Expansions in eigenfunctions of a two-parameter system of differential equations. *Quaestiones Math.*, 10(2):135–152.

[Faierman, 1987] Faierman, M. (1987). Expansions in eigenfunctions of a two-parameter system of differential equations. II. *Quaestiones Math.*, 10(3):217–249.

[Faierman, 1991a] Faierman, M. (1991a). *Two-parameter eigenvalue problems in Differential Equations*. Pitman Research Notes in Mathematics No. 205. Longmans, London.

[Faierman, 1991b] Faierman, M. (1991b). *Two-parameter eigenvalue problems in ordinary differential equations*, volume 205 of *Pitman Research Notes in Mathematics Series*. Longman Scientific & Technical, Harlow.

[Faierman and Mennicken, 2005] Faierman, M. and Mennicken, R. (2005). A non-standard multiparameter eigenvalue problem in ordinary differential equations. *Math. Nachr.*, 278(12–13):1550–1560.

[Faierman et al., 2008] Faierman, M., Möller, M., and Watson, B. A. (2008). Completeness theorems for a non-standard two-parameter eigenvalue problem. *Integral Equations Operator Theory*, 60(1):37–52.

[Faierman and Roach, 1988a] Faierman, M. and Roach, G. F. (1988a). Eigenfunction expansions for a two-parameter system of differential equations. *Quaestiones Math.*, 12(1):65–99.

[Faierman and Roach, 1988b] Faierman, M. and Roach, G. F. (1988b). Full and partial-range eigenfunction expansions for a multiparameter system of differential equations. *Appl. Anal.*, 28(1):15–37.

[Faierman and Roach, 1989] Faierman, M. and Roach, G. F. (1989). Eigenfunction expansions associated with a multiparameter system of differential equations. *Differential Integral Equations*, 2(1):45–56.

[Falckenberg, 1915] Falckenberg, H. (1915). Zur Theorie der Kreisbogenpolygone. *Math. Ann.*, 77:65–80.

[Falckenberg, 1917] Falckenberg, H. (1917). Zur Theorie der Kreisbogenpolygone. II. *Math. Ann.*, 78:234–256.

[Falckenberg, H. Hilb, E., 1916] Falckenberg, H. and Hilb, E. (1916). Die Anzal der Nullstellen der Hankelschen Funktionen. *Gött. Nachr.*, pages 190–196.

[Farenick, 1986] Farenick, D. (1986). *Gersgorin theory and determinantal operators*. PhD thesis, University of Calgary, Alberta, Canada. (M.Sc. dissertation).

[Farenick and Browne, 1987] Farenick, D. R. and Browne, P. J. (1987). Geršgorin theory and the definiteness of determinantal operators. *Proc. Roy. Soc. Edinburgh Sect. A*, 106(1–2):1–10.

[Fleige, 1999] Fleige, A. (1999). Non-semibounded sesquilinear forms and left-indefinite Sturm-Liouville problems. *Integral Equations Operator Theory*, 33(1):20–33.

[Fleige, 2008] Fleige, A. (2008). The Riesz basis property of an indefinite Sturm-Liouville problem with a non-odd weight function. *Integral Equations Operator Theory*, 60(2):237–246.

[Fulton et al., 2008a] Fulton, C., Pearson, D., and Pruess, S. (2008a). Efficient calculation of spectral density functions for specific classes of singular Sturm-Liouville problems. *J. Comput. Appl. Math.*, 212(2):150–178.

[Fulton et al., 2008b] Fulton, C., Pearson, D., and Pruess, S. (2008b). New characterizations of spectral density functions for singular Sturm-Liouville problems. *J. Comput. Appl. Math.*, 212(2):194–213.

[Fulton, 1977] Fulton, C. T. (1977). Two-point boundary value problems with eigenvalue parameter contained in the boundary conditions. *Proc. Roy. Soc. Edinburgh Sect. A*, 77(3-4):293–308.

[Gadzhiev, 1982] Gadzhiev, G. A. (1982). Multiparameter spectral theory of differential operators. In *Investigation of linear operators and their applications*, pages 64–75. Azerbaĭdzhan. Gos. Univ., Baku.

[Gadzhiev, 1985] Gadzhiev, G. A. (1985). A multitime equation and its reduction to a multiparameter spectral problem. *Dokl. Akad. Nauk SSSR*, 285(3):530–533.

[Gadzhiev, 1989] Gadzhiev, G. A. (1989). *Mnogoparametricheskaya spektralnaya teoriya v gilbertovom prostranstve*. Azerbaĭdzhan. Gos. Univ., Baku.

[Gadžiev, 1979] Gadžiev, G. A. (1979). Theorem on the completeness of a system of eigenvectors of a two-parameter Dirac system. *Azerbaĭdzhan. Gos. Univ. Uchen. Zap.*, 2:70–75.

[Greguš et al., 1971] Greguš, M., Neuman, F., and Arscott, F. M. (1971). Three-point boundary value problems in differential equations. *J. London Math. Soc. (2)*, 3:429–436.

[Grunenfelder and Košir, 1996a] Grunenfelder, L. and Košir, T. (1996a). An algebraic approach to multiparameter spectral theory. *Trans. Amer. Math. Soc.*, 348(8):2983–2998.

[Grunenfelder and Košir, 1996b] Grunenfelder, L. and Košir, T. (1996b). Coalgebras and spectral theory in one and several parameters. In *Recent developments in operator theory and its applications (Winnipeg, MB, 1994)*, volume 87 of *Oper. Theory Adv. Appl.*, pages 177–192. Birkhäuser, Basel.

[Grunenfelder and Košir, 1997] Grunenfelder, L. and Košir, T. (1997). Koszul cohomology for finite families of comodule maps and applications. *Comm. Algebra*, 25(2):459–479.

[Grunenfelder and Košir, 1998] Grunenfelder, L. and Košir, T. (1998). Geometric aspects of multiparameter spectral theory. *Trans. Amer. Math. Soc.*, 350(6):2525–2546.

[Guseĭnov, 1980] Guseĭnov, G. Š. (1980). Eigenfunction expansions for multiparameter differential and difference equations with periodic coefficients. *Dokl. Akad. Nauk SSSR*, 253(5):1040–1044.

[Hartman, 2002] Hartman, P. (2002). *Ordinary differential equations*, volume 38 of *Classics in Applied Mathematics*. Society for Industrial and Applied Mathematics (SIAM), Philadelphia, PA. Corrected reprint of the second (1982) edition [Birkhäuser, Boston, MA; MR0658490 (83e:34002)], With a foreword by Peter Bates.

[Haupt, 1910] Haupt, O. (1910). Bemerkungen über Oszillationstheoreme. *Gött. Nachr.*, 85–86.

[Haupt, 1911] Haupt, O. (1911). *Untersuchungen über Oszillationstheoreme.* Diss. Würzburg. Leipzig: B. Z. Teubner. 50 S. 8°.

[Haupt, 1912] Haupt, O. (1912). *Über die Entwicklung einer willkürlichen Funktion nach den Eigenfunktionen des Turbulenzproblems.* Münch. Ber. 1912.

[Haupt, 1914] Haupt, O. (1914). Über eine Methode zum Beweise von Oszillationstheoremen. *Math. Ann.*, 76:67–104.

[Haupt, 1918] Haupt, O. (1918). Über lineare homogene Differentialgleichungen zweiter Ordnung mit periodischen Koeffizienten. *Math. Ann.*, 79:278–285.

[Haupt, 1929] Haupt, O. (1929). Über ein Oszillationstheorem. *Sitzungsberichte Phys.-Med. Sozietfit Erlangen*, 61:203–206.

[Haupt and Hilb, 1923] Haupt, O. and Hilb, E. (1923). Oszillationstheoreme oberhalb der Stieltjesschen Grenze. *Math. Ann.*, 89(1–2):130–146.

[Haupt and Hilb, 1924] Haupt, O. and Hilb, E. (1924). Über Greensche Randbedingungen. *Math. Ann.*, 92(1–2):95–103.

[Heine, 1961] Heine, E. (1961). *Handbuch der Kugelfunctionen. Theorie und Anwendungen. Band I, II, 1881.* Zweite umgearbeitete und vermehrte Auflage. Thesaurus Mathematicae, No. 1. Physica-Verlag, Würzburg, 1961.

[Hilb, 1906] Hilb, E. (1906). Die Reihenentwicklungen der Potentialtheorie. *Math. Ann.*, 63:38–53.

[Hilb, 1907] Hilb, E. (1907). Eine Erweiterung des *Klein*schen Oszillationstheorems. *[J] Deutsche Math.-Ver.*, 16:279–285.

[Hilb, 1908] Hilb, E. (1908). Über Kleinsche Theoreme in der Theorie der linearen Differentialgleichungen. *Math. Ann.*, 66(2):215–257.

[Hilb, 1909] Hilb, E. (1909). Über Integraldarstellung willkürlicher Funktionen. *Math. Ann.*, 66:1–66.

[Hilb, 1911] Hilb, E. (1911). Über Reihenentwicklungen nach den Eigenfunktionen linearer Differentialgleichungen 2ter Ordnung. *Math. Ann.*, 71(1):76–87.

[Hilb, E., 1909] Hilb, E. (1909). Über Kleinsche Theoreme in der Theorie der linearen Differentialgleichungen (2. Mitt). *Math. Ann.*, 68:24–74.

[Hilbert, 1912] Hilbert, D. (1912). *Grundzüge einer allgemeinen Theorie der linearen Integralgleichungen.* Leipzig und Berlin: B. G. Teubner. XXVI u. 282 S. gr. 8°. (Fortschr. d. math. Wissensch. in Monographien hrsgb. von *O. Blumenthal*, Heft 3.).

[Hille, 1969] Hille, E. (1969). *Lectures on ordinary differential equations.* Addison-Wesley Publ. Co., Reading, Mass.-London-Don Mills, Ont.

[Hinton, 2005] Hinton, D. B. (2005). Sturm's 1836 oscillation results: Evolution of the theory. In *Sturm-Liouville Theory,-Past and Present,* W. Amrein et. al., eds, pages 1–27. Birkhauser, Basel.

[Hochstenbach and Plestenjak, 2003] Hochstenbach, M. E. and Plestenjak, B. (2003). Backward error, condition numbers, and pseudospectra for the multiparameter eigenvalue problem. *Linear Algebra Appl.,* 375:63–81.

[Hosseinabady, 2003] Hosseinabady, A. (2003). Some properties of the eigenfunctions of Sturm-Liouville problem with two turning points. *Iran. J. Sci. Tech.,* 27(A2):381–388.

[Howe, 1971] Howe, A. (1971). Klein's oscillation theorem for periodic boundary conditions. *Canad. J. Math.,* 23:699–703.

[Ince, 1956] Ince, E. L. (1956). *Ordinary Differential Equations.* Dover Publications, New York.

[Isaev, 1977] Isaev, G. (1977). Numerical range and spectrum of holomorphic operator-functions of several complex variables. *[J] Usp. Mat. Nauk,* 32(6):253–254.

[Isaev, 1982] Isaev, G. (1982). *Selected questions of multiparameter spectral theory.* PhD thesis, Steklov Institute, Moscow. (Abstract of dissertation).

[Isaev, 1975] Isaev, G. A. (1975). The numerical range of operator pencils and multiple completeness in the sense of M. V. Keldyš. *Funkcional. Anal. i Priložen.,* 9(1):31–34.

[Isaev, 1976a] Isaev, G. A. (1976a). On multiparameter spectral theory. *Dokl. Akad. Nauk SSSR,* 229(2):284–287.

[Isaev, 1976b] Isaev, G. A. (1976b). Separation of the numerical range of multiparameter spectral problems. *Izv. Akad. Nauk Azerbaǐdžan. SSR Ser. Fiz.-Tehn. Mat. Nauk,* 5:9–13.

[Isaev, 1979] Isaev, G. A. (1979). Questions in the theory of selfadjoint multiparameter problems. In *Spectral theory of operators (Proc. Second All-Union Summer Math. School, Baku, 1975) (Russian),* pages 87–102. "Èlm," Baku.

[Isaev, 1980a] Isaev, G. A. (1980a). Introduction to a general multiparameter spectral theory. In *Spectral theory of operators, No. 3 (Russian),* pages 142–201. "Èlm," Baku.

[Isaev, 1980b] Isaev, G. A. (1980b). The root elements of multiparameter spectral problems. *Dokl. Akad. Nauk SSSR,* 250(3):544–547.

[Isaev, 1981a] Isaev, G. A. (1981a). Expansion in eigenfunctions of selfadjoint singular multiparameter differential operators. *Dokl. Akad. Nauk SSSR*, 260(4):786–790.

[Isaev, 1981b] Isaev, G. A. (1981b). On the theory of indices of the defect of multiparameter differential operators of Sturm-Liouville type. *Dokl. Akad. Nauk SSSR*, 261(4):788–791.

[Isaev, 1983] Isaev, G. A. (1983). Genetic operators and multiparameter spectral problems. *Dokl. Akad. Nauk SSSR*, 268(4):785–788.

[Isaev, 1985] Isaev, H. (1985). Lectures on multiparameter spectral theory. Technical report, University of Calgary.

[Isaev, 1986a] Isaev, G. A. (1986a). Quadratic integrability of the product of solutions of Sturm-Liouville multiparametric equations. In *Spectral theory of operators and its applications, No. 7 (Russian)*, pages 109–143. "Èlm," Baku.

[Isaev, 1986b] Isaev, G. A. (1986b). Singular multiparameter differential operators. Expansion theorems. *Mat. Sb. (N.S.)*, 131(173)(1):52–72, 127.

[Isaev, 1991] Isaev, H. (1991). Glimpses of multiparameter spectral theory. In *Ordinary and partial differential equations, Vol. III (Dundee, 1990)*, volume 254 of *Pitman Res. Notes Math. Ser.*, pages 106–125. Longman Sci. Tech., Harlow.

[Isaev and Allahverdiev, 1980] Isaev, G. A. and Allahverdiev, B. P. (1980). Oscillation theorems for multiparameter spectral problems connected with second-order differential equations. In *Spectral theory of operators, No. 3 (Russian)*, pages 202–221. "Èlm," Baku.

[Isaev and Faĭnshteĭn, 1987] Isaev, G. A. and Faĭnshteĭn, A. S. (1987). The Taylor spectrum and multiparameter spectral theory for systems of operators. *Dokl. Akad. Nauk SSSR*, 297(1):30–34.

[Isaev and Kuliev, 1986] Isaev, G. A. and Kuliev, T. Y. (1986). Analytic properties of the boundary of the numerical range of operators in a Hilbert space. *Dokl. Akad. Nauk SSSR*, 291(5):1050–1053.

[Ismagilov and Kostyuchenko, 2008] Ismagilov, R. S. and Kostyuchenko, A. G. (2008). On the asymptotics of the spectrum of a nonsemibounded vector Sturm-Liouville operator. *Funktsional. Anal. i Prilozhen.*, 42(2):11–22, 95.

[Ji, 1991a] Ji, X. (1991a). An iterative method for the numerical solution of two-parameter eigenvalue problems. *Int. J. Comput. Math.*, 41(1–2):91–98.

[Ji, 1991b] Ji, X. Z. (1991b). Numerical solution of joint eigenpairs of a family of commutative matrices. *Appl. Math. Lett.*, 4(3):57–60.

[Jodayree Akbarfam and Mingarelli, 2005] Jodayree Akbarfam, A. and Mingarelli, A. B. (2005). Duality for an indefinite inverse Sturm-Liouville problem. *J. Math. Anal. Appl.*, 312(2):435–463.

[Källström and Sleeman, 1974] Källström, A. and Sleeman, B. D. (1974). A multi-parameter Sturm-Liouville problem. In *Ordinary and partial differential equations (Proc. Conf., Univ. Dundee, Dundee, 1974)*, pages 394–401. Lecture Notes in Math., Vol. 415. Springer, Berlin.

[Källström and Sleeman, 1975a] Källström, A. and Sleeman, B. D. (1974-1975a). An abstract relation for multiparameter eigenvalue problems. *Proc. Roy. Soc. Edinburgh Sect. A*, 74:135–143 (1976).

[Källström and Sleeman, 1975b] Källström, A. and Sleeman, B. D. (1974-1975b). A left definite multiparameter eigenvalue problem in ordinary differential equations. *Proc. Roy. Soc. Edinburgh Sect. A*, 74:145–155 (1976).

[Källström and Sleeman, 1976] Källström, A. and Sleeman, B. D. (1976). Solvability of a linear operator system. *J. Math. Anal. Appl.*, 55(3):785–793.

[Källström and Sleeman, 1977] Källström, A. and Sleeman, B. D. (1977). Multiparameter spectral theory. *Ark. Mat.*, 15(1):93–99.

[Källström and Sleeman, 1985] Källström, A. and Sleeman, B. D. (1985). Joint spectra for commuting operators. *Proc. Edinburgh Math. Soc. (2)*, 28(2):233–248.

[Kaper et al., 1984] Kaper, H. G., Kwong, M. K., and Zettl, A. (1984). Regularizing transformations for certain singular Sturm-Liouville boundary value problems. *SIAM J. Math. Anal.*, 15(5):957–963.

[Karabash and Kostenko, 2008] Karabash, I. and Kostenko, A. (2008). Indefinite Sturm-Liouville operators with the singular critical point zero. *Proc. Roy. Soc. Edinburgh Sect. A*, 138(4):801–820.

[Khazanov, 1994] Khazanov, V. B. (1994). Multiparameter eigenvalue problem: Jordan semilattices of vectors. *Zap. Nauchn. Sem. S.-Peterburg. Otdel. Mat. Inst. Steklov. (POMI)*, 219(Chisl. Metody i Voprosy Organ. Vychisl. 10):213–220, 224.

[Klein, 1881a] Klein, F. (1881a). Über Körper, welche von confocalen Flächen zweiten Grades begrenzt sind. *Math. Ann.*, 18:521–539.

[Klein, 1881b] Klein, F. (1881b). Über Laméschen Functionen. *Math. Ann.*, 18:512–520.

[Klein, 1890] Klein, F. (1890). Zur Theorie der allgemeinen Laméschen Functionen. *Gött. Nachr.*, 1890(4):540–549.

[Klein, 1891] Klein, F. (1891). Über Normirung der linearen Differentialgleinchungen zweiter Ordnung. *Math. Ann.*, 38:144–152.

[Klein, 1892] Klein, F. (1892). Über den Hermítéschen Fall der Laméschen Differentialgleichung. *Math. Ann.*, 40(1):125–129.

[Klein, 1894] Klein, F. (1894). Autographirte Vorlesungshefte. *Math. Ann.*, 45(1):140–152.

[Klein, 1895] Klein, F. (1895). Autographirte Vorlesungshefte. *Math. Ann.*, 46:77–90.

[Klein, 1907a] Klein, F. (1907a). Bemerkungen zur Theorie der linearen Differentialgleichungen zweiter Ordnung. *Math. Ann.*, 64(2):175–196.

[Klein, 1907b] Klein, F. (1907b). *Über den Zusammenhang zwischen dem sogenannten Oszillationstheorem der linearen Differentialgleichungen und dem Fundamentaltheorem der automorphen Funktionen.* Deutsche Math.-Ver. 16, 537 .

[Klein, 1922] Klein, F. (1922). *Anschauliche Geometrie. Substitutionsgruppen und Gleichungstheorie. Zur Mathematischen Physik.* Gesammelte Abhandlungen, Bd. 2. hg. V. R. Fricke/ H. Vermeil, Berlin.

[Klein, 1980] Klein, F. (1980). Zum Oszillationstheorem jenseits der Stieltjesschen Grenze (1894). In *Gesammelte mathematische Abhandlungen*, volume 2, pages 597–600. Springer Verlag, Auflage: 1 (Mai 1998), Berlin, New York. (Klein's Collected Works).

[Kong and Kong, 2008a] Kong, L. and Kong, Q. (2008a). Right-indefinite half-linear sturm-liouville problems. *Computers and Mathematics with Applications*, 55(11):2554–2564.

[Kong and Kong, 2008b] Kong, L. and Kong, Q. (2008b). Right-indefinite half-linear Sturm-Liouville problems. *Comput. Math. Appl.*, 55(11):2554–2564.

[Kong et al., 2001] Kong, Q., Wu, H., and Zettl, A. (2001). Left-definite Sturm-Liouville problems. *J. Differential Equations*, 177(1):1–26.

[Kong et al., 2008] Kong, Q., Wu, H., and Zettl, A. (2008). Limits of Sturm-Liouville eigenvalues when the interval shrinks to an end point. *Proc. Roy. Soc. Edinburgh Sect. A*, 138(2):323–338.

[Konstantinov, 1994] Konstantinov, A. Y. (1994). Eigenfunction expansion of right definite multiparameter problems. In *Mathematical results in quantum mechanics (Blossin, 1993)*, volume 70 of *Oper. Theory Adv. Appl.*, pages 343–346. Birkhäuser, Basel.

[Konstantinov, 1995] Konstantinov, A. Y. (1995). Nonuniformly definite multiparameter spectral problems. *Ukraïn. Mat. Zh.*, 47(5):659–670.

[Konstantinov and Stadnyuk, 1993] Konstantinov, A. Y. and Stadnyuk, A. G. (1993). Eigenfunction expansion of symmetric multiparameter problems. *Ukraïn. Mat. Zh.*, 45(8):1166–1169.

[Koornwinder, 1980] Koornwinder, T. H. (1980). A precise definition of separation of variables. In *Geometrical approaches to differential equations (Proc. Fourth Scheveningen Conf., Scheveningen, 1979)*, volume 810 of *Lecture Notes in Math.*, pages 240–263. Springer, Berlin.

[Korotyaev, 2003] Korotyaev, E. (2003). Periodic "weighted" operators. *J. Differential Equations*, 189(2):461–486.

[Košir, 1994] Košir, T. (1994). Finite-dimensional multiparameter spectral theory: the nonderogatory case. In *Proceedings of the 3rd ILAS Conference (Pensacola, FL, 1993)*, volume 212/213, pages 45–70.

[Košir, 2003] Košir, T. (2003). The Cayley-Hamilton theorem and inverse problems for multiparameter systems. *Linear Algebra Appl.*, 367:155–163.

[Košir, 2004] Košir, T. (2004). Root vectors for geometrically simple multiparameter eigenvalues. *Integral Equations Operator Theory*, 48(3):365–396.

[Košir and Plestenjak, 2002] Košir, T. and Plestenjak, B. (2002). On stability of invariant subspaces of commuting matrices. *Linear Algebra Appl.*, 342:133–147.

[Krüger and Teschl, 2008] Krüger, H. and Teschl, G. (2008). Relative oscillation theory for Sturm-Liouville operators extended. *J. Funct. Anal.*, 254(6):1702–1720.

[Kublanovskaya and Khazanov, 2006] Kublanovskaya, V. N. and Khazanov, V. B. (2006). On the solution of inverse eigenvalue problems for parametric matrices. *Zap. Nauchn. Sem. S.-Peterburg. Otdel. Mat. Inst. Steklov. (POMI)*, 334(Chisl. Metody i Vopr. Organ. Vychisl. 19):174–192, 272.

[Lamé, 1837] Lamé, G. (1837). Mémoire sur les surfaces isothermes dans les corps solides homogènes en équilibre de température. *J. Maths. Pures Appl.*, 2:147–183.

[Lamé, 1854] Lamé, G. (1854). Mémoire sur l'équilibre et l'élasticité des enveloppes sphériques. *J. Maths. Pures Appl.*, 19:51–87.

[Lamé, 1859] Lamé, G. (1859). *Leçons sur les coordonnées curvilignes et leurs diverses applications*. Mallet-Bachelier, Paris.

[Lawther, 2006] Lawther, R. (2006). Erratum: "On the straightness of eigenvalue interactions" [Comput. Mech. Online First, 2005]. *Comput. Mech.*, 37(4):362–368.

[Levinson, 1949] Levinson, N. (1949). Criteria for the limit-point case for second order linear differential operators. *Časopis Pěst. Mat. Fys.*, 74:17–20.

[Li and Liu, 1996] Li, X. P. and Liu, X. G. (1996). The problem of joint eigenvalues of pencils of real symmetric matrices. *Sichuan Daxue Xuebao*, 33(5):494–498.

[Liu, 2008a] Liu, C.-S. (2008a). A Lie-group shooting method for computing eigenvalues and eigenfunctions of Sturm-Liouville problems. *CMES Comput. Model. Eng. Sci.*, 26(3):157–168.

[Liu, 2008b] Liu, C.-S. (2008b). Solving an inverse Sturm-Liouville problem by a Lie-group method. *Bound. Value Probl.*, pages Art. ID 749865, 18.

[Liu and Ma, 2007] Liu, X. G. and Ma, H. Y. (2007). Sensitivity analysis for multiparameter eigenvalue problems. *J. Ocean Univ. China Nat. Sci.*, 37(5):837–840.

[Liu and Xu, 2008] Liu, X. G. and Xu, L. H. (2008). The Crawford number of definite multiparameter systems. *J. Ocean Univ. China Nat. Sci.*, 38(4):681–684.

[Liu, 2008c] Liu, Y. (2008c). On Sturm-Liouville boundary value problems for second-order nonlinear functional finite difference equations. *J. Comput. Appl. Math.*, 216(2):523–533.

[Loud, 1975] Loud, W. S. (1975). Stability regions for Hill's equation. *J. Differential Equations*, 19(2):226–241.

[Lützen and Mingarelli, 2008] Lützen, J. and Mingarelli, A. B. (2008). Charles François Sturm and Differential Equations. In *The Complete Works of Charles François Sturm*, J.-C. Pont, ed., pages 25–45. Birkhauser, Basel.

[Lyashenko, 2000] Lyashenko, B. M. (2000). Mathematical models of the two-center problem in quantum mechanics. *Vīsn. Kiïv. Unīv. Ser. Fīz.-Mat. Nauki*, 3:263–268.

[Ma, 1972a] Ma, S. S.-P. (1972a). *Boundary value problems for a matrix-differential system in two parameters.* PhD thesis, University of Toronto, Department of Mathematics, Toronto, Ontario, Canada, M5S 1A1.

[Ma, 1972b] Ma, S. S.-P. (1972b). *Boundary value problems for a matrix differential system in two parameters.* PhD thesis, University of Toronto, Department of Mathematics, Sidney Smith Hall. (Abstract of dissertation).

[Magnus and Winkler, 1966] Magnus, W. and Winkler, S. (1966). *Hill's equation.* Interscience Tracts in Pure and Applied Mathematics, No. 20. Interscience Publishers John Wiley & Sons New York–London–Sydney.

[Makin, 2008] Makin, A. S. (2008). Characterization of the spectrum of regular boundary value problems for the Sturm-Liouville operator. *Differ. Uravn.*, 44(3):329–335, 429.

[Marasi and Jodayree Akbarfam, 2007] Marasi, H. R. and Jodayree Akbarfam, A. (2007). On the canonical solution of indefinite problem with m turning points of even order. *J. Math. Anal. Appl.*, 332(2):1071–1086.

[Marasi et al., 2008] Marasi, H. R., Jodayree Akbarfam, A., and Saei, F. D. (2008). Asymptotic form of the solution of Sturm-Liouville problem with m turning points. *Int. J. Pure Appl. Math.*, 44(5):691–702.

[Marletta and Zettl, 2004] Marletta, M. and Zettl, A. (2004). Spectral exactness and spectral inclusion for singular left-definite Sturm-Liouville problems. *Results Math.*, 45(3–4):299–308.

[May and Easton, 1996] May, R. L. and Easton, A. K., editors (1996). *Computational techniques and applications: CTAC95*, River Edge, NJ. World Scientific Publishing Co. Inc.

[McGhee, 1983] McGhee, D. F. (1982-1983). Multiparameter problems and joint spectra. *Proc. Roy. Soc. Edinburgh Sect. A*, 93(1–2):129–135.

[McGhee and Picard, 1988] McGhee, D. F. and Picard, R. H. (1988). *Cordes' two-parameter spectral representation theory*, volume 177 of *Pitman Research Notes in Mathematics Series*. Longman Scientific & Technical, Harlow.

[McGhee and Roach, 1982] McGhee, D. F. and Roach, G. F. (1981–1982). The spectrum of multiparameter problems in Hilbert space. *Proc. Roy. Soc. Edinburgh Sect. A*, 91(1–2):31–42.

[McGhee and Sallam, 1988] McGhee, D. F. and Sallam, M. H. (1988). Simple eigenvalues and bifurcation for a multiparameter problem. *Proc. Edinburgh Math. Soc. (2)*, 31(1):77–88.

[McLachlan, 1947] McLachlan, N. W. (1947). *Theory and Application of Mathieu Functions*. Oxford, at the Clarenden Press.

[Meixner and Schäfke, 1954] Meixner, J. and Schäfke, F. W. (1954). *Mathieusche Funktionen und Sphäroidfunktionen mit Anwendungen auf physikalische und technische Probleme*. Die Grundlehren der mathematischen Wissenschaften in Einzeldarstellungen mit besonderer Berücksichtigung der Anwendungsgebiete, Band LXXI. Springer-Verlag, Berlin.

[Miller, 1988] Miller, Jr., W. (1988). Mechanisms for variable separation in partial differential equations and their relationship to group theory. In *Symmetries and nonlinear phenomena (Paipa, 1988)*, volume 9 of *CIF Ser.*, pages 188–221. World Sci. Publ., Teaneck, NJ.

[Mingarelli, 1983] Mingarelli, A. B. (1983). *Volterra-Stieltjes integral equations and generalized ordinary differential expressions*, volume 989 of *Lecture Notes in Mathematics*. Springer-Verlag, Berlin.

[Mingarelli, 1986] Mingarelli, A. B. (1986). A survey of the regular weighted Sturm-Liouville problem—the nondefinite case. In *International workshop on applied differential equations (Beijing, 1985)*, pages 109–137. World Sci. Publishing, Singapore.

[Mingarelli, 1994] Mingarelli, A. B. (1994). A class of maps in an algebra with indefinite metric. *Proc. Amer. Math. Soc.*, 121(4):1177–1183.

[Mingarelli, 2004] Mingarelli, A. B. (2004). Characterizing degenerate Sturm-Liouville problems. *Electron. J. Differential Equations*, pages No. 130, 8 pp. (electronic).

[Mingarelli, 2005] Mingarelli, A. B. (2005). A glimpse into the life and times of F. V. Atkinson. *Math. Nachr.*, 278(12–13):1364–1387.

[Mingarelli and Halvorsen, 1988] Mingarelli, A. B. and Halvorsen, S. G. (1988). *Nonoscillation domains of differential equations with two parameters*, volume 1338 of *Lecture Notes in Mathematics*. Springer-Verlag, Berlin.

[Mirzoev, 2008] Mirzoev, K. A. (2008). On the singular multiparameter Sturm-Liouville problem. *Dokl. Akad. Nauk*, 421(1):24–28.

[Morse and Feshbach, 1953] Morse, P. M. and Feshbach, H. (1953). *Methods of theoretical physics. 2 volumes.* McGraw-Hill Book Co., Inc., New York.

[Müller, 1982] Müller, R. E. (1982). Discretization of multiparameter eigenvalue problems. *Numer. Math.*, 40(3):319–328.

[Neamaty and Sazgar, 2008] Neamaty, A. and Sazgar, E. A. (2008). The Neumann conditions for Sturm-Liouville problems with turning points. *Int. J. Contemp. Math. Sci.*, 3(9–12):551–559.

[Neuman, 1991] Neuman, F. (1991). *Global properties of linear ordinary differential equations*, volume 52 of *Mathematics and its Applications (East European Series)*. Kluwer Academic Publishers Group, Dordrecht.

[Niessen, 1966] Niessen, H.-D. (1966). Algebraische Untersuchungen über separierbare Operatoren. *Math. Z.*, 94:328–348.

[Niessen and Zettl, 1992] Niessen, H.-D. and Zettl, A. (1992). Singular Sturm-Liouville problems: the Friedrichs extension and comparison of eigenvalues. *Proc. London Math. Soc. (3)*, 64(3):545–578.

[Pell, 1922] Pell, A. J. (1922). Linear equations with two parameters. *Trans. Amer. Math. Soc.*, 23(2):198–211.

[Perera and Shivaji, 2008] Perera, K. and Shivaji, R. (2008). Positive solutions of multiparameter semipositone p-Laplacian problems. *J. Math. Anal. Appl.*, 338(2):1397–1400.

[Plestenjak, 1998] Plestenjak, B. (1998). A numerical algorithm for computing a basis for the root subspace at a nonderogatory eigenvalue of a multiparameter system. *Linear Algebra Appl.*, 285(1–3):257–276.

[Plestenjak, 2001] Plestenjak, B. (2001). A continuation method for a weakly elliptic two-parameter eigenvalue problem. *IMA J. Numer. Anal.*, 21(1):199–216.

[Podlevs'kiĭ, 2006] Podlevs'kiĭ, B. M. (2006). A numerical algorithm for solving linear multiparameter eigenvalue problems. *Mat. Metodi Fiz.-Mekh. Polya*, 49(2):86–89.

[Pokornyĭ et al., 2008a] Pokornyĭ, Y. V., Zvereva, M. B., and Shabrov, S. A. (2008a). On an extension of the Sturm-Liouville oscillation theory to problems with impulse parameters. *Ukraïn. Mat. Zh.*, 60(1):95–99.

[Pokornyĭ et al., 2008b] Pokornyĭ, Y. V., Zvereva, M. B., and Shabrov, S. A. (2008b). Sturm-Liouville oscillation theory for impulsive problems. *Uspekhi Mat. Nauk*, 63(1(379)):111–154.

[Reid, 1971] Reid, W. T. (1971). *Ordinary differential equations*. John Wiley & Sons Inc., New York.

[Reid, 1980] Reid, W. T. (1980). *Sturmian theory for ordinary differential equations*, volume 31 of *Applied Mathematical Sciences*. Springer-Verlag, New York. With a preface by John Burns.

[Richardson, 1910] Richardson, R. G. D. (1910). Das Jacobische Kriterium der Variationsrechnung und die Oszillationseigenschaften linearer Differentialgleichungen 2. Ordnung (i). *Math. Ann.*, 68(2):279–304.

[Richardson, 1911] Richardson, R. G. D. (1911). Das Jacobische Kriterium der Variationsrechnung und die Oszillationseigenschaften linearer Differentialgleichungen 2. Ordnung (ii). *Math. Ann.*, 71(2):214–232.

[Richardson, 1912] Richardson, R. G. D. (1912). Theorems of oscillation for two linear differential equations of the second order with two parameters. *Trans. Amer. Math. Soc.*, 13(1):22–34.

[Richardson, 1913] Richardson, R. G. D. (1913). Über die notwendigen und hinreichenden Bedingungen für das Bestehen eines Kleinschen Oszillationstheorems. *Math. Ann.*, 73(2):289–304. (correction Vol. 74, p.312).

[Richardson, 1918] Richardson, R. G. D. (1918). Contributions to the study of oscillation properties of the solutions of linear differential equations of the Second Order. *Amer. J. Math.*, 40(3):283–316.

[RIMS, 1996] RIMS, editor (1996). *Hembun mondai to sono shūhen (Japanese)*, Kyoto. Kyoto University Research Institute for Mathematical Sciences. Sūrikaisekikenkyūsho Kōkyūroku No. 973 (1996).

[Roach, 1984] Roach, G. e. (1984). *University of Strathclyde seminars in applied mathematical analysis: Multiparameter problems*. Shiva Mathematics Series, 8. Nantwich, Cheshire: Shiva Publishing Limited. VIII, 101 p. hbk.

[Roach, 1974a] Roach, G. F. (1974a). A classification and reduction of general multiparameter problems. Technical report, Georg-August-Universität Göttingen.

[Roach, 1974b] Roach, G. F. (1974b). Transform theorems for multiparameter problems in Hilbert space. Technical report, Georg-August-Universität Göttingen.

[Roach, 1976] Roach, G. F. (1976). A Fredholm theory for multiparametric problems. *Nieuw Arch. Wisk. (3)*, 24(1):49–76.

[Roach, 1977] Roach, G. F. (1977). Variational methods for multiparametric eigenvalue problems. In *Numerik und Anwendungen von Eigenwertaufgaben und Verzweigungsproblemen (Tagung, Math. Forschungsinst., Oberwolfach, 1976)*, pages 119–132. Internat. Schriftenreihe zur Numer. Math., Vol. 38. Birkhäuser, Basel.

[Roach, 1982] Roach, G. F. (1982). Symmetry groups and multiparameter eigenvalue problems. *Zeszyty Nauk. Politech. Łódz. Mat.*, 14:21–39.

[Roach, 1983] Roach, G. F. (1983). On a class of nonhomogeneous multiparameter problems. In *Dynamical problems in mathematical physics (Oberwolfach, 1982)*, volume 26 of *Methoden Verfahren Math. Phys.*, pages 183–193. Lang, Frankfurt am Main.

[Roach and Sleeman, 1976] Roach, G. F. and Sleeman, B. D. (1976). Generalized multiparameter spectral theory. In *Function theoretic methods for partial differential equations (Proc. Internat. Sympos., Darmstadt, 1976)*, pages 394–411. Lecture Notes in Math., Vol. 561. Springer, Berlin.

[Roach and Sleeman, 1978b] Roach, G. F. and Sleeman, B. D. (1977–1978b). On the spectral theory of operator bundles. *Applicable Anal.*, 7(1):1–14.

[Roach and Sleeman, 1978a] Roach, G. F. and Sleeman, B. D. (1978a). Coupled operator systems and multiparameter spectral theory. *Proc. Roy. Soc. Edinburgh Sect. A*, 80(1–2):23–34.

[Roach and Sleeman, 1979] Roach, G. F. and Sleeman, B. D. (1979). On the spectral theory of operator bundles. II. *Applicable Anal.*, 9(1):29–36.

[Rynne, 1988a] Rynne, B. P. (1988a). Multiparameter spectral theory and Taylor's joint spectrum in Hilbert space. *Proc. Edinburgh Math. Soc. (2)*, 31(1):127–144.

[Rynne, 1988b] Rynne, B. P. (1988b). Multiparameter spectral theory of singular differential operators. *Proc. Edinburgh Math. Soc. (2)*, 31(1):49–66.

[Rynne, 1988c] Rynne, B. P. (1988c). Perturbation theory of multiparameter eigenvalue problems. *J. Math. Anal. Appl.*, 131(2):465–485.

[Rynne, 1990] Rynne, B. P. (1990). Uniform convergence of multiparameter eigenfunction expansions. *J. Math. Anal. Appl.*, 147(2):340–350.

[Rynne, 1995] Rynne, B. P. (1995). The structure of the zero sets of non-linear mappings near generic multiparameter eigenvalues. *J. Math. Anal. Appl.*, 194(1):147–173.

[Saei et al., 2008] Saei, F. D., Jodayree Akbarfam, A., and Marasi, H. R. (2008). Higher-order asymptotic distributions of non-definite Sturm-Liouville problem. *Int. J. Pure Appl. Math.*, 44(5):661–671.

[Sallam, 1994] Sallam, M. H. (1994). Generalised simple eigenvalues and bifurcation for a linked multiparameter eigenvalue problem. *Bull. Calcutta Math. Soc.*, 86(3):227–232.

[Sallam, 1996] Sallam, M. H. (1996). Generalized simple eigenvalues and bifurcation for a linked multiparameter eigenvalue problem. *Kyungpook Math. J.*, 36(2):229–236.

[Schäfke and Schneider, 1966a] Schäfke, F. W. and Schneider, A. (1965/1966a). *S*-hermitesche Rand-Eigenwertprobleme. I. *Math. Ann.*, 162:9–26.

[Schäfke and Schneider, 1966b] Schäfke, F. W. and Schneider, A. (1966b). *S*-hermitesche Rand-Eigenwertprobleme. II. *Math. Ann.*, 165:236–260.

[Schäfke and Schneider, 1968] Schäfke, F. W. and Schneider, A. (1968). *S*-hermitesche Rand-Eigenwertprobleme. III. *Math. Ann.*, 177:67–94.

[Schäfke and Volkmer, 1989] Schäfke, R. and Volkmer, H. (1989). Bounds for the eigenfunctions of multiparameter Sturm-Liouville systems. *Asymptotic Anal.*, 2(2):139–159.

[Seeger and Lay, 1990] Seeger, A. and Lay, W. e. (1990). *Centennial Workshop on Heun's Equation*. Max Planck Institut für Metallforschung, Institut für Physik, Stuttgart, Germany.

[Seyranian and Mailybaev, 2003] Seyranian, A. P. and Mailybaev, A. A. (2003). *Multiparameter stability theory with mechanical applications*, volume 13 of *Series on Stability, Vibration and Control of Systems. Series A: Textbooks, Monographs and Treatises*. World Scientific Publishing Co. Inc., River Edge, NJ.

[Shibata, 1996a] Shibata, T. (1996a). Nonlinear multiparameter eigenvalue problems. *Sūrikaisekikenkyūsho Kōkyūroku*, 973:40–52. Variational problems and related topics (Japanese) (Kyoto, 1996).

[Shibata, 1996b] Shibata, T. (1996b). Spectral asymptotics and bifurcation for nonlinear multiparameter elliptic eigenvalue problems. *Bull. Belg. Math. Soc. Simon Stevin*, 3(5):501–515.

[Shibata, 1997a] Shibata, T. (1997a). Multiparameter variational eigenvalue problems with indefinite nonlinearity. *Canad. J. Math.*, 49(5):1066–1088.

[Shibata, 1997b] Shibata, T. (1997b). Nonlinear eigenvalue problems with several parameters. *Sūrikaisekikenkyūsho Kōkyūroku*, 984:176–190. Structure of functional equations and mathematical methods (Japanese) (Kyoto, 1996).

[Shibata, 1997c] Shibata, T. (1997c). Nonlinear multiparameter eigenvalue problems on general level sets. *Nonlinear Anal.*, 29(7):823–838.

[Shibata, 1997d] Shibata, T. (1997d). Spectral asymptotics of nonlinear multiparameter Sturm-Liouville problems. *Differential Integral Equations*, 10(4):625–648.

[Shibata, 1997e] Shibata, T. (1997e). Two-parameter nonlinear eigenvalue problems. In *Nonlinear waves (Sapporo, 1995)*, volume 10 of *GAKUTO Internat. Ser. Math. Sci. Appl.*, pages 425–430. Gakkōtosho, Tokyo.

[Shibata, 1997f] Shibata, T. (1997f). Variational method in nonlinear two-parameter eigenvalue problems. *Boll. Un. Mat. Ital. B (7)*, 11(3):539–562.

[Shibata, 1997g] Shibata, T. (1997g). Variational methods for nonlinear multiparameter elliptic eigenvalue problems. *Nonlinearity*, 10(5):1319–1329.

[Shibata, 1998a] Shibata, T. (1998a). Asymptotic profiles of variational eigenvalues of two-parameter non-linear Sturm-Liouville problems. *Math. Methods Appl. Sci.*, 21(18):1619–1635.

[Shibata, 1998b] Shibata, T. (1998b). Nonlinear multiparameter Sturm-Liouville problems. *Asymptot. Anal.*, 18(3–4):173–192.

[Shibata, 1998c] Shibata, T. (1998c). Two-parameter nonlinear Sturm-Liouville problems. *Proc. Edinburgh Math. Soc. (2)*, 41(2):225–245.

[Shibata, 1999] Shibata, T. (1999). Spectral asymptotics for nonlinear multiparameter problems with indefinite nonlinearities. *Czechoslovak Math. J.*, 49(124)(2):317–340.

[Shibata, 2001a] Shibata, T. (2001a). Asymptotic behavior of eigenvalues for two-parameter perturbed elliptic sine-Gordon type equations. *Results Math.*, 39(1–2):155–168.

[Shibata, 2001b] Shibata, T. (2001b). Spectral asymptotics of nonlinear elliptic two-parameter problems. *Nonlinear Anal.*, 43(1):75–90.

[Shibata, 2003a] Shibata, T. (2003a). Asymptotic formulas and critical exponents for two-parameter nonlinear eigenvalue problems. *Abstr. Appl. Anal.*, 11:671–684.

[Shibata, 2003b] Shibata, T. (2003b). Precise asymptotic formulas for variational eigencurves of semilinear two-parameter elliptic eigenvalue problems. *Ann. Mat. Pura Appl. (4)*, 182(2):211–229.

[Shibata, 2004] Shibata, T. (2004). Asymptotic expansion of the variational eigencurve for two-parameter simple pendulum-type equations. *Nonlinear Anal.*, 58(3–4):425–440.

[Shibata, 2005] Shibata, T. (2005). Precise asymptotic properties of solutions to two-parameter elliptic eigenvalue problems in a ball. *Adv. Nonlinear Stud.*, 5(1):33–56.

[Shibata, 2006a] Shibata, T. (2006a). Boundary layer and variational eigencurve in two-parameter single pendulum type equations. *Commun. Pure Appl. Anal.*, 5(1):147–154.

[Shibata, 2006b] Shibata, T. (2006b). The relationship between two-parameter perturbed sine-Gordon type equation and nonlinear Klein-Gordon equation. *Indiana Univ. Math. J.*, 55(5):1557–1572.

[Shibata, 2007a] Shibata, T. (2007a). Asymptotic behavior of solutions to the perturbed simple pendulum problems with two parameters. *Math. Nachr.*, 280(4):451–462.

[Shibata, 2007b] Shibata, T. (2007b). Precise asymptotic shapes of solutions to nonlinear two-parameter problems. *Results Math.*, 50(3–4):259–273.

[Shonkwiler, 1975] Shonkwiler, R. (1975). Multiparameter resolvents. *J. Math. Anal. Appl.*, 52(2):344–350.

[Sini, 2003] Sini, M. (2003). Some uniqueness results of discontinuous coefficients for the one-dimensional inverse spectral problem. *Inverse Problems*, 19(4):871–894.

[Sini, 2004] Sini, M. (2004). On the one-dimensional Gelfand and Borg-Levinson spectral problems for discontinuous coefficients. *Inverse Problems*, 20(5):1371–1386.

[Sleeman, 1980] Sleeman, B. (1980). Multiparameter periodic differential equations. In Everitt, W. N., editor, *Ordinary and Partial Differential Equations: Proceedings of the Fifth Conference Held at Dundee, Scotland, March 29–31, 1978*, volume 827, pages 229–250.

[Sleeman, 1971b] Sleeman, B. D. (1970-1971b). The two parameter Sturm-Liouville problem for ordinary differential equations. *Proc. Roy. Soc. Edinburgh Sect. A*, 69:139–148.

[Sleeman, 1971a] Sleeman, B. D. (1971a). Multi-parameter eigenvalue problems in ordinary differential equations. *Bul. Inst. Politehn. Iași (N.S.)*, 17(21)(3–4, sect. I):51–60.

[Sleeman, 1972a] Sleeman, B. D. (1972a). Multi-parameter eigenvalue problems and k-linear operators. In *Conference on the Theory of Ordinary and Partial Differential Equations (Univ. Dundee, Dundee, 1972)*, pages 347–353. Lecture Notes in Math., Vol. 280. Springer, Berlin.

[Sleeman, 1972b] Sleeman, B. D. (1972b). The two-parameter Sturm-Liouville problem for ordinary differential equations. II. *Proc. Amer. Math. Soc.*, 34:165–170.

[Sleeman, 1973a] Sleeman, B. D. (1973a). Completeness and expansion theorems for a two-parameter eigenvalue problem in ordinary differential equations using variational principles. *J. London Math. Soc. (2)*, 6:705–712.

[Sleeman, 1973b] Sleeman, B. D. (1973b). Singular linear differential operators with many parameters. *Proc. Roy. Soc. Edinburgh Sect. A*, 71(part 3):199–232.

[Sleeman, 1975] Sleeman, B. D. (1975). Left-definite multiparameter eigenvalue problems. In *Spectral theory and differential equations (Proc. Sympos., Dundee, 1974; dedicated to Konrad Jörgens)*, pages 307–321. Lecture Notes in Math., Vol. 448. Springer, Berlin.

[Sleeman, 1978a] Sleeman, B. D. (1978a). *Multiparameter spectral theory in Hilbert space*, volume 22 of *Research Notes in Mathematics*. Pitman (Advanced Publishing Program), Boston, Mass.

[Sleeman, 1978b] Sleeman, B. D. (1978b). Multiparameter spectral theory in Hilbert space. *J. Math. Anal. Appl.*, 65(3):511–530.

[Sleeman, 1979] Sleeman, B. D. (1979). Klein oscillation theorems for multiparameter eigenvalue problems in ordinary differential equations. *Nieuw Arch. Wisk. (3)*, 27(3):341–362.

[Sleeman, 2008] Sleeman, B. D. (2008). Multiparameter spectral theory and separation of variables. *J. Phys. A*, 41(1):015209, 20.

[Sleeman et al., 1984] Sleeman, B. D., Smith, P. D., and Wright, G. P. (1984). Doubly-periodic Floquet theory. *Proc. Roy. Soc. London Ser. A*, 391(1800):125–147.

[Stäckel, P., 1891] Stäckel, P. (1891). Über die Integration der Hamilton-Jacobi'schen Differentialgleichung mittels Separation der Variabeln. Habilitationsschrift, Halle, 26 pp.

[Stieltjes, 1885] Stieltjes, T. J. (1885). On some polynomials verifying a linear differential equation of the second order and on the theory of Lamé functions. (Sur certains polynômes qui vérifient une équation différentielle linéaire du second ordre et sur la théorie des fonctions de Lamé.). *Acta Math.*, 6:321–326.

[Sturm, 1829] Sturm, C. (1829). Extrait d'un mémoire sur l'intégration d'un système d'équations différentielles linéaires. *Bull. Sci. Math. Astr. Phys.*, 12:315–322.

[Sturm, 1836] Sturm, C. (1836). Mémoire sur les équations différentielles linéaires du second ordre. *J. Maths. Pures Appl.*, 1(1):106–186.

[Sun and Ge, 2008] Sun, B. and Ge, W. (2008). Existence and iteration of positive solutions to a class of Sturm-Liouville-like p-Laplacian boundary value problems. *Nonlinear Anal.*, 69(4):1454–1461.

[Szegő, 1975] Szegő, G. (1975). *Orthogonal polynomials*. American Mathematical Society, Providence, R.I., fourth edition. American Mathematical Society, Colloquium Publications, Vol. XXIII, Reprint of the (1939) edition.

[Taylor, 1970] Taylor, J. L. (1970). A joint spectrum for several commuting operators. *J. Functional Analysis*, 6:172–191.

[Thomson, 1873] Thomson, W. and Tait, P. G. (1873). *The Elements of Natural Philosophy*. Vol. II. Oxford at the Clarendon Press [Reprinted by the University of Michigan Library (2001)].

[Thomson, 1863] Thomson, W. (1863). Dynamical problems regarding elastic spheroidal shells and spheroids of incompressible liquid. *Phil. Trans. Roy. Soc. London*, 153:583–616.

[Titchmarsh, 1958] Titchmarsh, E. C. (1958). *Eigenfunction expansions associated with second-order differential equations. Vol. 2*. Oxford, at the Clarendon Press.

[Titchmarsh, 1962] Titchmarsh, E. C. (1962). *Eigenfunction expansions associated with second-order differential equations. Part I*. Second Edition. Clarendon Press, Oxford.

[Tumanov, 2000] Tumanov, S. N. (2000). Asymptotics of the eigenvalues of an indefinite Sturm-Liouville problem with two turning points. *Uspekhi Mat. Nauk*, 55(5(335)):179–180. English translation: Russian Math. Surveys 55 (2000), no. 5, 1007–1008.

[Tumanov, 2001] Tumanov, S. N. (2001). Asymptotic formulas for the real eigenvalues of the Sturm-Liouville problem with two turning points. *Izv. Ross. Akad. Nauk Ser. Mat.*, 65(5):153–166.

[Turyn, 1980] Turyn, L. (1980). Sturm-Liouville problems with several parameters. *J. Differential Equations*, 38(2):239–259.

[Turyn, 1983] Turyn, L. (1983). Perturbation of linked eigenvalue problems. *Nonlinear Anal.*, 7(1):35–40.

[Turyn, 1986] Turyn, L. (1986). Bifurcation without mixed mode solutions. In *Multiparameter bifurcation theory (Arcata, Calif., 1985)*, volume 56 of *Contemp. Math.*, pages 335–341. Amer. Math. Soc., Providence, RI.

[Verma, 1989] Verma, R. U. (1989). Multiparameter spectral theory of a separating operator system. *Appl. Math. Lett.*, 2(4):391–393.

[Vladimirov et al., 2002] Vladimirov, A. A., Griniv, R. O., and Shkalikov, A. A. (2002). Spectral analysis of periodic differential matrices of mixed order. *Tr. Mosk. Mat. Obs.*, 63:45–86.

[Volkmer, 1982] Volkmer, H. (1982). On multiparameter theory. *J. Math. Anal. Appl.*, 86(1):44–53.

[Volkmer, 1984a] Volkmer, H. (1984a). Eigenvector expansions in multiparameter eigenvalue problems. Multiparameter problems, Univ. Strathclyde Semin. appl. math. Anal., Shiva Math. Ser. 8, 93–101.

[Volkmer, 1984b] Volkmer, H. (1984b). On the completeness of eigenvectors of right definite multiparameter problems. *Proc. Roy. Soc. Edinburgh Sect. A*, 96(1–2):69–78.

[Volkmer, 1986] Volkmer, H. (1986). On the minimal eigenvalue of a positive definite operator determinant. *Proc. Roy. Soc. Edinburgh Sect. A*, 103(3–4):201–208.

[Volkmer, 1987] Volkmer, H. (1987). On an expansion theorem of F. V. Atkinson and P. Binding. *SIAM J. Math. Anal.*, 18(6):1669–1680.

[Volkmer, 1988] Volkmer, H. (1988). *Multiparameter eigenvalue problems and expansion theorems*, volume 1356 of *Lecture Notes in Mathematics*. Springer-Verlag, Berlin.

[Volkmer, 1996] Volkmer, H. (1996). Asymptotic spectrum of multiparameter eigenvalue problems. *Proc. Edinburgh Math. Soc. (2)*, 39(1):119–132.

[Volkmer, 2004] Volkmer, H. (2004). Matrix Riccati equations and matrix Sturm-Liouville problems. *J. Differential Equations*, 197(1):26–44.

[Walter, 1973] Walter, J. (1973). Regular eigenvalue problems with eigenvalue parameter in the boundary condition. *Math. Z.*, 133:301–312.

[Wang and Sun, 2008a] Wang, G. X. and Sun, J. (2008a). Inequalities among eigenvalues of left-definite and definite Sturm-Liouville problems. *Math. Practice Theory*, 38(1):163–168.

[Wang and Sun, 2008b] Wang, G. X. and Sun, J. (2008b). Properties of eigenvalues of a class of discontinuous Sturm-Liouville problems. *J. Inn. Mong. Norm. Univ. Nat. Sci.*, 37(3):291–295.

[Wei and Fu, 2008] Wei, G. and Fu, S. (2008). Left-definite spaces of singular Sturm-Liouville problems. *J. Math. Anal. Appl.*, 345(1):420–430.

[Weyl, 1910a] Weyl, H. (1910a). Über gewöhnliche Differentialgleichungen mit Singularitäten und die zugehörigen Entwicklungen willkürlicher Funktionen. *Math. Ann.*, 68(2):220–269.

[Weyl, 1910b] Weyl, H. (1910b). Über gewöhnliche lineare Differentialgleichungen mit singulären Stellen und ihre Eigenfunktionen (2. Note). *Nachr. Ges, Wiss. Göttingen*, 1910:442–467.

[Whittaker and Watson, 1996] Whittaker, E. T. and Watson, G. N. (1996). *A course of modern analysis*. Cambridge Mathematical Library. Cambridge University Press, Cambridge. Reprint of the fourth (1927) edition.

[Yang, 2008] Yang, Z. (2008). Existence of nontrivial solutions for a nonlinear Sturm-Liouville problem with integral boundary conditions. *Nonlinear Anal.*, 68(1):216–225.

[Yoshikawa, 1910a] Yoshikawa, J. (1910a).   Dreiparametrige Randwertaufgaben. *Gött. Nachr.*, 1910:563–585.

[Yoshikawa, 1910b] Yoshikawa, J. (1910b). Ein zweiparametriges Oszillationstheorem. *Gött. Nachr.*, 1910:586–594.

[Zettl, 2005] Zettl, A. (2005). *Sturm-Liouville theory*, volume 121 of *Mathematical Surveys and Monographs*. American Mathematical Society, Providence, RI.

[Zhidkov, 2008] Zhidkov, P. (2008). On an inverse eigenvalue problem for a semilinear Sturm-Liouville operator. *Nonlinear Anal.*, 68(3):639–644.

Index